WATER-WORKS

THE ARCHITECTURE AND ENGINEERING OF THE NEW YORK CITY WATER SUPPLY

PATRICK
66080
66080

WATER-WORKS

THE ARCHITECTURE AND ENGINEERING OF THE NEW YORK CITY WATER SUPPLY

KEVIN BONE, Editor

GINA POLLARA, Associate Editor

PAUL DEPPE, Archive Project Associate

Essays By

ALBERT F. APPLETON

KEVIN BONE

PETER H. GLEICK

GERARD KOEPPEL

With Excerpts from *City Water Tunnel #3* by

MARTY POTTENGER

The Irwin S. Chanin School of Architecture of The Cooper Union

THE MONACELLI PRESS

First published in the United States of America in 2006 by The Monacelli Press, Inc.
611 Broadway, New York, New York 10012

Library of Congress Cataloging-in-Publication Data

Water-works : the architecture and engineering of the New York City water supply / Kevin Bone, editor ; Gina Pollara, associate editor ; Paul Deppe,archive project associate ; essays by Albert F. Appleton, Kevin Bone, Peter H. Gleik, Gerard Koeppel ; with excerpts from City Water Tunnel #3 by Marty Pottenger.
p. cm.
Includes bibliographical references and index.
ISBN-13: 978-1-58093-176-2 (hardcover)
ISBN-10: 1-58093-176-6 (hardcover)
1. Waterworks--New York (State)--New York--History. 2. Water-supply--New York (State)--New York--History. I. Bone, Kevin. II. Pollara, Gina. III. Appleton, Albert F.
TD225.N5W377 2006
628.109747'1--dc22

2006024663

Designed by Abigail Sturges

Printed and bound in Hong Kong

This publication was made possible with support from the American Council of Engineering Companies of New York, Bone/Levine Architects, the Graham Foundation for Advanced Studies in the Fine Arts, The Lucius N. Littauer Foundation, Malcolm Pirnie, Inc., the New York State Laborers' Union Local 147, and the New York State Laborers'-Employers' Cooperation & Education Trust.

FRONTISPIECE *City Tunnel #2. Bronze section valve for a shaft on the truck ready for transport from storage yard in Long Island City, Queens. The lower Manhattan skyline appears beneath the Queensboro Bridge. November 6, 1931.*

DEDICATION *Catskill Aqueduct. April 5, 1912.*

For

JOHN Q. HEJDUK

1929–2000

Dean of The Irwin S. Chanin School of Architecture

for his insights into the richness of these materials,
for his encouragement throughout the project,
and for his unwavering dedication to architectural education.

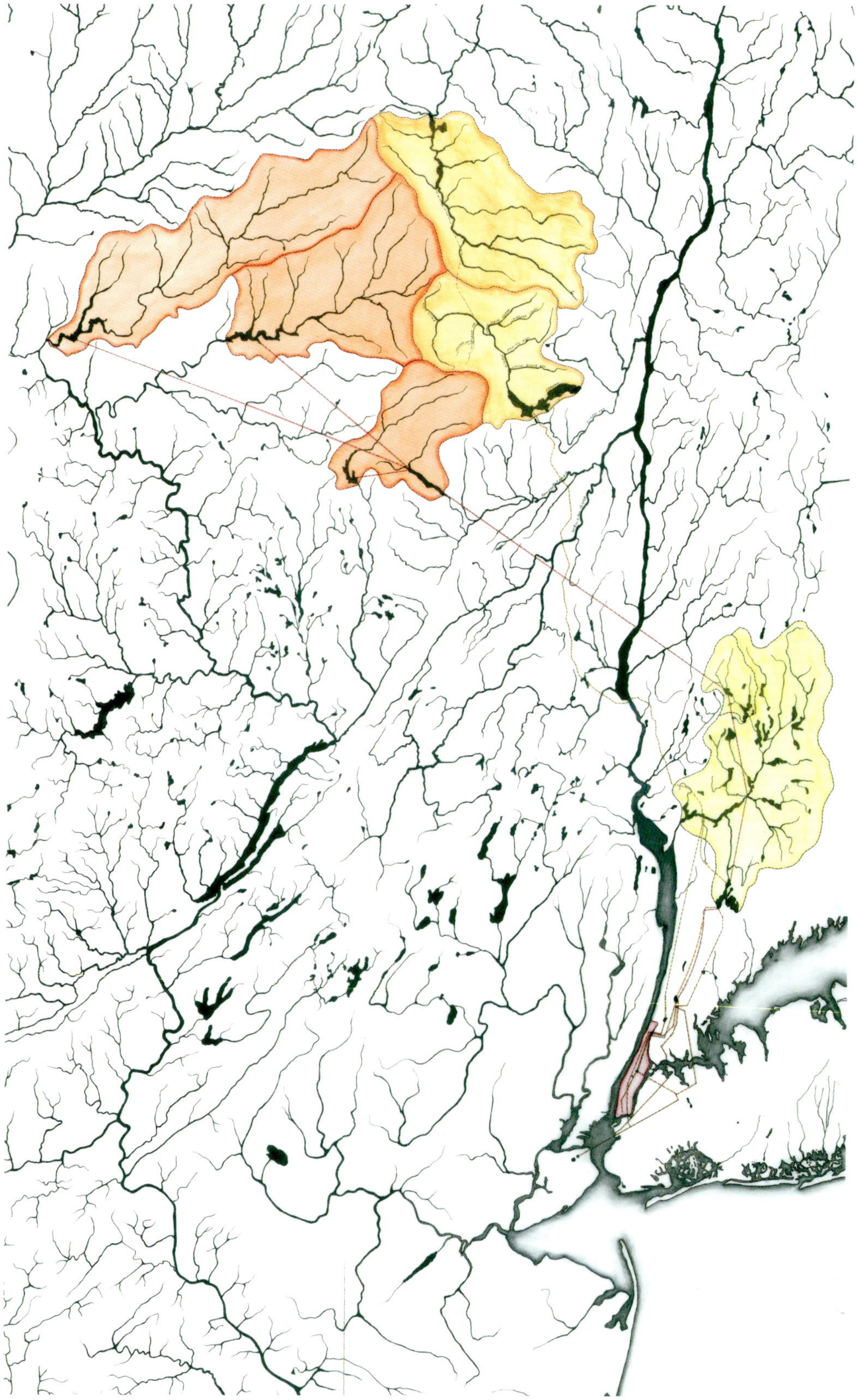

New York City Water Supply Watersheds, Rivers, and Streams.

CONTENTS

FOREWORD

GEORGE CAMPBELL JR.

Water-Works: The Architecture and Engineering of the New York City Water Supply captures one of the most prodigious public works projects in history. The end product of a ten-year research effort led by Kevin Bone and his collaborators at The Irwin S. Chanin School of Architecture at The Cooper Union for the Advancement of Science and Art, this book documents the development of the systems that have brought water to New York City residents since colonial times. It exhibits the historical accomplishments of major waterworks projects—unanticipated breakthroughs in technology and singular architectural wonders—as well as the intellectual challenges and human suffering that were an inherent part of the creation of the water supply system.

These images of city infrastructure leave no doubt about the central role of the visual arts in public works. Successful engineering and architectural projects have an absolute and critical dependence on the graphic precision and elegance intrinsic to the skills of well-trained artists. As with other components of the built environment, public works incorporate the history, culture, values, social attributes, and artistic sensibilities of their communities. This is unequivocally affirmed in the exquisite drawings, the beautiful watercolor renderings, and the illuminating photographic record presented in this book.

As important as supplying water to cities has been in the past, it looms even larger in the future. Within the next fifteen years the majority of the world's population will reside in cities. While enormous resources have been poured into water delivery over the millennia, industrialization and other kinds of human intervention have rendered previously reliable sources of water unfit to drink. Tragically, it is estimated that 1.2 billion people today lack access to potable drinking water. As growing municipalities struggle to meet the needs of surging populations, *Water-Works* can serve as a reminder of the difficulties and responsibilities we face in bringing clean water to inhabitants of the urban environment.

The work of the Cooper Union team follows a long-standing intellectual tradition at the institution. Beginning with its founding in 1859, the college adopted urbanism as a central unifying theme in its unique mix of academic and professional curricula. Its programs offer degrees in precisely those disciplines that are critical to creating a habitable, sustainable urban infrastructure—architecture, art, and engineering. The institution's founder, industrialist Peter Cooper, not only established the school's tradition of social engagement that encourages a commitment to civic responsibility, he was himself one of the city's most involved citizens. As a member of the Common Council, he participated in the early development of the water supply system. In 1829 he served on the Special Committee on Water, and in 1840 he was a member of the Joint Special Committee on the Croton Aqueduct.

Cooper's dedication to New York City has continued through the institution's graduates for over a century. All students at the college are recipients of a full tuition scholarship as mandated by Peter Cooper. These beneficiaries of his generous legacy include many who have made powerful contributions to the public works of New York City. Among those whose names are associated with the city's infrastructure are John Reidel and Patrick Quilty, both engineering graduates from the class of 1898, who respectively became chief engineer of the City of New York and commissioner, Department of Water Supply, Gas and Electricity. Alexander Lurkis, class of 1930, also took on the responsibility of commissioner, Department of Water Supply, Gas and Electricity. Samuel Sigman, class of 1938, was chief of Waterworks Construction, and Angelika Forndran, class of 1973, was chief of Engineering and Scientific Services, both in the New York City Department of Environmental Protection. George Fox, class of

The Foundation Building of The Cooper Union. Photograph taken in preparation for City Tunnel #1 shaft construction. November 18, 1910.

1940, was president of Grow Tunneling Corporation, the organization responsible for the planning, design, and management of major public works related to water supply, including extensive construction work on City Tunnel #3.

Kevin Bone must be counted among those representing The Cooper Union who have made exceptional contributions to New York City. The publication of this book is a major milestone in his commitment to the preservation of the city's archives, documenting the history of what historian Kenneth T. Jackson has described as the "best water system that any people has built since the Romans 2,000 years ago."[1]

Water-Works is the result of work that began in the mid-1990s when Bone and his academic associates examined the archived materials produced during the construction of the Croton, Catskill, and Delaware water systems and discovered that during 150 years of administrative change and reorganization, the archives had become orphaned and homeless, abandoned and forgotten.

Impelled by an ongoing commitment to urban research, in 1994 The Cooper Union forged a collaboration with the Department of Environmental Protection, the current custodian of the archives, to study, organize, document, and preserve the hundreds of thousands of documents. The Cooper Union team pulled together a group of dedicated architects, engineers, and students to carry out the project. One of the highlights of their effort was the critically acclaimed exhibition "Water-Works: The Architecture and Engineering of the New York City Water Supply" at The Cooper Union's Arthur A. Houghton Jr. Gallery in 2001. It was curated by Kevin Bone with Gina Pollara and Paul Deppe, both architects and Cooper graduates who were among the first to work on the project.

The archive that documents and illuminates the building of the New York water supply system has been preserved for posterity and is now accessible to those whose work it is to care for the system. Indeed, the pages that follow access the archives' bounty for scholars and general public alike. This book is important for all of the reasons that the preservation and recounting of history are important and necessary. But it is also of great consequence because *Water-Works* celebrates what is, above all, a story of unparalleled human achievement.

PREFACE

KEVIN BONE

There are many important reasons why The Cooper Union embarked on the archival recovery project for the Department of Environmental Protection (DEP), upon which this book is based. The records that describe the building of the New York City water supply are vital for the ongoing management and maintenance of the system. They also comprise a critically important body of historic information describing everything from the social conditions and labor practices of the time to the streetscapes of New York City. The work needed to be done, and The Cooper Union was uniquely positioned to help.

My own interests were not primarily the preservation of records or historic data, important as these endeavors are. I came to this project as an architect looking for insights into the making of architecture. During the 1990s, I taught a building technology class at The Irwin S. Chanin School of Architecture of The Cooper Union. I often assigned problems that called for the investigation of the industrial and civil engineering heritage of New York City, as these structures, unencumbered by style and ornament, adeptly revealed issues of material and method. I sent my students to find the drawings that described works such as bridges, towers, tunnels, roadways, and industrial facilities. In the pre–9/11 environment, the offices of engineers, both public and private, were incredibly open and supportive. Students often came back to school with copies of magnificent drawings and photographs describing the indigenous works of the industrial and civil engineering world. The efforts to organize the waterworks archive were for me a logical continuation of these investigations.

Of all the engineering materials that I have seen, no body of civil engineering compares to that accomplished during the building of the New York City Water Supply System. Since 1836, New York City has continuously designed and built structures that collect, store, and deliver water [Fig. 1]. The city waterworks is widely regarded as one of the best urban water supplies in the world, and the civil engineering is without equal, representing a range of design and construction innovations whose formal clarity and structural logic are both technically inspired and artistically beautiful. The earlier waterworks, dating from the mid-nineteenth century, are masonry structures reminiscent of ancient Roman engineering [Fig. 2]. By the time the Ashokan Reservoir was under construction in 1907, the works had become decisively modern in spirit, demonstrating the structural and plastic possibilities of reinforced concrete. These later structures, with their sparse and precise forms, elegantly reflect the utility of the program and the demands of modern engineering [Fig. 3].

Infrastructure and civil engineering have always offered lessons for architects, but not all lessons are about advances in building methods. The elemental forms and pure geometry of the waterworks are architectural in their own right, exploring conditions of space, order, volume, void, light, and shadow. Their depictions are often mysterious and poetic, marked by spatial contrasts of plasticity, contour, and profile. The sculpted earthworks, with their rational and fluid forms, are profoundly beautiful and reveal unintentional architectural relationships: concurrences of structure and spatial conditions that can be translated into other building types. By stepping outside the conventional boundaries of the field, as architects have done throughout the rise of modernism, we discovered an exquisite source of insight and inspiration.

About 1997, several years into the project, I went to see Dean John Hejduk to update him on our team's efforts. I carried a box of prints that Paul Deppe and Gina Pollara had organized. There were several hundred pages, each of which contained sixteen to twenty small images and a written description. The material was

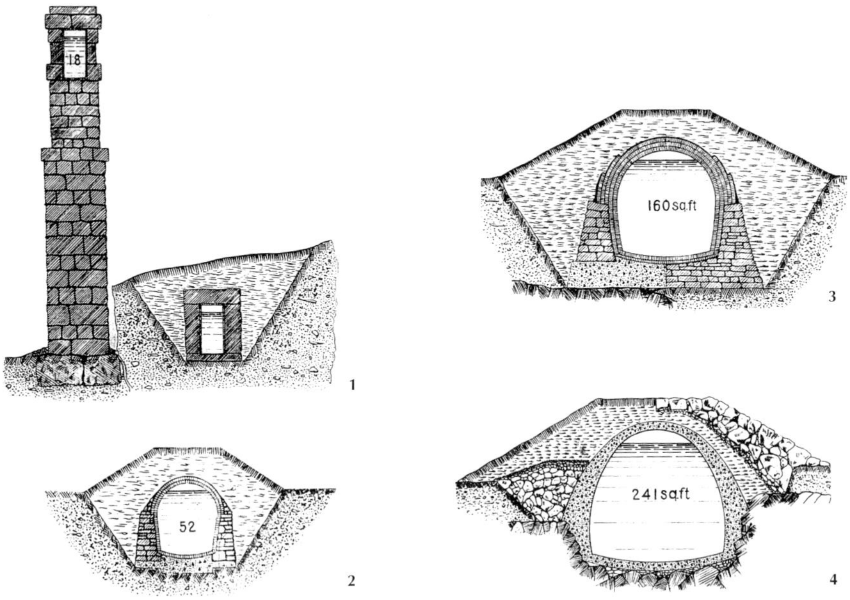

FIG. 1. *Comparative square-footage diagrams for cross sectional areas of (1) Roman aqueduct Aqua Claudia; (2) Old Croton Aqueduct; (3) New Croton Aqueduct; (4) Catskill Aqueduct. December 31, 1909.*

FIG. 2. *Aqueduct near the Villa of Quintili. Rome, Italy.*

FIG. 3. *Industrial vestiges from mining operation. Gold Hill Mesa. Colorado Springs, Colorado. 1973.*

grouped according to structure: towers, tunnels, chambers, dams, and embankments, among others [Figs. 4, 5, 6, 7, 8]. Until this time Hejduk had seen little of the specifics of the project, and I expected him to just glance at the materials for a sense of its scope. Instead, he went through the plates for hours, gazing at all the images and reinforcing our view that we had fallen upon a body of material whose value to architects could not be underestimated. He repeated dozens of times in his deep Bronx accent, "Will you look at this! Do you see that?" Often, when I am reviewing these materials, I ask myself whether I really have looked closely and seen all that is there. This standard of deliberation, Hejduk's standard, is one that we have tried to maintain throughout the process of making this book, and we have sought in every way to present and preserve the quality of the original materials.

As we continued to delve into the archival collection, I came to appreciate the materials more and more, not just for the abstractions of space, form, and structure that they presented but also for what they described. The drawings, photographs, and documents are a portrait of a magnificent human enterprise: an 170-year undertaking that created a system vital to the survival of New York City. This great system is far more than the monumental structural works that store and deliver water. It represents a synthesis of engineering and ecology; an integration of design, construction, and public determination. Cultivated through generations of research and civic discourse, the evolution of the system follows the rise of modern construction and the advent of environmental protection. This system, in all its manifestations, provides what many take for granted, and what is increasingly in short supply around the world: a high-quality domestic water supply.

The fact that New York City has continued its commitment to increase and improve the water supply remains one of the city's great attributes, but it has not always gone smoothly. There have been periods when its residents have suffered dearly for the lack of good sources of water. The description of the building of the system, and comments on the modern state of water, is told here in five essays.

In "The Rise to Croton," Gerard Koeppel describes the city's early struggles with disease, fire, and filth. The pure and abundant water that New Yorkers now enjoy was in notoriously short supply during the first 150 years of city life. Early residents of lower Manhattan took their water from shallow wells, but the availability of clean ground water did not last long. The expanding population, living and working in a densely inhabited environment, failed to recognize the relationship between surface contamination and sub-surface water quality, and by 1730 many of the city's public and private wells were seriously contaminated. Water-borne contagious diseases such as yellow fever, typhus, and cholera killed significant portions of the population and threatened the future of the city. The deterioration of the water quality at the southern end of Manhattan drove people north in search of unpolluted streams and potential wells. By 1833 New York City sought to develop sources of water outside the local supplies. The Croton River system in Westchester County was the first area to be tapped. A modest masonry dam was built near the mouth of the Croton River, and a fully enclosed masonry aqueduct began supplying fresh, clean water to Manhattan in 1842. The dam and the aqueduct, along with two receiving reservoirs in Manhattan and the High Bridge across the Harlem River, constituted the main components of what became known as the Old Croton system.

This system did not meet the city's water requirements for long, and continuous building efforts were required to realize the three "legs" of the modern system, described in essays on the New Croton, the Catskill, and the Delaware systems. After the completion of the Old Croton system, use of water soared. The rapidly increasing demand led to the expansion of the Croton system, and by 1911 a total of twelve reservoirs and three controlled lakes utilizing every available tributary within the Croton watershed had been built. The Croton system was completed when a second, larger aqueduct was added in 1885 and a magnificent new dam was finished in 1907. The New Croton system supplies 10 percent of the city's total water needs, up to 250 million gallons per day (mgd), and remains in service today.

After the consolidation of Greater New York City in 1898, water consumption increased at a staggering rate, and by 1899, even with the expanded Croton system capacity, water shortages still existed. The Catskill Mountain water sources were the only reasonable alternative. Over local opposition, the state passed laws that would allow New York City to take water from the upstate mountains. The legislation also provided for the establishment of the Board of Water Supply, the agency charged with the task of engineering and building the system that would deliver the Catskill water. Over the next twenty-two years, five reservoirs and 130 miles of primary aqueduct were designed and built under the direction of this agency. The Catskill system added 571 square miles of watershed, which provides up to 650 mgd to the city. Despite this bounty, supply was still not meeting the demands of the expanding population, and in 1923 the Board of Water Supply began to examine the western Catskill region for additional sources. Because these waters drain into the interstate Delaware River, water rights negotiations with both Pennsylvania and New Jersey were required. The Delaware portion of the city waterworks became the largest and most productive of the water sources. Work started in 1937 and by its completion in 1965, the Delaware system consisted of 1,015 square miles of watershed, four major storage reservoirs, and 159 miles of pressure tunnels, including the 84-mile-long Delaware Aqueduct, with the capacity to deliver 850 mgd.

Since the completion of the Delaware system, the city has pursued three primary

FIG. 4. *Catskill Aqueduct. Kensico Reservoir. Screen chamber. June 5, 1912.*

FIG. 5. *Delaware Aqueduct. Rondout Reservoir. Effluent chamber. July 21, 1949.*

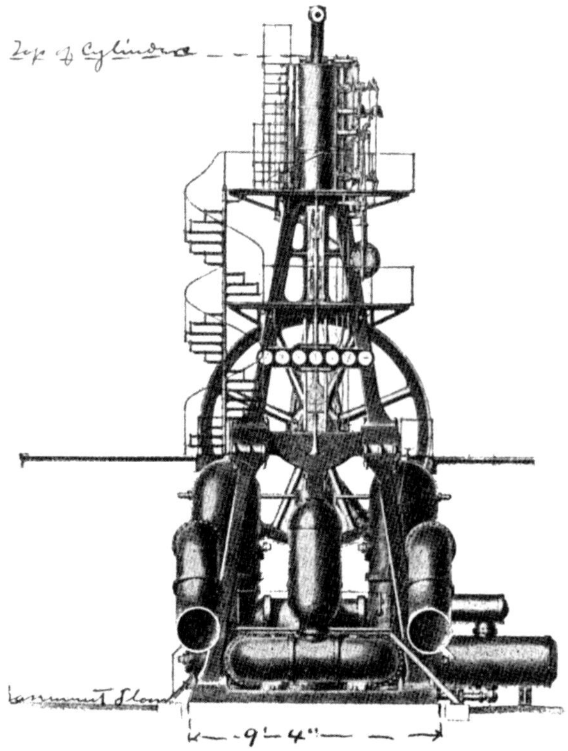

FIG. 6. *Pumping apparatus from pump house on Long Island. Elevation.*

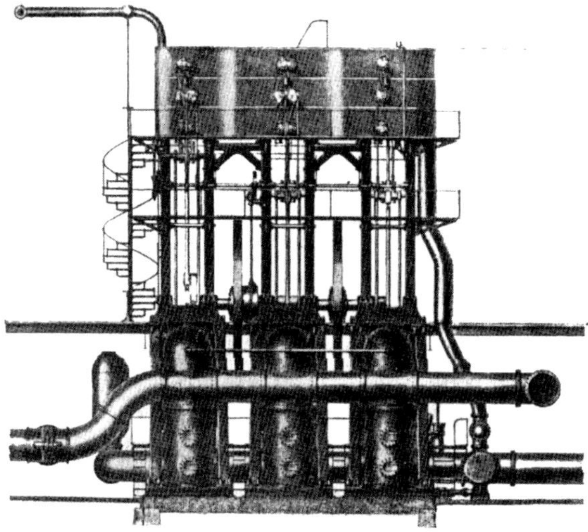

FIG. 7. *Pumping apparatus from pump house on Long Island. Side Elevation.*

programs related to water supply: improved local delivery, protection of the watersheds, and conservation through reduction of waste. These initiatives are described in former DEP Commissioner Albert Appleton's essay, "The Post-Delaware Era." City Tunnel #3 and its related structures, which are scheduled for completion in 2020, will provide new distribution routes to the boroughs. To protect the watersheds and maintain water quality, the city initiated negotiations with those upstate communities whose activities most impacted water quality. In the early 1990s, a Memorandum of Understanding was drafted describing how water quality is to be protected in the watersheds. Finally, efforts to implement effective conservation led to revised city building codes that reduced use through the installation of low-flow water fixtures and universal metering.

The resources required for building the water supply system have far exceeded all other urban infrastructure initiatives in New York City, including the building of the subway system. The cost in dollars and in human life has been great, but unwavering civic commitment has insured there is enough water for the city. With diligence, vigorous care, conservation, and improvement, fresh clean water will continue to be available to New York City. The rest of the world is not so fortunate and in many countries the future of urban water resources is extremely uncertain. In "The Future of Water," Dr. Peter Gleick looks at the state of global water, and the concerns of the future.

The summer of 1999 was dry. The Cooper Union archive team was hunkered down in the midtown warehouse that houses the DEP archival records. The United States Geological Survey indicated that 80 percent of all streams and rivers in the Mid-Atlantic areas had flow rates far below normal and that 20 percent of the water courses were at all-time lows. Fish kills were widespread. Crop failures devastated 30 to 40 percent of the average annual yield to local farmers. Scientists suggested that we were long overdue for a cataclysmic drought, and indeed, it was the third worst drought in the region in one hundred years. Others argued that global warming had raised average temperatures. Regional restrictions on water use were imposed from Maryland to western New York. Residential lawn watering was prohibited and fountains and swimming pools emptied. New York City was one step away from calling for a drought emergency. The grass lawns of the parks were brown and many of the trees had begun to lose their leaves early.

But the upstate New York City reservoirs had water. Life in New York City went on, generally unimpacted by the diminishing supply of fresh water available from nature. Pressure was good, and basic domestic service functioned. New York had its water. And that is something those of us who enjoy it can never forget, or forget to be thankful for.

FIG. 8. *Catskill System. Hill View Reservoir. Three waterways from the uptake chamber. October 28, 1913.*

INTRODUCTION

GINA POLLARA

This book is the culmination of an archival rescue mission that took place from 1994 to 2003. The Cooper Union team first learned of this vast trove of documents during an earlier and related archival effort on materials produced by the Department of Docks (now called the Ports and Trades Collection and housed at the Municipal Archives of the City of New York). There were reports of an archive belonging to the Department of Environmental Protection (DEP) that contained materials similar to those in the waterfront collection, specifically hand-drafted architectural and engineering drawings executed in ink and wash on watercolor paper or drafting cloth dating back to the mid-1830s that detailed the design and construction of the city's water supply system.

Ken Cobb, director of the Municipal Archives from 1990 to 2005, introduced us to the DEP staff in 1994. We were lucky to encounter DEP officials who saw the great opportunity in our interest and allowed us access to their warehouse on Manhattan's east side where the documents were stored. Most of the ground floor is a garage—the facility is the DEP's Bureau of Water and Sewer Operations, the division responsible for the maintenance and repair of water mains, valves, hydrants, and sewers—but 9,600 square feet of the second floor was being used as a document storehouse. At that time we had limited knowledge of the immensity of the water supply system and were unaware of the size and complexity of the DEP's holdings. What we found buried in that space—packed almost to the ceiling with boxes, vertical files, flat files, discarded furniture, and broken equipment—was a body of valuable architectural and engineering documents that details the development and construction of New York City's enormous water supply system.

The archive contains approximately one hundred thousand drawings in various media, fifty thousand printed photographs, and three thousand lantern slides as well as two hundred glass-plate and twenty thousand film negatives. It also includes countless technical engineering books and journals, construction contracts, legal documents, personnel records, and work logs. These materials, which constitute some of the agency's oldest records, chronicle the history of the system, document and record the physical characteristics and hydrological data of the watersheds, describe the construction and performance of the many structures that store and deliver water, and depict visions and plans that were never realized. The records are the photographic and drawn representation of the huge network of dams, reservoirs, tunnels, bridges, shafts, chambers, valves, and pipes that channels water to the city from as far as 125 miles away. As architects, we marveled at the precision and clarity of the technical drawings and the artistic aesthetic they exemplified.

In spite of the disorder, the archive was still relevant: many of these documents were in active use by DEP engineers and outside consultants. Since there was no staff and no obvious record arrangement, engineers had to search through drawer after drawer of materials, often without success. Some drew maps of the space and made basic inventories of various file cabinets in order to later locate groups of drawings, but the facility offered little security for these records, and no document-handling protocols were in place to safeguard the collection. It was clear there was a critical need to organize and protect these materials so that the information they contained could be made available without further damage to the documents.

In the summer of 1995, a pilot program to organize the DEP collection was launched. The Cooper Union provided structural and financial support for the project. The ultimate goal was to create a permanent, fully functioning archive dedicated to the care and preservation of New York City's essential water supply

FIG. 1. *Board of Water Supply engineers. Front row, left to right: William H. Burr, John R. Freeman, and F. P. Stearns. Back row: Thomas Hassett (Secretary), J. Waldo Smith (Chief Engineer). 1907.*

FIG. 2. *Board of Water Supply drafting room. c. 1910.*

records. Both the former president of The Cooper Union John Jay Iselin and the late dean of The Irwin S. Chanin School of Architecture John Hejduk championed the effort and allocated the institution's financial and intellectual resources, support that has been continued by the current president, George Campbell Jr., and dean, Anthony Vidler. During the 1990s a collective of academic institutions, referred to as the Consortium, had a unique arrangement with the city through the Mayor's Office of Construction (MOC). This partnership promoted educational exchange and offered opportunities for students to acquire professional experience. Rather than hire an expensive team of consultants to provide a short-term service, MOC outsourced the work to a member of the Consortium, whose students would execute it under academic supervision at a significant cost savings to the city. It was through this mechanism that The Cooper Union received its first contract with the city to organize the archive.

To bring some basic order to the warehouse and to facilitate an initial assessment, the entire layout of the archive was reorganized with the help of teams of dedicated Cooper Union students. The space was measured, floor plans drawn, and every box, file cabinet, shelf, and flat file

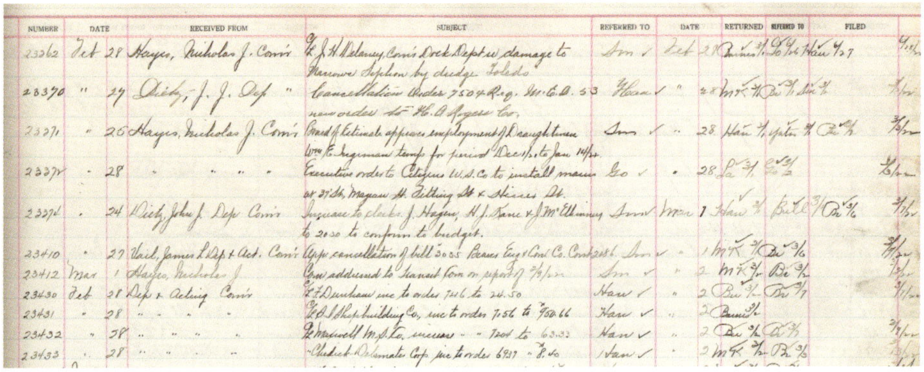

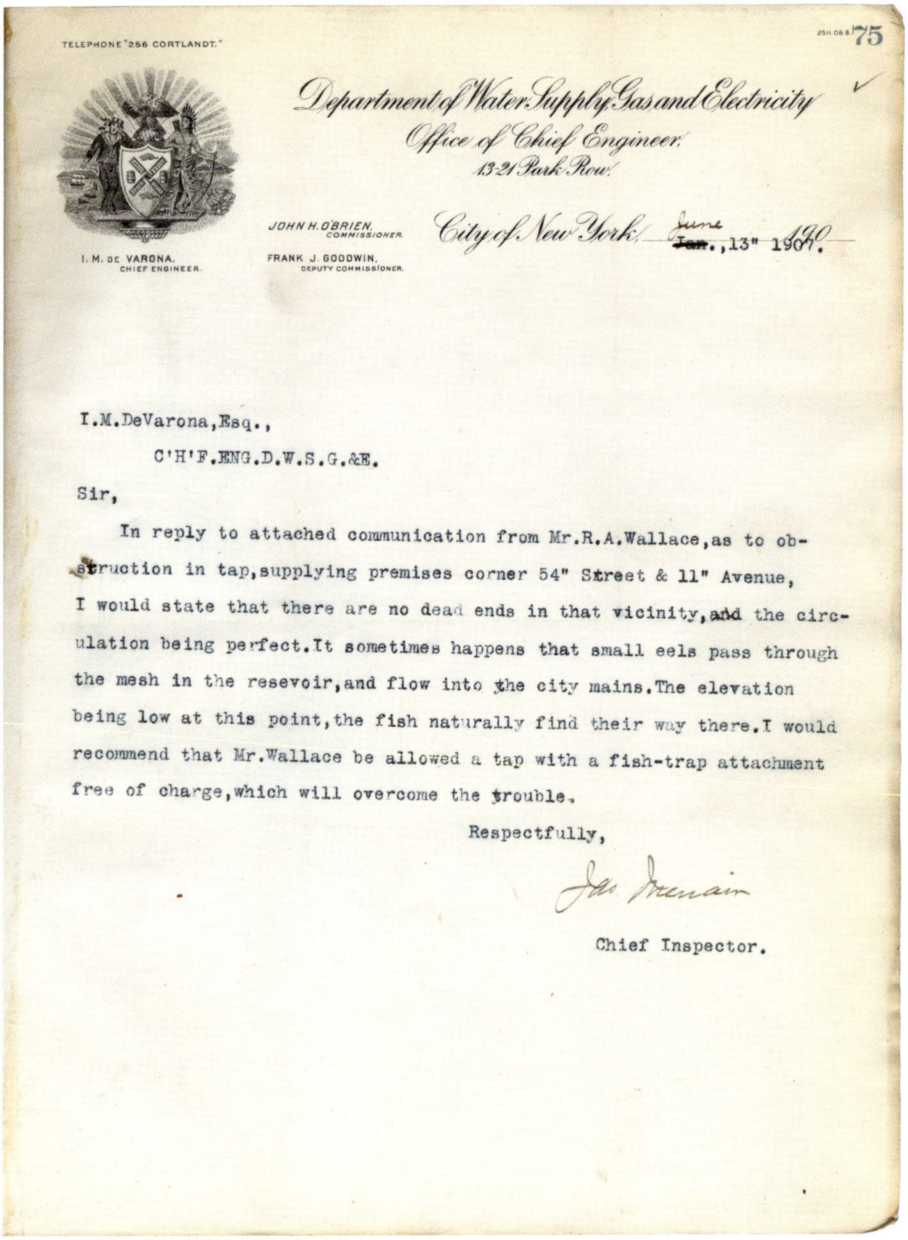

TELEPHONE "256 CORTLANDT."

Department of Water Supply, Gas and Electricity
Office of Chief Engineer,
13-21 Park Row.

JOHN H. O'BRIEN, COMMISSIONER.
FRANK J. GOODWIN, DEPUTY COMMISSIONER.
I. M. DE VARONA, CHIEF ENGINEER.

City of New York, June 13" 1907.

I.M.DeVarona,Esq.,
C'H'F.ENG.D.W.S.G.&E.

Sir,

In reply to attached communication from Mr.R.A.Wallace,as to obstruction in tap,supplying premises corner 54" Street & 11" Avenue, I would state that there are no dead ends in that vicinity,and the circulation being perfect.It sometimes happens that small eels pass through the mesh in the resevoir,and flow into the city mains.The elevation being low at this point,the fish naturally find their way there.I would recommend that Mr.Wallace be allowed a tap with a fish-trap attachment free of charge,which will overcome the trouble.

Respectfully,

Jas. Mena[illegible]

Chief Inspector.

was numbered and examined. The photographs, drawings, and books that had been mixed with other records were extracted, and a general inventory was developed that became the basis of a searchable index. Since the archive facility lacked the environmental controls necessary to protect the materials, window air-conditioning units were installed in two small side rooms. These also provided some respite from the extreme heat and dust in the main space. After this slightly more stable environment was created, the most fragile drawings, photographs, and books were relocated to prevent further deterioration. In retrospect, the amount of work accomplished that first summer was astounding. The dust, the dirt, and the heat were intense. There were long days when little progress seemed to be made and it was difficult not to feel overwhelmed by the magnitude of what needed to be done. Then an exquisite drawing or a hand-colored lantern slide would be unearthed and the purpose of the project reaffirmed.

As we slowly sorted through the collection, the core of what had once been a highly organized and fastidiously maintained drawing and filing system emerged. This central method of arrangement was instituted by the Board of Water Supply (BWS), the agency created by an act of legislation in 1905 to plan, design, and construct the water supply system. The BWS employed some of the foremost engineers of their day, and many worked for the organization for their entire careers [Figs. 1, 2]. The caliber and consistency of staff fostered a collective institutional memory that ensured a high level of professionalism, stability, and reliability through decades of construction. The building of such a large system required a tight hierarchical structure and a meticulous uniformity that is evident throughout the collection. The draftsmen adhered to strict

FIG. 3. *Ledger entry page recording correspondence received.*

FIG. 4. *Letter regarding obstructions in water faucet. July 13, 1907.*

formats and precise drawing standards, and we found multiple copies of the templates outlining these specifications. There are also examples of painstaking record-keeping. A specially constructed cabinet holds oversize ledger books into which each letter received by the department was registered by hand with details of recipient, date, content, and action taken [Fig. 3]. The ledgers have companion volumes into which the letters themselves are bound, along with the responses, all cross-referenced [Fig. 4]. When the BWS was dissolved in 1981, their records were moved from place to place before landing at the East Side warehouse. As additional file cabinets and boxes of documents accumulated over the years, the organization and provenance of the primary records were buried.

During the years of work on the archive that followed, an annual student summer internship program provided the staff necessary to advance the project. In addition to organizing the documents at the main storage facility, materials from various other DEP offices and facilities, including Shaft 17 in upstate New York and the Jerome Avenue Pumping Station in the Bronx, were identified and accessioned into the collection. Photographic materials were rehoused in archival boxes and folders, and a separate photography archive room was created. Similarly, a library was established for the substantial collection of historic and technical books, construction contracts, and annual reports. The most fragile drawings were rehoused in acid-free folders and moved to a separate rare-drawing room furnished with new flat files, and shelving was installed to house the thousands of boxed records. As the archive environment was stabilized, staff within the agency came forward with other valuable materials they had been safeguarding.

FIG. 5. *Catskill System. Gilboa Dam under construction. Town of Gilboa is on upstream side of dam. October 1, 1925.*

FIG. 6. *Catskill System. Schoharie Reservoir filling behind Gilboa Dam. November 20, 1926.*

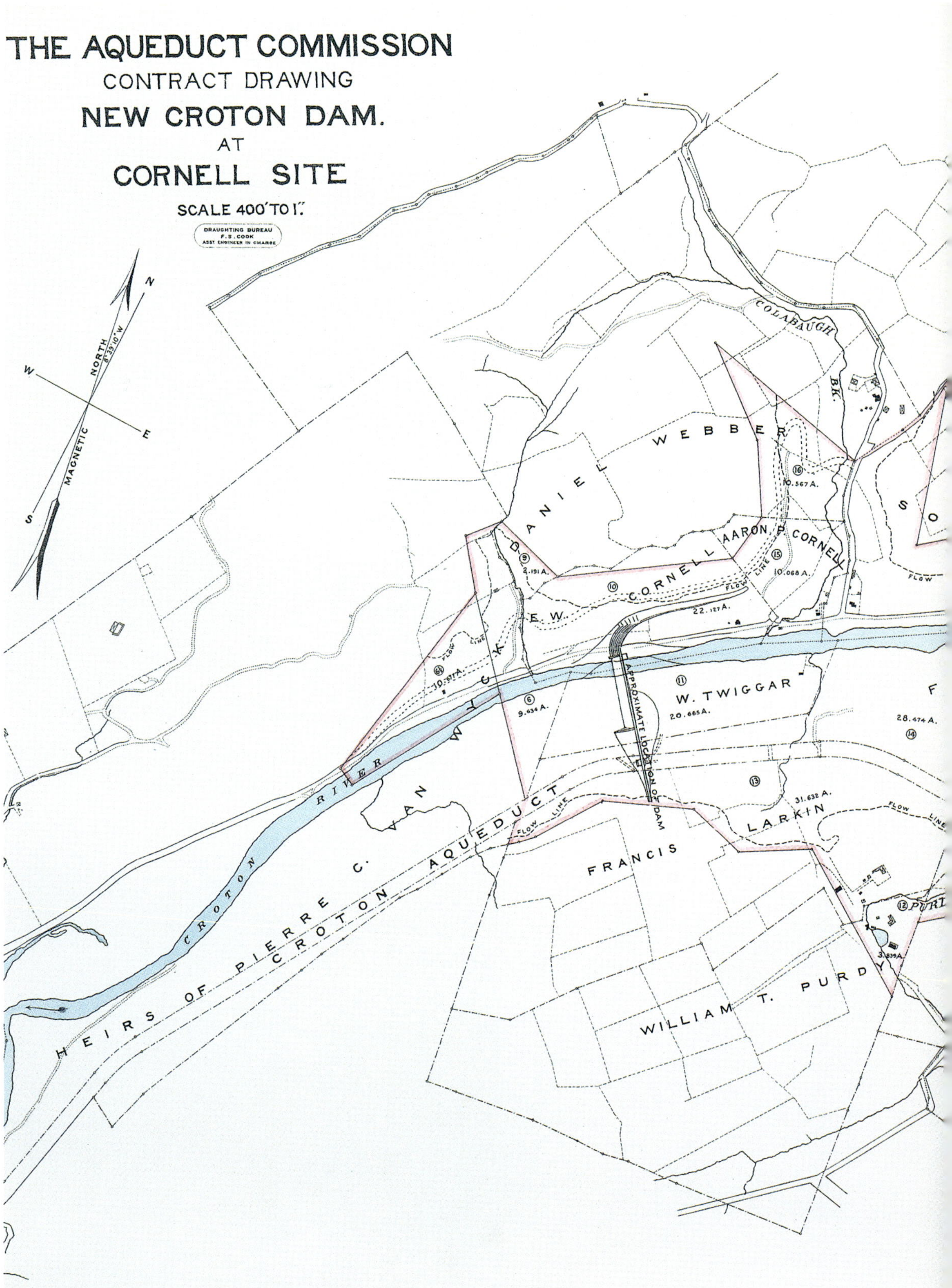

FIG. 7. *New Croton Dam. Site plan showing private parcels of land in area flooded for reservoir. May 2, 1892.*

SHEET №2
HUNTER'S BK.
ISAAC LOSEE
GEORGE F. TEED
WEBB
FLOW LINE
CALEB McCORD
9.537 A.
WALTER TWIGGAR
8.032 A.
LEONARD
CHADEYNE
42.349 A.
ORSER
24.910 A.
20.996 A.
85.666 A.
LARKIN
24.665 A.
JOSEPH BAILEY
0.505 A.
10.324 A.
CALEB McCORD
35.197 A.
BRADY J.
BAILEY'S BK.
JOSEPH BAILEY
COURTLAND
OF YORKTOWN
A. Fteley
CHIEF ENGINEER
MAY 2ND 1892.

FIG. 8. *Camp Kensico. Street view of workers' quarters. Westchester County, New York. July 5, 1910.*

FIG. 9. *Camp Hill View. Mess hall. October 26, 1910.*

FIG. 10. *Board of Water Supply police. Catskill Aqueduct. Peekskill, New York, Division. March 30, 1908.*

Throughout the long course of this project we were sustained by the scope and quality of the materials, and by the evident dedication of those who devoted their lives to this monumental effort. From antiquity to the present, the need to supply water to cities has stimulated some of the world's greatest engineering achievements, as well as some of its greatest sacrifices. Lists of workers who had died during construction were often included at the end of publications that celebrated the completion of a particular structure or milestone in the project. These necrologies were reminders that this was dangerous work, and that the sandhogs—construction workers who build exclusively underground—did not just undertake a job they provided a service to the greater good, a commitment to future generations of city dwellers and to the city itself.

In addition to the workers who dedicated their lives to building this system, thousands of people were displaced when whole towns were flooded to create the reservoirs [Figs. 5, 6]. Cemeteries were exhumed and relocated. Lists of land takings (and the relative pittances paid for them) underscore the personal sacrifice of thousands of families and individuals whose livelihoods were destroyed by the city's thirst [Fig. 7]. For many upstate New Yorkers, this is still a bitter memory.

The archive is a rich collection and should become the source of many studies. The photographs record the history of building technology and the continued innovation in construction methods, from horse and wagon to steam-powered machinery to the latest tunnel-boring equipment. They also document the towns and villages before demolition. The lantern slide collection contains beautiful images of old Roman aqueducts. The assortment of letters alone provides a glimpse into the graphics and typography of the 1910s and 1920s. There is also evidence of an entire culture that evolved from the thousands of people needed to build the system. Artifacts and ephemera depict what was a fully evolved community with all the complexities, trivialities, and prejudices of any society. Towns

erected to house the workers and their families included day schools for the children, night schools for the laborers, stores, hospitals, dental clinics, and dining halls [Figs. 8, 9]. The DEP maintained (as it does to this day) its own police force [Fig. 10]. A bimonthly newspaper was established—first the *Catskill Water Supply News*, then the *Delaware Water Supply News*—where engineers and workers published accounts of construction techniques and detailed advances in the development and use of materials such as bronze, cast iron, and Portland cement, a key ingredient in making concrete. The newsletters also include BWS bowling league scores, as well as announcements of births, deaths, marriages, or poetry and songs about the water supply system [Fig. 11].

One of the reasons we undertook this project was to be able to bring these beautiful materials before the public eye, and in order to prepare some of the most fragile drawings for exhibition, we turned once again to the Municipal Archives. We were able to help broker a mutually beneficial Memorandum of Agreement between the MA and DEP for the conservation work. With care and artistry, MA staff restored and preserved a substantial part of the drawing collection.

In January 2001, we mounted an exhibition of some of these materials at The Cooper Union's Arthur A. Houghton Jr. Gallery [Fig. 12]. The exhibition attracted a wide range of architects, engineers, and other professionals, as well as historians, students, and New Yorkers interested in both the history of their city and its water supply system. Many of the active and retired sandhogs came with their families to share their stories and celebrate their achievements. The size and breadth of the audience emphasized the significance and relevance of this collection, but perhaps most importantly, it underscored the imperative of preserving the architectural and engineering history of the city of New York. It has been our privilege to make a small contribution toward this effort.

FIG. 11. *Board of Water Supply songbook. February 1912.*

FIG. 12. *Water-Works exhibition at The Cooper Union in the Arthur A. Houghton Jr. Gallery. January 8–February 10, 2001.*

GALILEO'S TEARS

The water man from Phoenix said: "You have to recognize we're a closed system—the earth is a closed system—three quarters of it covered with water, only a small percentage of that is drinkable . . . the rest brackish. The challenge is universal. Consequently, it being a closed system, we're drinking the same water that Napoleon drank. We're drinking the same water that Archimedes drank. We're drinking the same water that Galileo drank. So you just have to recognize that, put it into that context, and realize you can only do so much with it."

Excerpt from *City Water Tunnel #3*,
written and performed by Marty Pottenger

THE RISE TO CROTON

GERARD KOEPPEL

Twenty-first-century New York is arguably the world's capital city. But it was not always so. New York's primacy was not inevitable (nor is it inviolable). When Dutch New Amsterdam became English New York in 1664, the leading colonial city was Boston. Over the next hundred years, New York was outranked in population by two others: Philadelphia and Charles Town to the south. By the turn into the first full century of the new American nation, Philadelphia was the country's largest city, but its supremacy was draining away down the hundred river miles that separated it from ocean shipping lanes. In 1810 New York counted ninety-six thousand citizens, slightly more than Philadelphia. Ten years later, the Erie Canal was under construction, joining the Hudson River to the previously impenetrable continental interior. Two hundred years after its founding, the island settlement by the sea finally emerged as the trade center of the Western world.

In many ways, New York was prepared to take up its mantle as the nation's premier city. In 1811 it began stretching paved fingers up Manhattan from its jumbled southern tip. Replication of Dutch-made disorder—narrow winding streets haphazardly laid—would not do. A state commission led by Gouverneur Morris, a home-grown gentryman and a draftsman of the U.S. Constitution, decided that Manhattan should be a natureless grid of twelve avenues and 155 streets: two thousand similar rectilinear blocks from suburban North (now Houston) Street to the seeming limit of future migration, designated as 155th Street. Manhattan's hills—the Delaware Indian name "Manahata" is thought to mean "island of many hills"—were to be shoved into its valleys. Its many fresh streams, extensive marshes, scattered ponds, and bubbling springs were to be suffocated with pavement and basement floors. Rational order was to transform this continental shard into prized real estate. "[A] city is to be composed principally of the habitations of men," wrote the sure-handed Morris, "and . . . straight-sided and right-angled houses are the most cheap to build and the most convenient to live in."[1] Thus was Manhattan gridded for its glorious future.

As urban settlement pushed north from Houston Street to Washington Square to 21st Street and beyond, the political and social landscape adapted as well. Starting in the late 1790s, control of city and state government shifted from the Federalists (later reborn as Whigs and now Republicans) to the Democratic-Republicans (today's Democrats): from Hamiltonian elites to Jeffersonian merchants, from the manor-born to the Tammany served. In 1804 the state legislature, over conservative objections, significantly expanded city suffrage to all taxpayers. In 1830 the first major city charter revision in a hundred years created modern governance: the Common (now City) Council was split into two houses, each with legislative authority, and the mayor—appointed only since 1821 by the council instead of Albany—was severed from the council and given veto power, subject to majority override. Annual popular elections for mayor began in 1834. Mayoral control of proto-bureaucracy followed, and a city under executive management emerged. And none too soon.

From 1810 to 1840 New York's population more than tripled, adding an average of 7,200 new people a year, many of them poor Irish immigrants. New York turned from an overgrown colonial-age town into a dense industrial city of 330,000. And yet, on the cusp of world dominance, New York was still drinking water as it had in the days of burghers and bouweries: from wells that had always been distastefully hard and brackish but now were extraordinarily deficient and dangerously polluted.

FACING PAGE
FIG. 1. *Old Croton Aqueduct. Cross section. c. 1836.*
This is believed to be the oldest drawing in the DEP Archive.

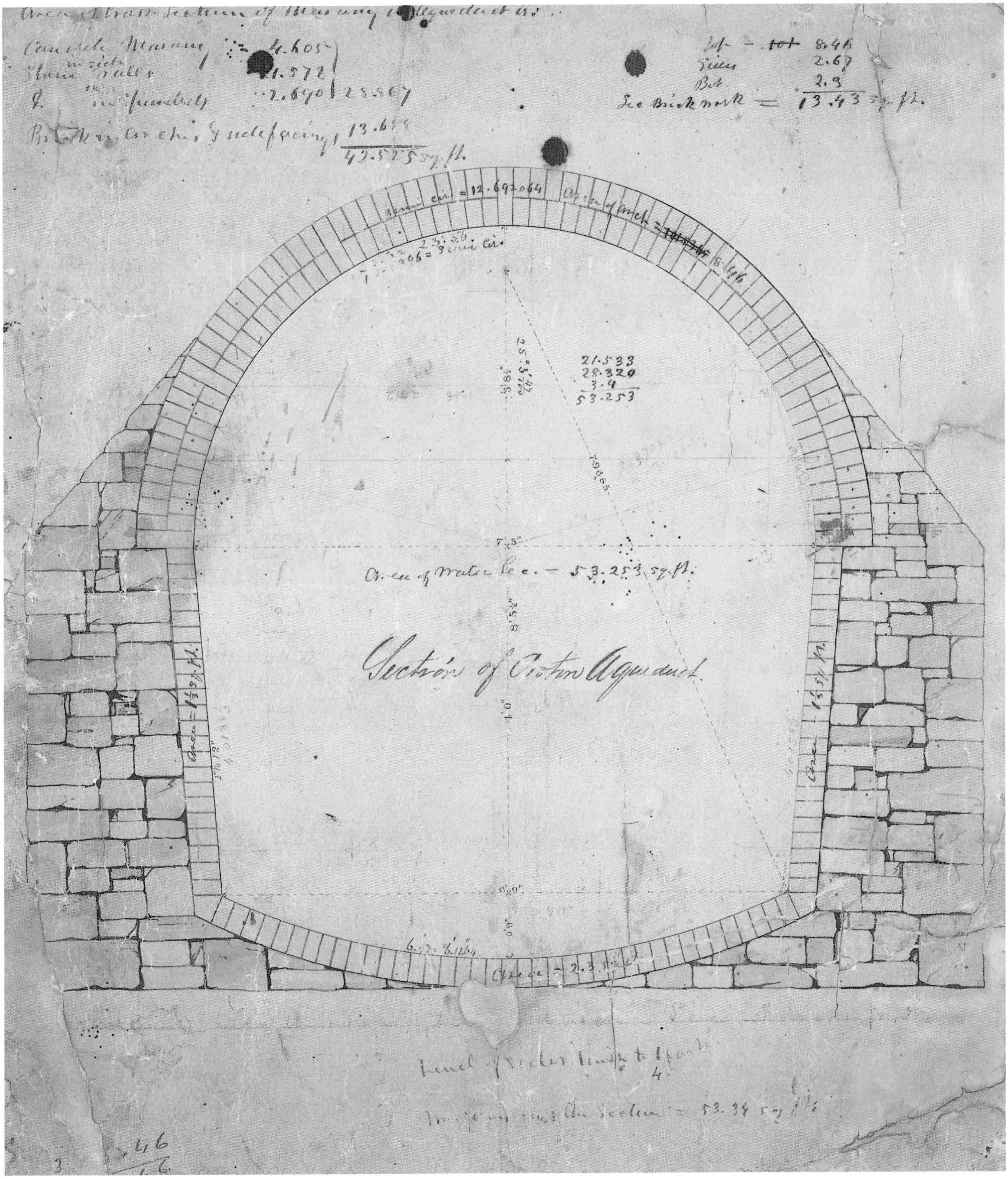
Section of Croton Aqueduct
Area of Water Sec. = 53.253 sq. ft.
21.533
28.320
3.4
53.253
13.43 sq. ft.

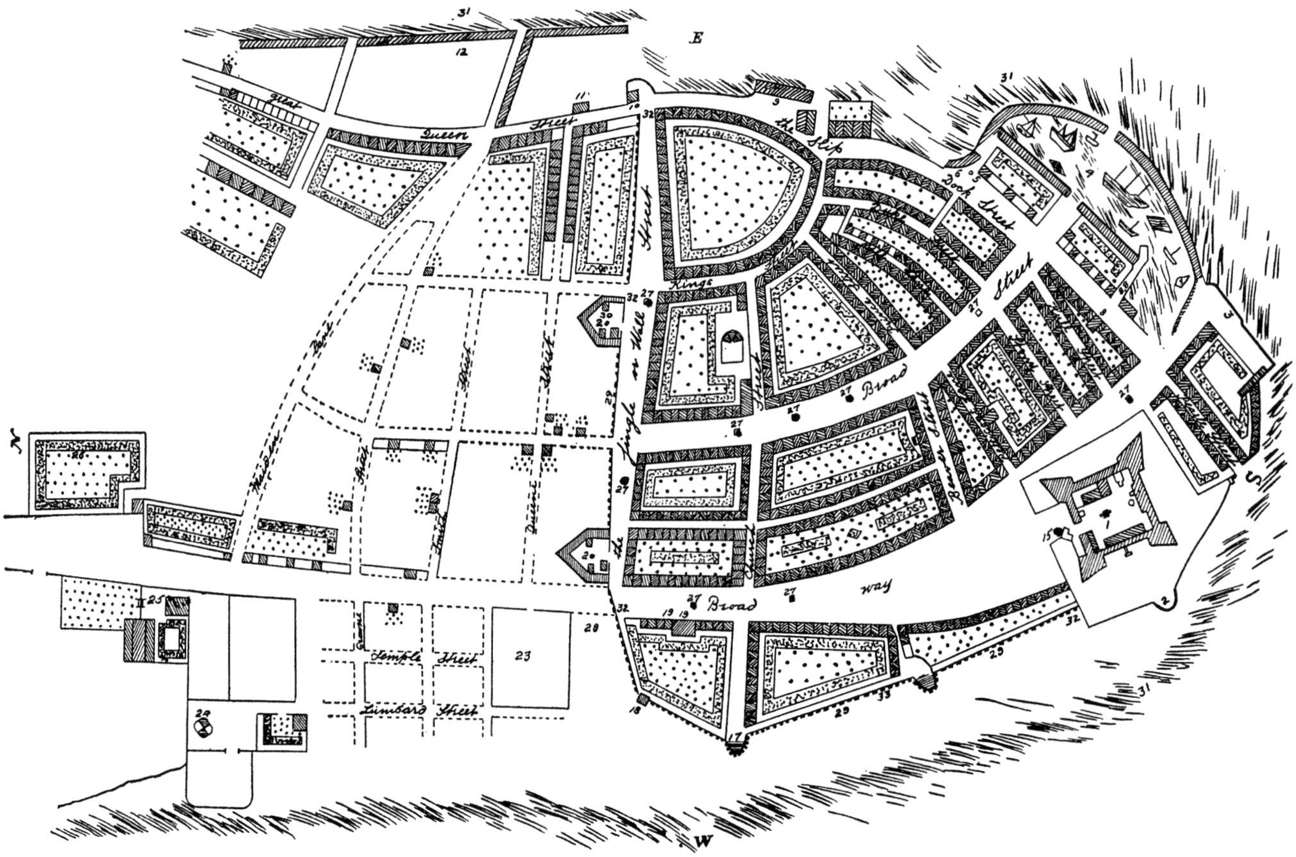

FIG. 2. *Plan of New York, 1695, showing nine public wells below Wall Street. Small dark squares indicate well locations.*

New York's eternal struggle with filth was noted as early as 1697 when Boston physician Benjamin Bullivant took offense to the "nasty & unregarded" streets.[2] Annapolis physician Alexander Hamilton (an intellectual bon vivant, not the later statesman) was the first to record New York's water trouble. "They have very bad water in the city," he observed in his private travel journal during the summer of 1744, after finding excellent and abundant well water in Philadelphia.[3] Swedish botanist Peter Kalm made New York's water a matter of public record a few years later. "There is no good water to be met with in the town itself," he wrote in the widely published account of his North American travels. "This want of good water lies heavy upon the horses of the strangers that come to this place; for they do not like to drink the water from the wells of the town."[4] English clergyman Andrew Burnaby passed the final verdict on colonial New York: "[It] is subject to one great inconvenience, which is the want of fresh water."[5] U.S. Supreme Court Justice and native New Yorker Brockholst Livingston remarked in 1810 on the bodily wastes seeping into wells public and private: "Inhabitants literally in their water [are] drinking a proportion of their own evacuations, as well as that of their horses, cows, dogs, cats and other putrid liquids so plentifully dispensed in the different yards, streets, and alleys."[6] Livingston might have added the noxious leechings from slaughterhouses, tanneries, hatteries, and other industrial waste deposited, along with organic waste, in yards and street gutters. Another generation later, Washington Irving lamented the lack of progress: "It is a pity that so rich and luxurious a city which lavishes countless thousands on curious wines, cannot afford itself wholesome water."[7]

Irving was writing in early August 1832 from the safety of a spa at Saratoga Springs. Having ravaged the world for six years, Asiatic cholera had made its inevitable way to New York, deemed (by former mayor and great diarist Philip Hone) to be a filthier host than any of the cholera-violated cities of Europe and Canada. No one knew that cholera was a bacillus transmitted in waste-polluted water. In 1832 no New Yorker made the medical connection between well water and the waste that infiltrated the wells from the streets and yard privies. Over 3,500 New Yorkers died, nearly one in sixty, the highest cholera death rate among the largest cities of the Western world. One hundred thousand New Yorkers fled, a desertion rivaled only by the patriotic flight from the English occupation during the Revolution. This was no way to run the leading city of the United States.

Wells had figured almost exclusively in the water fortunes and misfortunes of New York for its first two centuries. The Dutch town (confined within a fortified wall that became Wall Street) drank only from private wells, especially those used commercially by a half-dozen brewers. New Amsterdam's wells were shallow and few. The best were lined with wood and had a large bucket suspended from a long sweep. The lack of a well in the fur-trading settlement's fort at the foot of Broadway contributed to Governor Peter Stuyvesant's quick surrender to an English fleet in 1664. There was no way to water his besieged troops, Stuyvesant later explained to New Amsterdam's proprietor, the Dutch West India Company, which found the explanation "very strange."[8]

New York's first water shortage was resolved two years later, when the new English governor sunk a well in the fort "beyond the imagination of the Dutch" who had believed the location unsuited for a well.[9] It was the city's first public well. The second, sunk in 1671 in the rear yard of the Stadt Huys, a tavern turned into City Hall on the East River at Pearl Street and Coenties Slip, was the first stone-lined well. A deep wood-lined well dug in front of the fort in 1689 was fitted five years later with a pump, another first. By then the town boasted nine other stone-lined public wells. They were situated mid-street and systematically ordered, paid for, and maintained. Locations were designated by the Common Council, costs were split between the city and the given neighborhood, and residents were assessed on the basis of their proximity to a local well. Prominent residents were named (and compensated) as caretakers of their neighborhood wells, so that most of these first public wells became identified not by their locations but by the names of their caretakers. The well in Broadway just south of what is now Exchange Place, for example, was known as "Mr. (Francis) Rombout's Well," for the former mayor who owned much of the land along the west side of Broadway. A detailed city plan drawn in 1695 noted thirty landmarks, not the least of which were the public wells in Broadway, Broad Street, Wall Street, Bridge Street, and the fort [Fig. 2]. A hundred and forty years later, hundreds of public wells equipped with brass pumps had been shifted from the center of the streets to curbs and were maintained by the city.

During the generations of water deprivation, there were periods of hope. All revolved around the greatest geographic feature of lower Manhattan: the Fresh Water Pond. Spreading seventy suburban acres across what is now the courthouse neighborhood of Foley Square, the pond in its natural splendor was ringed by wooded hills and fed by clean subterranean springs. Its outlets spanned the island: marshy ground led southeast to a small stream opening onto the East River at what is now Catharine Street, and extensive marshes led northwest along what later became a canal (and then the street of that name), draining to the Hudson River. As the expanding city encroached on this area in the early 1800s, the Fresh Water came to be called the Collect [Figs. 3, 4], a corruption of the Dutch word for *pond*—*kolk*—and disparagingly emblematic of the pond's increasing collection of rubbish and the occasional murder victim. In the 1810s the pond was landfilled into real estate, with Centre Street roughly defining its longitudinal axis. The pond was never directly tapped for drinking water, but its prodigious springs were the source for three water supplies of varying inspiration and efficacy.

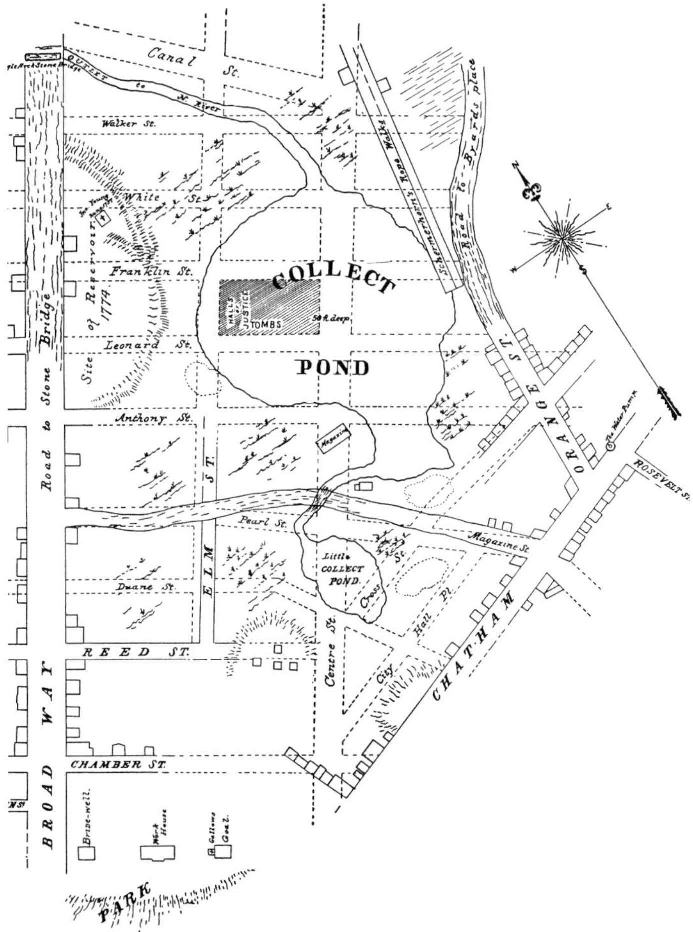

FIG. 3. *Vicinity of Collect Pond showing streets and significant landmarks before and after filling of pond in 1810s.*

FIG. 4. *Collect Pond. c. 1787.*

In the early 1740s, the Tea Water Pump started a six-decade run, ending when the city grew up around it and its sources grew polluted. The pump was located just east of the Fresh Water Pond on a country road that is now Park Row; the exact spot is beneath the grounds of the Chatham Towers apartment complex. The pump was owned by the Hardenbrooks, a family of prominent tanners, and leased to a succession of operators; the well tapped a spring that was for many years deemed the finest for New Yorkers' tea.

The pump got its start after the so-called Great Negro Conspiracy of 1741. The conspiracy was supposedly hatched by slaves and suspect whites at a private well near a Hudson River dock above Wall Street to which leading citizens' slaves were sent to fetch water. New York was ripe for intrigue, due to a combination of economic recession, strained colonial finances—England was warring with Spain—a severe winter, and lingering unsubstantiated rumors about a black plot to poison public wells. The 1741 conspiracy, "exposed" by the spectacular testimony of a sixteen-year-old white servant girl, called for the slaughter of all whites, the burning of the town, and rule by a junta of assorted tavernkeepers, slaves, and prostitutes. Hysterical justice claimed thirty-five lives (all but four black) at the stake and gallows, and dozens of blacks and whites were banished. A new law prohibited blacks from getting water from anywhere but the nearest neighborhood well.

The "conspiracy" turned the Hardenbrooks' suburban well into an entrepreneurial opportunity. Before long, the Tea Water Pump was a concession, its water delivered in casks on horse-drawn carts by city-regulated Tea Water men to customers who could afford the charge. By the early 1770s thousands of households were getting a few pails of water a day for forty-five shillings a year. As fine as the water was, its annual cost equaled a month's earnings for working people, who continued to use the inferior but free street wells. By the early 1800s, the quality of the Tea Water had fallen dramatically and it passed out of use as a regular supply. It was last seen in 1827, trickling from a pipe into a liquor store at what had become 126 Chatham Street.

The inequities and inefficiencies in the water supply of late colonial New York were addressed by Christopher Colles, the first individual clearly identified with New York's water fortunes. Colles hailed from a prominent Anglo-Irish family. Particularly notable is his nephew Abraham Colles, the great surgeon and medical educator of early-nineteenth-century Dublin. Christopher Colles's successes, however, were few. "Had I been brought up to the trade of hatter," he lamented late in a long and ultimately impoverished life, "people would begin to come into the world without heads."[10]

In 1774 Colles proposed New York's first piped water supply, a public enterprise he would both build and superintend. On the outskirts of town he would create a waterworks featuring a deep well, a steam engine, and a large reservoir. A network of hollowed log pipes would be laid through the main streets. Although increasingly distracted by rising revolutionary fervor, the Common Council agreed to the idea and began issuing promissory notes to finance the estimated £18,000 ($45,000) cost [Figs. 5, 6]. That was six times the city's annual revenues and far more than its accumulated debt. The "New-York Water Works" notes were the first paper currency issued by an American city.

Over the next two years, Colles constructed his works on high ground purchased by the city just west of the Fresh Water Pond, at what is now Broadway between Franklin and White Streets. The well was thirty feet wide and twenty-eight feet deep; Samuel Bard, the city's leading scientist, tested its water and deemed it superior even to the Tea Water. Next to the well, Colles built a reservoir that was 165 feet square, with a capacity of two million gallons. Twelve-foot-high vertical interior walls of brick (or stone) were supported by exterior sloping earthen embankments.

The centerpiece of the works was the steam engine that pumped water from the well into the reservoir. At the time there was no working steam engine in America. Twenty years earlier, a large and powerful Newcomen-type engine had been imported in parts from England, where steam engines had long been in use to pump out mines; the imported engine was put into service at the famous Schuyler copper mine in New Jersey but was destroyed by fire early in 1773. Later that year, the newly immigrated Colles designed and built the first American steam engine for a Philadelphia distillery. It was cheaply made and barely worked, but it inspired design improvements for his New York engine. Its major components were cast at the New York Air Furnace, a prominent foundry at the site now occupied by the Woolworth Building.

The completed engine was an immediate and fascinating attraction. Few New Yorkers had seen a steam engine in operation. Workers raised a flag on a high pole at the waterworks before tests. Large crowds flocked up from the town to experience this belching, hissing, clanking harbinger of the industrial future. These tests began in March 1776. Six months later, British forces occupied the city. Most New Yorkers, including the patriotic Colles, fled. The New-York Water Works, a "dangerous" example of American ingenuity, was destroyed by British troops. The log pipe distribution network was never laid.

Colles's waterworks (like all of his engineering and scientific projects over the next four decades) amounted to very little, but they inspired many other would-be water purveyors in the years after the revolution. The Common Council, cautious after the failed first effort and harried by the claims of many Colles contractors, refused to sanction any of the numerous plans similar to his. None advanced beyond written proposals. This changed in 1798.

Yellow fever had been a recurring warm-weather plague in New York since 1702, when the mysterious ailment was called a "malignant distemper" and officially blamed on "our manifold sins immorality & profaneness."[11] By the mid-1700s the disease had taken its modern name from the jaundice its victims suffered, but the suspected causative agent had advanced only to miasmas of bad air rising from swamps and foul standing water. Not until the early 1900s was yellow fever conclusively understood as an acute viral disease transmitted by infected mosquitoes. Eighteenth-century New York, with its surrounding swamps and filthy unsewered streets, was a notorious mosquito breeding ground.

In the summer of 1798, two thousand of the city's sixty thousand residents died from mosquito bites they believed to be nothing more than a nuisance. The wisest survivors recognized that filth must be involved. "The present sickness will subside and soon be forgotten," warned editor and public health advocate Noah Webster, "and men will proceed in the same round of folly and vice . . . piling together buildings, accumulating filth, and destroying fresh air, and preparing new and more abundant materials for pestilence, which will continue to assume greater virulence and to prove more destructive to human life, in proportion to the magnitude of our cities."[12]

Webster's newspapers advocated heavily for New York's cleansing. "Get water into the city," demanded one correspondent. "Take the matter into consideration, and resolve every man for himself, to leave no stone unturned to have this grand object of watering carried thro," urged another.

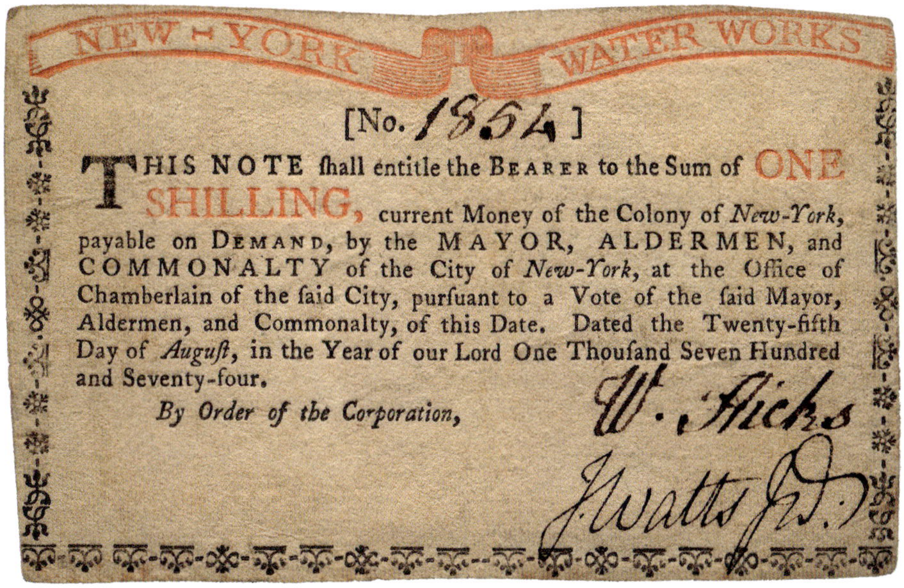
NEW-YORK WATER WORKS

[No. 1854]

THIS NOTE ſhall entitle the BEARER to the Sum of ONE SHILLING, current Money of the Colony of *New-York*, payable on DEMAND, by the MAYOR, ALDERMEN, and COMMONALTY of the City of *New-York*, at the Office of Chamberlain of the ſaid City, purſuant to a Vote of the ſaid Mayor, Aldermen, and Commonalty, of this Date. Dated the Twenty-fifth Day of *Auguſt*, in the Year of our Lord One Thouſand Seven Hundred and Seventy-four.

By Order of the Corporation,

FIG. 5. *Promissory note for the waterworks of 1774–76. Obverse.*

FIG. 6. *Promissory note for the waterworks of 1774–76. Reverse.*

"Then New-York will be as famous as old Rome was, and the other cities may learn from us how to do clean things."[13]

Fever-ravaged New Yorkers were already talking about a cleansing plan offered in July by a Westchester doctor named Joseph Browne. Little did they know what a dirty business it would become.

A surgeon and author of an earlier treatise on yellow fever, Browne proposed that a private company provide water to New York from the pristine Bronx River, which flowed through southernmost Westchester (now part of the Bronx) opposite upper Manhattan. The river would be dammed some fourteen miles from the city and diverted into an open channel to the Harlem River. There, a river-powered pumpworks would raise the Bronx water to a height sufficient for its passage over an embankment across the Harlem River. From upper Manhattan, the water would flow in wooden pipes down to a receiving reservoir five miles north of town, and from there to a distributing reservoir in the city proper. Browne put the cost at $200,000, half of that for a twenty-mile network of distribution pipe. As proposed by Browne, the public-spirited company would guarantee a minimum of 300,000 gallons of water a day; after household allowances of 30 daily gallons, at an average charge of $10 a year, excess supply would be available to the city for street cleaning and firefighting. For investors, Browne calculated a 13 percent return in ten years.

Browne's was the first proposal to water New York from off its island. In that, it was historic. Otherwise, as subsequent developments showed, it was nothing more than a front for the larger designs of Browne's confederate, brother-in-law Aaron Burr.

Finally propelled to action by yellow fever, the Common Council endorsed Browne's Bronx plan but insisted that the city itself, rather than a profit-minded company, do it. The council communicated this desire to its delegation of senators and assemblymen in Albany, where the city's water needs would be decided by the state legislature, meeting in early 1799. The legislature and the council were controlled (though not for much longer) by Federalists; the city's Assembly delegation, however, was headed by Democratic-Republican Burr, who had other ideas.

By the time Burr had completed an extraordinary sequence of manipulations over the ensuing weeks, the legislature chartered a private company as historic as the plan that prompted it. The highlight of Burr's efforts was enlisting political archrival Alexander Hamilton to convince fellow Federalists that the annually elected and perpetually underfunded city government was inadequate to the task of building and managing a water supply. Why Hamilton agreed to get his city's government to change its mind in favor of a private company is an enduring mystery, but Burr was a master at convincing opponents where their best interests lay, and they were invariably his own. In April 1799, "an act for supplying the city of New-York with pure and wholesome water" became law. The Manhattan Company was created, ostensibly to supply New York with water based on Browne's plan, with Burr at the head of the board of directors.

The company was capitalized at $2 million, an unprecedented cash base in those days of few and strictly limited corporations. An obscure clause Burr slipped into the charter late in legislative debate allowed the company to use surplus funds from its water operations for any legal purpose. This was practically revolutionary and soon revealed the true purpose of Browne's water plan and Burr's exploitation of it. Five months after it was created and many months before it supplied any water, the Manhattan Company opened a bank. It immediately thrived as a source of economic and political power for Democratic-Republicans, quickly outpacing the influence of the city's two other banks—the Bank of New York and the local branch of the Bank of the United States—both Federalist-dominated institutions created by Alexander Hamilton. Burr's banking coup is known today as JPMorganChase. To hoodwinked Federalists who had unwittingly supported Burr's endeavor as a water company, the bank was "a greater pestilence than the Yellow Fever."[14] But the water business was even worse.

In order to do banking, the Manhattan Company, under the terms of its charter, had to supply water. It endeavored to do this in as capital-conserving a manner as possible. Instead of throwing money and limited engineering knowledge at the distant Bronx River, the company opted for a

FIG. 7. *The Manhattan Company's Chambers Street Reservoir. 1825.*

waterworks just south of the Fresh Water Pond, by then universally and derisively known as the Collect. In his Bronx proposal of 1798, Joseph Browne had called the Collect a "large stagnating filthy pond" and doubted that the city would "ever seriously think of forcing the inhabitants to drink [its] disgusting water."[15] In 1799, Joseph Browne, appointed at his brother-in-law's behest as the Manhattan Company's superintendent, endeavored to do just that.

Browne oversaw construction of the waterworks between Chambers and Reade Streets. It was built on the Colles model of a quarter-century earlier (the aging Colles actually gave it his blessing) but of dramatically less imagination. A large well was dug, but its water was not submitted to scientific testing. On the brink of the steam age, the company opted for a horse-operated pump to raise the water, initially into a round iron tank 41 feet across and 15 feet deep, set on a 23-foot-high stone foundation. Plans for a grand reservoir were gradually scaled back. Superintendent Browne's initial proposal for a great stone octagon holding a million gallons was replaced by the plan of architect John McComb Jr. (the future builder of today's City Hall) for a 250,000 gallon reservoir. Ultimately, the company board approved a reservoir, likely designed by City Hall codesigner Joseph Mangin, holding a mere 100,000 gallons. New York was then a rapidly expanding city of nearly that many people. The Manhattan Company could not adequately supply them from its tank and its meager reservoir.

The reservoir was not unattractive. It was built of brick and stone on a lot with a fifty-foot front on Chambers Street. Rising twenty feet, the gracefully sloping facade featured a portico of Doric columns surmounted by a statue of Oceanus, recumbent and pouring water from an urn [Fig. 7]. The reservoir was something of a landmark until it was torn down shortly after the Croton Aqueduct came on line in 1842. (The Surrogate's Court building now stands on the site.) The more utilitarian tank had a longer life. Enclosed in a four-story building, it was maintained for generations by the Manhattan Company for fear of losing its charter. During the widening of Centre Street in 1898, the building's eastern wall was torn down, drawing crowds to view the century-old iron tank, into which a bank employee reportedly pumped a bit of water every day, in service to the charter. The tank was tightly reentombed inside a new, curving wall until 1914, when tank and building were torn down and the land was leased to a developer [Figs. 8, 9]. The company's banking business proceeded without incident until the mid-1900s when a merger with the Chase National Bank and a national banking charter legally relieved Aaron Burr's corporate descendant of any water obligations.

After using horsepower to raise its water level, the company eventually moved into the steam age. The company managed to lay some twenty-five miles worth of distribution pipe, initially made of hollowed yellow pine logs of small bore and notoriously leaky. In 1810 the company wanted the city to chop down all its Lombardy poplar trees because their thirsty roots

FIG. 8. *Iron water tank of the Manhattan Company, Centre and Reade Streets, during dismantling. July 14, 1914.*

FIG. 9. *Radial arched masonry supports after removal of iron water tank. July 29, 1914.*

were clogging company pipe; the city responded with a tree protection ordinance. In the 1820s the company replaced some of its wooden pipes with cast iron mains.

For decades after its founding, the Manhattan Company's influential directors and lawyers, mayor and future governor DeWitt Clinton among them, thwarted numerous plans by the city or private interests to develop a proper water supply.

Wary of threats to its banking activities, the Manhattan Company effectively argued that it alone was entitled to supply New York with piped drinking water. In the meantime, these very pipes froze in the winter, clogged in the summer, rarely yielded ample water, and that water never deemed healthy or flavorful. In 1824 botanist David Hosack was among fifteen leading city physicians condemning company water as "highly injurious" to New Yorkers' health.[16] Several years later, a lyrical correspondent to the *New York Evening Journal* derided the "poisonous nature of the pernicious Manhattan water . . . the unpalatableness of this abominable fluid."[17]

If New Yorkers suffered for years with Manhattan Company water, the company disposed of its founder promptly. Having dipped into its bank for $120,000 in unsecured loans—nearly as much as the company had invested in its water operations—Burr was removed from the board in 1802 (when he was Jefferson's vice president). Brother-in-law Joseph Browne was shown the door shortly thereafter. The company's practice of ripping up streets to lay pipe and the bitter negotiations with the city to pay for their repair helped topple Browne from concurrent service as company superintendent and the city's street commissioner. This was certainly one of the era's great conflicts of civic interest.

Yellow fever preyed on New Yorkers for many summers after the Manhattan Company was founded to combat it. Four hundred people died in 1822, the last epidemic. By then, the Collect and its swampy outlets had been filled in for development, serendipitously eliminating a prominent mosquito breeding ground. Ten years later, New York would be ravaged by cholera, the disease that finally prompted a new solution to its water troubles.

In the meantime, the city expanded in length and width, through the island's broad girth between Grand and 14th Streets. Settled areas increased, as did the wood and other combustible structures crowded into them. Increasingly they caught on fire and progressively there was less water to save them. Major fires devastated residential and commercial neighborhoods in 1804, 1811, and 1816; the city suffered an average of twenty fires a year during that period. Spectacular blazes consumed the historic Park Theatre (now the site of J&R Music World) in 1820 and the Bowery Theatre (in today's Chinatown opposite the Confucius Plaza Apartments) in 1828.

It was fire that inspired the city's first public waterworks: not a bold aqueduct from a distant river but a modest works on suburban 13th Street. Begun in 1830, the system, with well, steam-engine, tank, and distributing pipes, is largely forgotten today because the water was undrinkable and used only for firefighting. The 13th Street Reservoir was, however, a great step forward, at least geographically. After the Tea Water Pump and the Colles and Manhattan Company works, the 13th Street Reservoir system was New York's first effort to provide water from somewhere other than the subterranean sources of the Collect Pond. That the pond itself had been filled in nearly twenty years earlier indicates how slowly New York started crawling out from under the Manhattan Company's unfortunate water monopoly.

The 13th Street Reservoir system was the inspiration of Common Council fire committee chairman Samuel Stevens, "Alderman Sam," as he was known, for his long council service. Questioned during council debate about the prospects for water under newly laid 13th Street, Stevens said: "Give us the tank and pipes, and we [will] engage to fill them, if we have to carry the water in quart bottles."[18]

The works were built in 1829 on empty lots purchased by the city on the south side of 13th Street just east of upper Bowery (now Fourth Avenue) [Fig. 10]. Inside an octagonal stone structure, a cast-iron tank, over 40 feet in diameter and 20 feet high, was placed on a 27-foot-high stone base [Fig. 11]. Beside the 250,000 gallon tank, a twelve-horsepower steam engine was installed to raise water from a remarkable well. Dug and blasted over a three-year period, the well eventually reached a depth of 112 feet, mostly through solid rock. The main chamber was 16 feet across, but two broad lateral galleries near the bottom of the well each extended 75 feet, giving a total capacity of 175,000 gallons. The subterranean blasting killed at least three workers, the first examples of that special breed now known as sandhogs to die in the service of New York's public water supply.

By 1833 roughly 12 miles of cast iron mains, with inner diameters from 6 to 20 inches, had been laid. Major north-south lines ran along Hudson Street (and lower 8th Avenue), Broadway, the Bowery, East Broadway, Pearl Street, and William Street. Shorter crossing lines were laid under Waverly Street, Spring Street, Stanton Street, Delancey Street, Grand Street, Canal Street, and Chambers Street. In all, some 150 street hydrants made firefighting water available in portions of all twelve wards out to the city limits at 21st Street.

The partially completed works were officially opened in April 1831. A potentially significant fire several weeks later was put out with 13th Street water, the "most practical evidence of the certain success of this enterprize."[19] Rising 100 feet above what was already high ground, the handsome building presented "a very picturesque object to boats passing through both the East and North [now Hudson] Rivers."[20] This first landmark of the city's public water supply was short-lived. The works were shut down with the advent of the Croton water supply system in 1842, and the distinctive octagonal building and tank were torn down in the early 1850s. The location retains its heritage: for nearly a century, 108 East 13th Street has been a firehouse.

The 13th Street Reservoir system prevented much death and destruction by

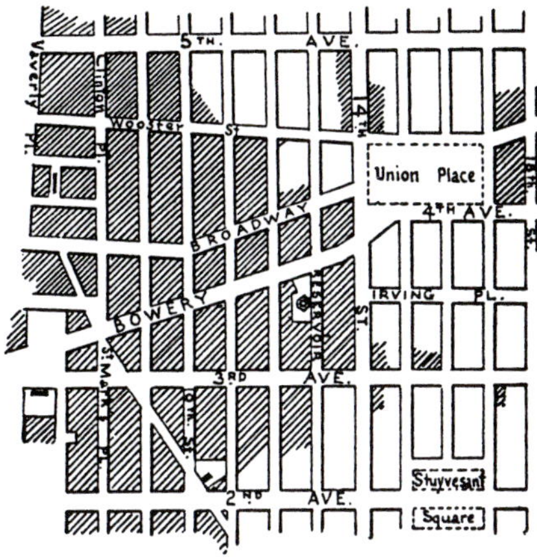

FIG. 10. *Map showing the location of the 13th Street Reservoir between Bowery and Third Avenue in Manhattan.*

FIG. 11. *The octagonal stone structure housing the 13th Street Reservoir.*

fire, but it was unsuited to protecting New Yorkers from epidemic disease, as the cholera epidemic of 1832 quickly proved. Water from local wells, the Manhattan Company, and up-island springs (purveyed, as was the Tea Water, by cart at considerable expense) were insufficient or likely harmful to cholera patients, and no 13th Street hydrants were tapped for health purposes.

The terror of cholera exposed the deficient visions of Stevens, Burr, and Colles before them. The island did not have the water to quench New Yorkers' thirst or cleanse them of disease. The first person to recognize and act on this fact was Myndert Van Schaick, a New Yorker of old Dutch stock. His name appears on no memorials except the New York Marble Cemetery vault where his remains share space with assorted relatives of the Clinton and Hone clans. For the man who brought Croton watershed water to New York City and made both famous for excellent water, his obscurity seems undeserved.

While thousands of prominent citizens died (including Magdelen Bristed, the eldest daughter of John Jacob Astor) or fled, Alderman Van Schaick stayed and witnessed death scenes that horrified him. He (among others) realized that the suffering had something to do with filth and water. Whereas a generation earlier Aaron Burr had manipulated city and state for personal and political benefit after a devastating epidemic, Van Schaick put his efforts to public manipulations for enduring social benefit.

As a Democratic alderman and then a state senator from 1832 to 1835, Van Schaick shaped policy and drafted legislation, guiding its passage and implementation. In 1832 he arranged for a report by DeWitt Clinton Jr., the late governor's son and a rising engineer, which called for the city to tap the distant Croton River.

The Croton River was not unknown to New York water planners. In the early 1820s, the Common Council had hired Erie Canal engineer Canvass White to explore the possibilities of tapping the Bronx, Byram, or Saw Mill rivers; White in turn had hired Westchester surveyor George W. Cartwright to examine the distant Croton River to see if it might be led down to any of these rivers. Cartwright gauged the Croton at 20 million gallons a day—a third of its true capacity—and White discounted it as a tributary source; he also dismissed the Croton as an independent source due to what he perceived to be insurmountable engineering issues. White's report of 1824, like so many before, was filed away and forgotten, until engineer Clinton gave Croton a second look.

The Croton River rises in the low rolling hills of Putnam County and flows in a southwesterly direction, merging with the Hudson River about 40 miles north of New York City. Three main rivulets and many minor streams drain a 375-square-mile area blessed with dependable rainfall (48 to 50 inches per year) and abundant ground water that bubbles up in numerous small springs, brooks, and marshes. Many of the local lakes are fed from underground streams.

Clinton's report proposed that the water be taken directly from the Croton River through a diversion weir to be built near Pine's Bridge, in the lower stretch of the Croton valley about five miles upriver from the confluence with the Hudson River. From the diversion it would head south to New York, flowing in an open canal, much like a simple agricultural irrigation ditch:

> The elevation at Pine's Bridge, by Mr. Cartwright's measurement, is 183 feet above tide. I would propose at this point to sink the bottom of the works below the bed of the stream, to avoid the risk of a dam, and more fully command the whole volume of the water if necessary. Sluices and gates should

be provided and other contrivances to prevent any impurities from the stream passing into the works . . . The elevation of the Croton at Pines Bridge being 183 feet, and the bottom of the work being sunk 6 feet below the bed of the river, it leaves 177 feet; and if the line from that point should descend uniformly 1 1/2 half feet in the mile . . . I have strong confidence in the practicability of delivering it [the water to New York] at 138 feet above tide.[21]

The canal would head west, then south along the southern banks of the Croton River, continuing along the steep and irregular terrain of the east bank of the Hudson. The route would take the water over the valley of Sleepy Hollow and other ravines that drain into the Hudson, turn inland to the east, crossing the Saw Mill River above Yonkers, and then crossing the Harlem River near Macomb's Dam, in the vicinity of what is now Yankee Stadium. (An alternate proposal crossed the Saw Mill River near its mouth at Yonkers and continued directly south to the Harlem River, crossing at the northern extremity of Manhattan Island around Marble Hill at what is now the Broadway Bridge.)

Clinton's vision for the new aqueduct was the first to propose a bridge over the Harlem River, estimated at 138 feet high and 1,000 feet long. Despite objections made by his critics at the time of its release, Clinton's plan was specific, well organized, and based on sound engineering. The overall thoroughness and credibility of his report helped turned the tide of official opinion toward the Croton as the preferable source. Clinton closed his report by calling for minute and careful surveys to help establish and refine the aqueduct line and determine the actual construction requirements.

In 1833 Van Schaick's "pioneer law to the Croton" wisely shifted planning authority from the overmatched city government to five state-appointed commissioners.[22] Van Schaick named the panel, all prominent Manhattanites and good Democrats, led by former Mayor Stephen Allen, a sailmaker turned civic booster. At Van Schaick's behest, the commissioners named military and civil engineer David Bates Douglass to conduct surveys. The West Point graduate and professor had earlier been hired to design the first building for the new college on Washington Square (now New York University) of which Van Schaick was a founder.

While Clinton had proposed a straight diversion from the river, Douglass introduced the idea of building a dam and an associated storage reservoir. The Romans had utilized water storage facilities for their great aqueducts. That concept would come to prevail in New York. The Croton River was reliable, but there would certainly be sustained dry times when the river yield could prove inadequate. A dam could retain a reserve source and get the city through the inevitable droughts impacting the Croton basin. The position of the dam on the river would affect the capacity for collection and storage. If the dam was located too far upriver, the amount of water available in the watershed would be reduced, as many of the watershed streams would enter the river downstream of the dam. But if the dam was located too far downstream the elevation at the reservoir would be too low and could limit the effective ability to deliver and distribute the water by gravity. A balance that would provide an advantage in establishing proper flow would have to be sought.

Van Schaick's 1834 legislation, "An Act to Provide for Supplying the City of New York with Pure and Wholesome Water," gave the commissioners complete authority over all aspects of aqueduct construction and funding (through bond issues), pending approval of their final plan by the Common Council and a public referendum. "Was it not prudent," Van Schaick explained, when the successfully completed project reached its eventual $13 million cost—nearly double the outlay for the Erie Canal twenty years earlier—"that the people should be pledged for the payment of the debt by their own act and deed and by their favorite rule of the majority?"[23]

In April 1835, New Yorkers (that is, those adult males who were entitled to vote) went to the polls to give the Croton project their blessing. Opposition was fierce from spring water suppliers, bored-well drilling advocates, the Manhattan Company (via politically safe anonymous pamphlets), and tax-wary rich and poor alike. Westchester landowners (many of them city dwellers as well) protested the extraterritorial invasion of their peaceable county by a city grown too large to water itself. Despite robust anti-Croton electioneering, memories of the 1832 cholera had been revived by a 1834 recurrence, and the vote came down 17,330 to 5,963 in favor of the plan to claim the Croton's water. Only three of the city's fifteen wards had majorities against the bill: one comprising the formerly separate village of Greenwich which still had good wells and clear springs, and two in the sunken neighborhoods of what is now the Lower East Side, which had the worst wells and the least money but the greatest fear of change.

Lingering opposition went up in smoke eight months later. On the bitter night of December 16, a fire was touched off in a Merchant (now Beaver) Street warehouse. Fanned by a gale, it consumed seven hundred buildings from south and east of Wall Street and Broadway [Fig. 12]. The thermometer read below zero, but the intensity of the flames liquefied copper roofs. The 13th Street Reservoir system, depleted by fires earlier in the week, trickled and froze in its hydrants. Wells and scattered cisterns remained frozen solid. Manhattan Company water goes unmentioned in any account. Property losses from the Great Fire of 1835 were more than $20 million, roughly 10 percent of the city's property value. While the commercial center of the nation continued to smolder three days later, Stephen Allen was named to oversee the rebuilding effort. Allen was already managing the Croton project that would help protect the city from future disasters. But he was growing impatient with the slow pace of his chief engineer's study.

The quantity of water to be delivered was based upon two main assumptions: that the system would ultimately serve 450,000 people (at the time of planning, the population was 225,000), and that each user would require 20 gallons per day.

FIG. 12. *Burning of the Merchants' Exchange. December 16–17, 1835.*

Unlike the aqueducts of the ancient world that often passed through arid or semi-arid lands, the Croton Aqueduct would have to traverse a complex and craggy landform marked by springs, creeks, and deep-cut draws that flooded with every rainfall. To some degree the aqueduct could follow the contours of the land, but in many places ravines would have to be.crossed with built-up embankments or open bridges. These crossings would add substantially to the sheer volume of construction for the conduit. Each creek would be routed into a culvert of stone, some one hundred feet long and deep below the hydraulic line of the enclosed structure.

Douglass studied and suggested two alternate routes. He developed specific structural ideas for each of these routes but ultimately settled on a plan where water was to be collected at numerous places in the branches and tributaries of the Croton watershed and taken via iron pipes to a small one acre confluent reservoir at Mechanicsville. The elevation of this basin would be 270 feet above the Hudson River at low tide, a significantly higher point of origin than had previously been proposed. This plan confronted various topological obstacles in moving the water directly south through the high rocky ground that separated the Croton River drainage from the Saw Mill River basin. The route would have required deep cutting and hard-rock tunneling through the rocky ridges, but once through the high ground, the aqueduct could follow a shorter route down the gentle terrain of the Saw Mill River valley. The construction characteristics proposed by Douglass were very close to those ultimately used.

Douglass developed a horseshoe-shaped cross section that had an inverted arch at the base, outward battered masonry walls on the side, and a brick arch cover (or wood roof) over the channel [Fig. 13]. The larger aqueduct section would be four feet wide at the base arch, and the smaller section would be two feet wide at the base arch. Clinton's original plans for the aqueduct proposed an open-cut trench type conduit, but concerns about contamination, freezing and evaporation, and maintenance prompted the decision to enclose the entire aqueduct with a masonry arch. The most typical and characteristic cross section was a brick-lined stone trench with slightly battered

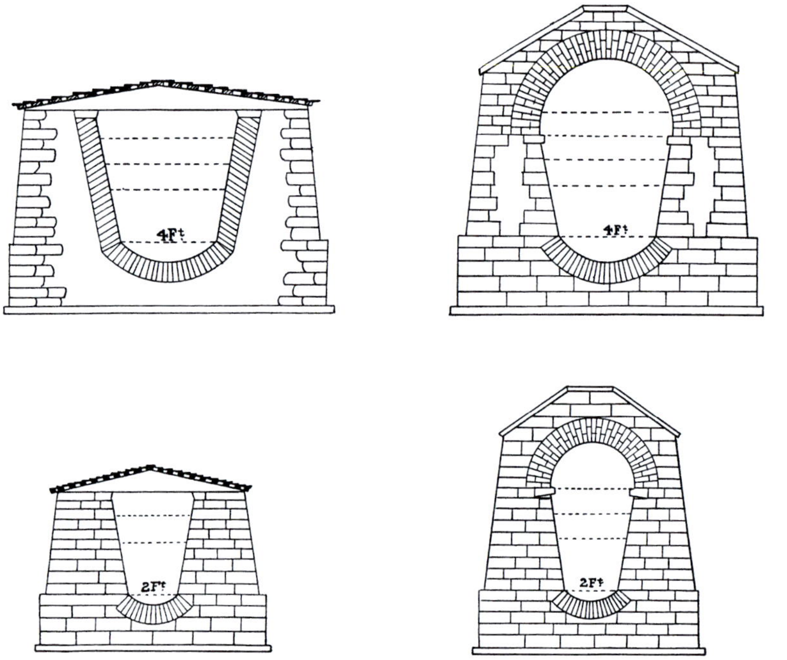

FIG. 13. *Various sections of the first Croton Aqueduct as proposed by Major David Bates Douglass. c. 1835.*

walls separated on the bottom by an inverted, nearly flat arch, and covered above by a semicircular brick arch, made of two thicknesses of brick, measuring 7 feet 8 inches in diameter. This masonry conduit would then be buried in an earthen berm, or would be set into a trenched cut and then covered. In addition to this typical construction, there were several masonry conduits developed to accommodate various geological conditions. These alternates included a stone arch cover and an irregular ovoid cross section that was used in rock tunneling [Figs. 14, 15].

By the fall of 1836, Douglass had completed four years of surveys and design but still hesitated on breaking ground. After increasingly uncivil quarreling, Allen replaced Douglass with John Jervis, an upstate farm boy turned valedictorian of the so-called Erie School of Engineering.

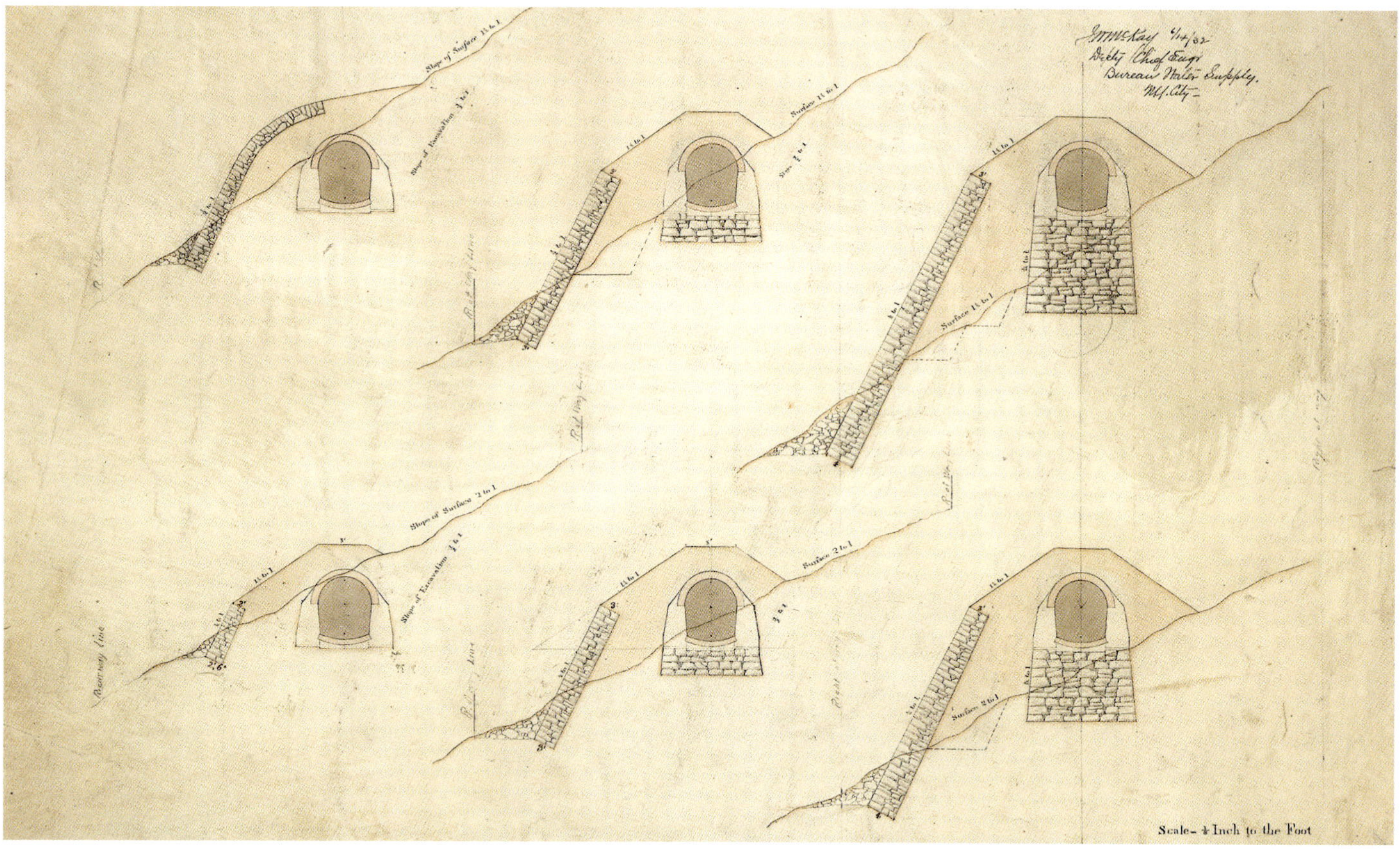

FIG. 14. *Old Croton Aqueduct. Cross sections showing different methods of construction on steep hillsides, inscribed: "This drawing probably made about 1836."*

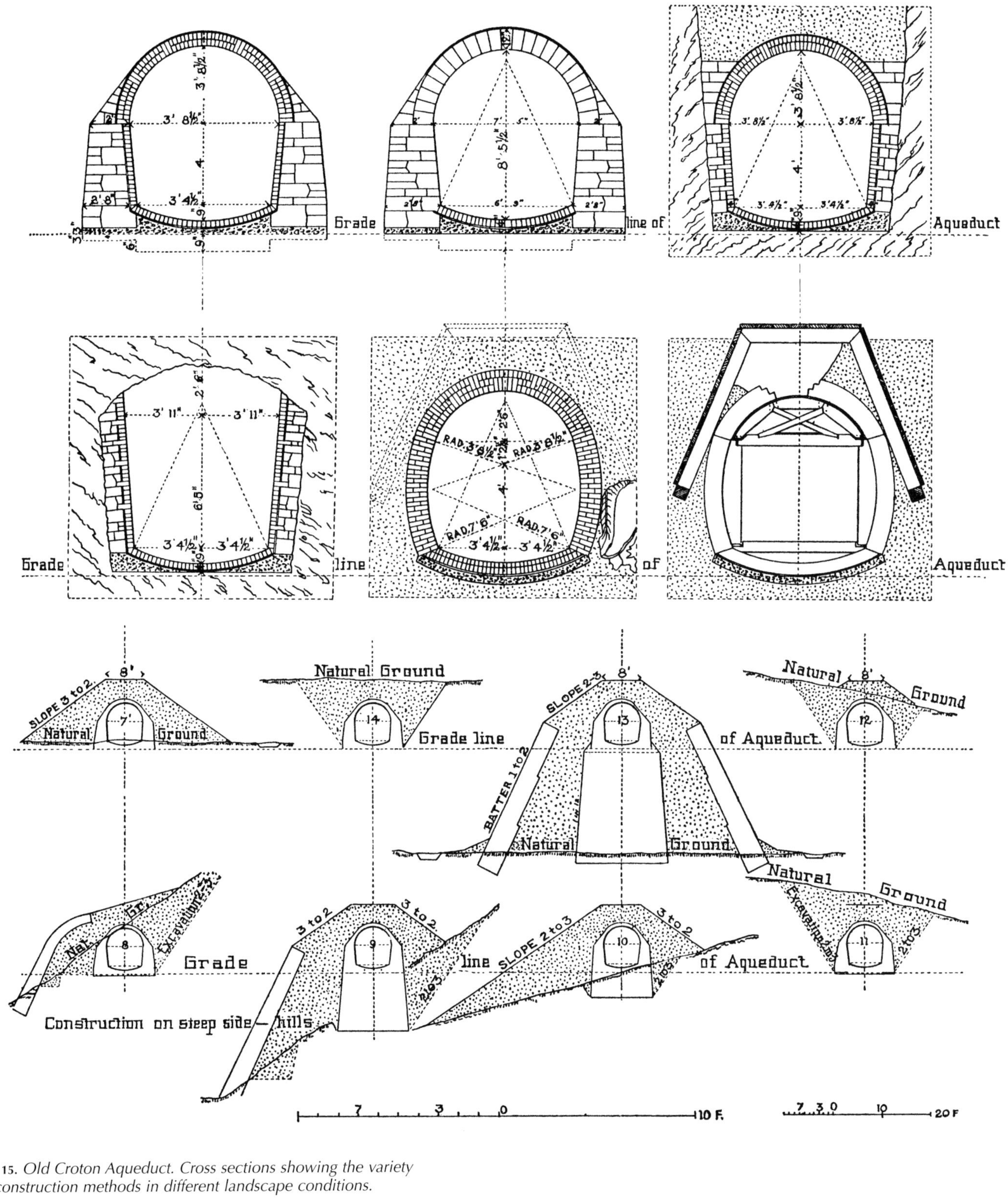

FIG. 15. *Old Croton Aqueduct. Cross sections showing the variety of construction methods in different landscape conditions.*

Whig politicians cried foul: Douglass was a Whig and Jervis a Democrat working on a project run by his political brethren. But Douglass was proving himself a conceptual engineer ("a ripe scholar . . . *and in theory,* well acquainted with the science of engineering" in Allen's public estimation),[24] while Jervis was a proven project engineer with a lengthening resume of railroad and canal work that began with his rise from Erie Canal axeman to chief. In the end, Douglass largely designed the Croton Aqueduct and Jervis built it, with some substantial design modifications and innovations along the way.

As conceived by Douglass and executed by Jervis, New York's great aqueduct started with a 50-foot-high dam 6 miles from the Croton River's junction with the Hudson, making a long natural reservoir of the rugged Croton River valley. Jervis shifted the dam location slightly downriver from where Douglass had staked it, but then paid the price. Heavy snow and rain in early January 1841 caused a flood that breached the mostly completed dam, killing one worker there and two men at a downstream wire mill destroyed by the rolling wall of water, ice, mud, boulders, and uprooted trees. Navigation on the lower Croton was permanently ruined; lawsuits continued for many years. But Jervis ingeniously redesigned the dam, with a stilling basin and a novel S-shaped face that minimized the destructive force of overflowing water and became a standard for American dams [Figs. 16, 17]. Jervis's dam survives, submerged for over a century now in the deeper reservoir created by the towering New Croton Dam.

The original plan for the dam was to have 25 percent of the length constructed as masonry, and the remaining 75 percent as earth embankment. As redesigned after the flood, the dam had an additional 180 feet of masonry. The river bottom was cleared of mud and boulders to reach a stable layer of riverbed. Timber cribs filled with boulders created the aprons upon which the dam rose. The cribs were a combination of elm, pine, and oak, with joints dovetailed and secured with wooden spikes of white oak (treenails), and capped with pine planks. This series of substructures was the armature of the dam. On these, rows of hydraulic stonework were keyed together to resist the great forces of restrained water. The face stone was granite. The profile of the dam allowed water to flow evenly over almost the entire upper surface and not through a restricted weir. On the downstream side of the dam, the stilling basin, or "dead water" pool, was created by a low secondary dam. The pool created a counterforce to the weight of the water backed up by the main dam and minimized the destructive force of the water spilling over the dam.

From the reservoir created by Jervis' dam, diverted river water began a 33-mile run by gravity to the Harlem River. Through the undulating hills and valleys of the lower Hudson Valley, Douglass had painstakingly determined the aqueduct's

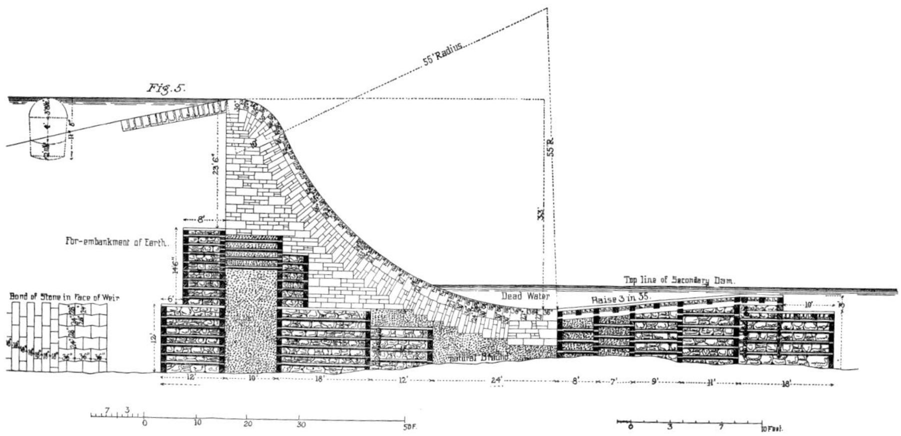

FIG. 16. *Old Croton Dam. Cross section.*

FIG. 17. *Water cresting over the Old Croton Dam. The gatehouse for the New Croton Aqueduct is on the hill at right. c. 1900.*

most efficient route to maintain the required descent of 13 inches per mile. In places, the aqueduct's underground course ran up to 30 feet deep in varied rocks and soils; in others, it passed in dozens of culverts and larger structures over rivers, streams, and hollows [Fig. 19]. Throughout, the basic dimensions were retained: the horseshoe-shaped conduit was just under 7½ feet wide and 8½ feet high. The bottom was a flat inverted arch and the straight sidewalls broadened slightly toward the rounded arch top. (In tunnels passing through rock, the conduit roof was formed by the natural rock; in tunnels deep through earth, the sidewalls were round.)

Jervis acknowledged Douglass' sensible route location and made only small refinements to his conduit dimensions and specifications, but Jervis designed most of the major aqueduct structures along the Westchester route, including the spectacular arched crossing of Sing Sing Kill, which is still standing in what is now Ossining, and the long embankment over the Mill (now Pocantico) River near Sleepy Hollow [Figs. 20, 21]. At the Harlem River, which separates the mainland from upper Manhattan, Jervis nearly met his match.

Douglass, with a taste for the magnificent, had suggested, but not designed, a grade-level crossing of the Harlem River on towering Roman arches, meeting Manhattan at what became 174th Street. At more than 1,400 feet long and 140 feet high, the scale of the bridge would have been unprecedented in America. Jervis, whose tastes ran toward the frugal, designed a pipe siphon on a relatively inexpensive low embankment, with one low arched passage for boats. This plan pleased the cost-conscious commissioners but quickly ran afoul of other interests.

Douglass defenders alleged that Jervis was afraid to build a great bridge. Real estate interests on both shores of the Harlem saw rising values in a high bridge. Shipping interests were outraged at the permanent obstruction to navigation posed by a low arch. And New Yorkers generally wondered why their fine aqueduct should come slinking onto the bold

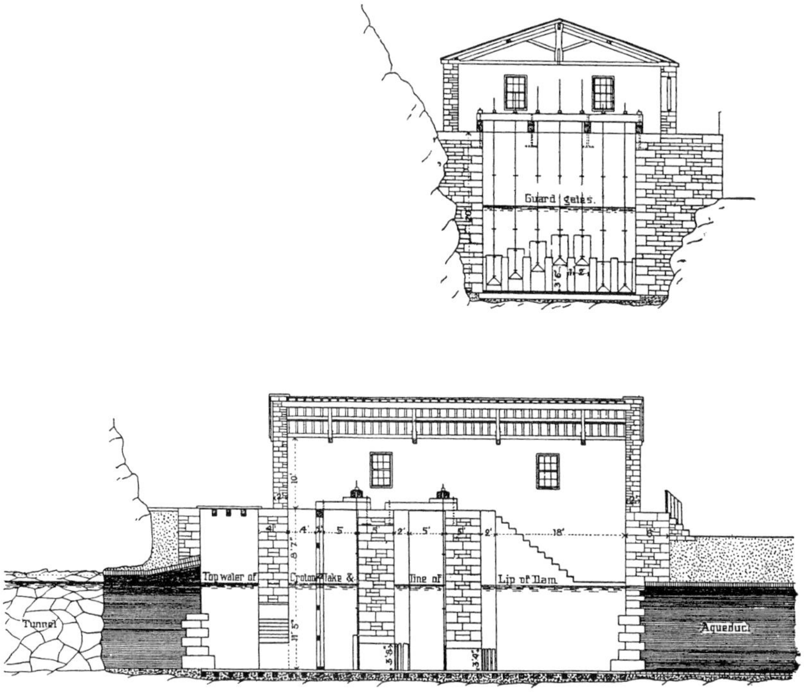

FIG. 18. *Old Croton Aqueduct. Inlet gatehouse.*

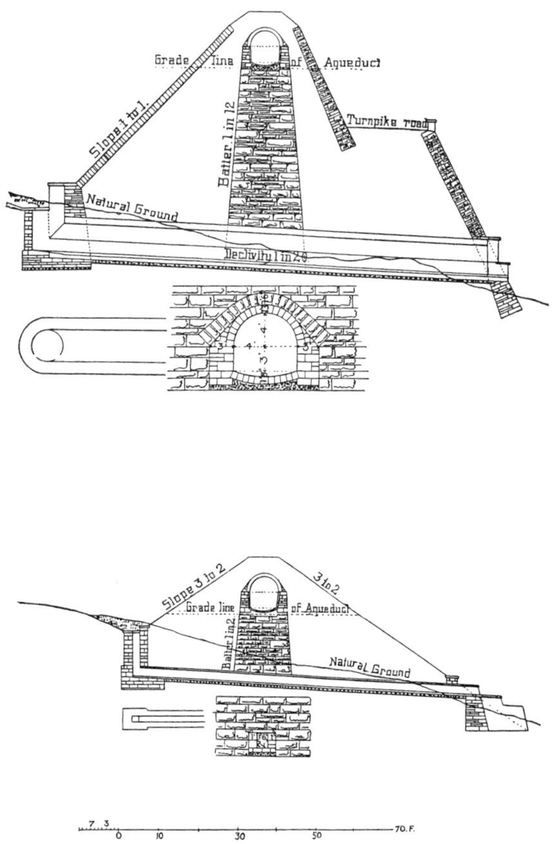

FIG. 19. *Old Croton Aqueduct. Cross sections of culverts.*

highlands of upper Manhattan in a style suited for lesser cities.

Passions became enflamed, especially those of Lewis Gouverneur Morris, grandnephew of the late city grid planner. In September 1838, the young heir to the Morrisania estate on the Harlem River led a strange naval assault on Macomb's Dam Bridge, which had been thrown across the Harlem two decades earlier without a state-mandated draw for river traffic. The dam portion of the wooden bridge, a mile downriver from the proposed Croton crossing, succumbed quickly to boatloads of axe-wielding mariners. The Harlem River was once again open for navigation, especially by Morris and other Harlem River estate owners, and the cause of Harlem navigation was dramatically improved: without Macomb's Dam Bridge, the argument for a river-blocking aqueduct embankment lost support (state courts eventually vindicated Morris for abating "a public nuisance;"[25] the third and current Macomb's Dam Bridge dates from 1895). The legislature soon ordered Jervis to choose between a high bridge or a tunnel.

Jervis knew that the effort to build the world's first underwater tunnel, Marc Brunel's Thames Tunnel in London, was only just coming to completion after fifteen years. The project had caused dozens of deaths and incapacitating injuries, and ran 400 percent over budget. Jervis offered a Harlem River tunnel plan but favored his plan for a multi-arched bridge with an Albany-mandated hundred-foot minimum clearance over the river: a high bridge but one that came in slightly below grade level using a cheaper pipe siphon instead of expensive additional masonry.

The legislature happily opted for the high bridge, a resolution not necessarily unwanted by the original low-bridge advocate: "It was natural," Jervis wrote in his memoirs, "that an engineer should incline to a work that would give prominence to professional character as a work of art."[26]

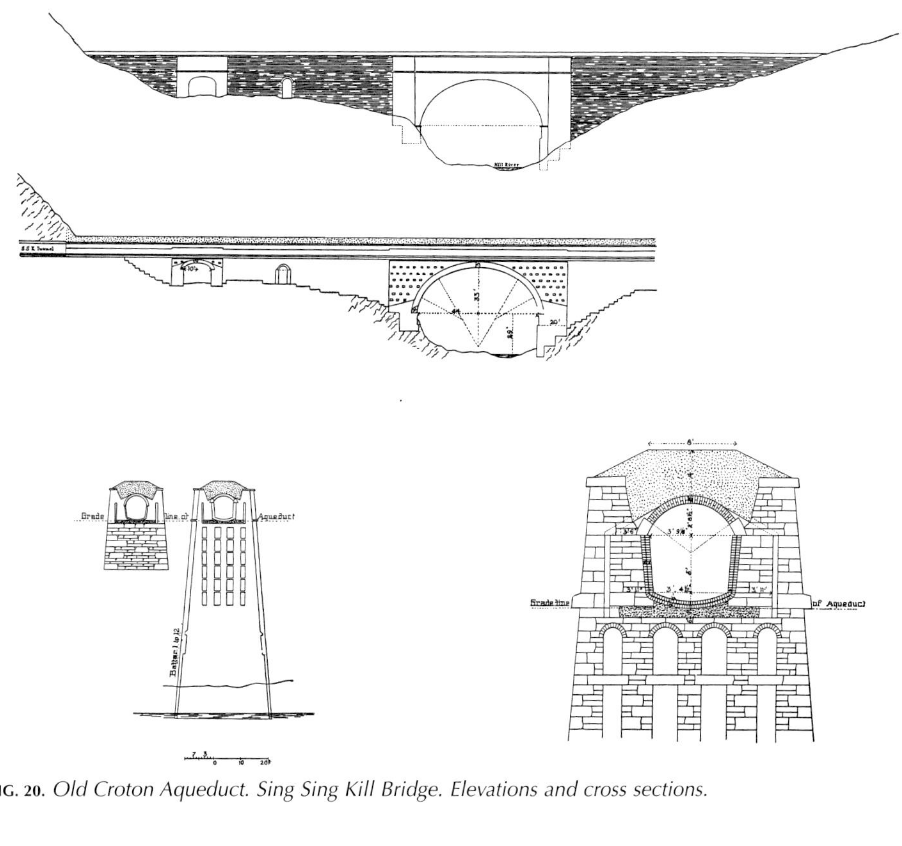

FIG. 20. *Old Croton Aqueduct. Sing Sing Kill Bridge. Elevations and cross sections.*

Not all on Jervis' staff agreed. Twenty-one-year-old assistant engineer Fayette Tower privately condemned a costly high bridge mandated "just for architectural beauty in a place where there is little necesity [sic] for it."[27] Tower soon learned the role of business and politics in engineering and went on to document the Croton Aqueduct in a book of exquisite engravings, including his panoramic view of the bridge that originally offended his youthful idealism.

Croton was built on the organizational model established for the Erie project twenty years earlier: a state-appointed commission hired the engineers who conducted the surveys, prepared the work for competitive bidding in small sections, and oversaw the contractors' progress. The construction had one hundred sections, each averaging less than a half-mile in length. As with the pioneering Erie work, the construction contracts for many sections were won by local farmers or landowners knowledgeable about their particular area, but a substantial number of sections were undertaken by associations of workers who had gained experience on canal projects elsewhere.

The lead contractor of the High Bridge, for example, was an aspiring capitalist named George Law. In the fifteen years after leaving the struggling upstate farm established by his Irish immigrant father, Law had risen from laborer on various American canal projects to contract winner on several small Pennsylvania canals. Having obtained some wealth and a wellborn wife, he moved to New York City to bid on Croton contracts. After winning the contract for the embankment over the Mill River (where he befriended a loquacious local whom he failed to recognize as Washington Irving), Law joined three other veteran Croton contractors to win the High Bridge job with a $755,000 bid. "[A] self-made man [who] worships his creator" and imbued with "a talent for making money out of other people,"[28] Law was soon a Fifth Avenue millionaire, a financier, and owner of steamship, ferry, and railroad lines ranging from New York to California and Panama. He was also an early challenger

FIG. 21. *Old Croton Aqueduct. Ossining Arch (Sing Sing Kill Bridge). July 9, 1913.*

for the White House in 1856 as a Know-Nothing Party candidate. Coarse and reactionary, Law was not popular among New York's established elite, but the Croton project had launched him into the city's broadening upper-middle class.

Most of the workers who labored for Law and other Croton contractors were fresh Irish immigrants. They had fled economic collapse in Ireland only to find the same crisis in America, where the Panic of 1837 (spawned by a burst bubble in speculative canal stocks) was spiraling into a six-year national depression. Croton, financed with $12 million in state bonds issues, which sold initially at impressive premiums but soon at deep discounts, was nevertheless a timely public works project that employed thousands of the otherwise jobless. In good seasons, unskilled Croton laborers happily earned a dollar a day. This was considerably more than the average wage for unskilled labor in the city, and relatively in line with the $75 monthly salary for first assistant engineers. In bad seasons, though, such as 1839 when the commissioners had to pay contractors in discounted city bonds instead of cash, daily wages for laborers dropped to seventy-five cents.

Regardless of the season, workers were always at risk of injury or death. Because workers were hired by contractors, and

FIG. 22. *High Bridge. Cross sections and partial elevation showing cofferdams and centering for arch construction. c. 1845.*

FIG. 23. *High Bridge. View from mainland, looking up Harlem River.*

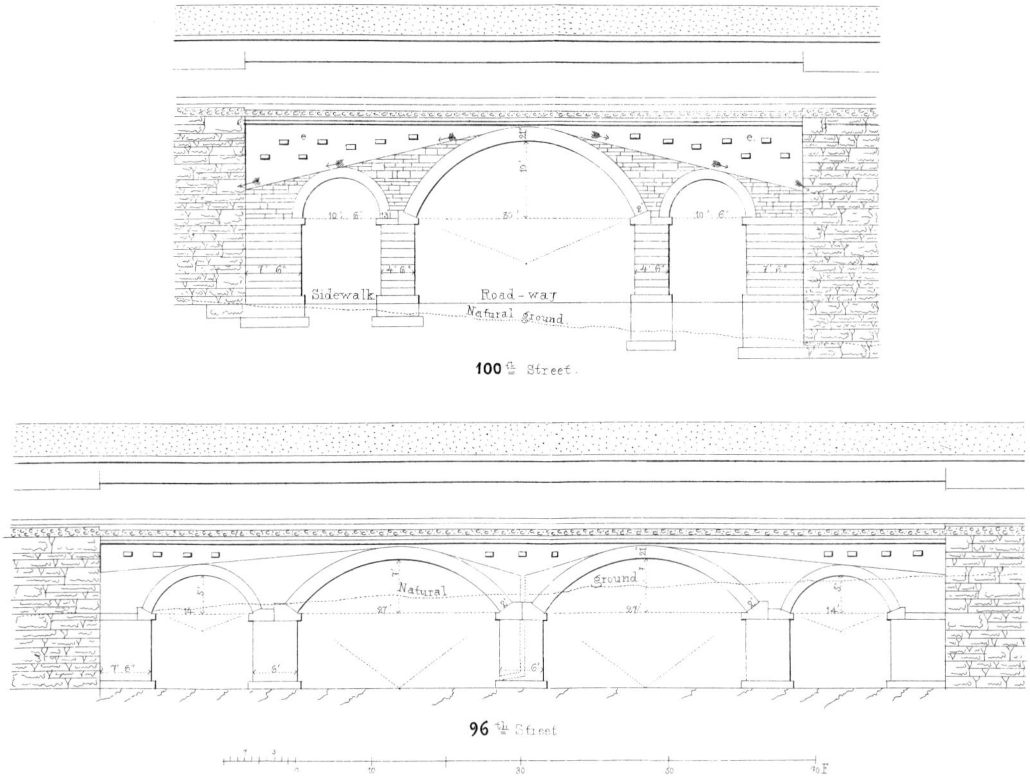

FIG. 24. *Sections of elevated aqueduct through the Clendening Valley showing arched passageways for roads and sidewalks.*

governmental oversight was generations in the future, there is no record of Croton casualties other than the occasional reporting in local papers of particularly gruesome incidents. The imprecise science of rock blasting, especially for the aqueduct's sixteen tunnels, earned the most ink and the least sympathy. On a section near Dobbs Ferry in January 1838, the "unfortunate" Patrick Carr was "negligent of the precaution of keeping a sufficient quantity of water in the hole, and when a part of the covering of the charge was removed, a hissing noise was heard, and in the twinkling of an eye" Carr was blown to bits.[29]

Unsafe on the job, Croton workers were never fully at peace with each other or their employers. Southern ("corkonians") and northern ("fardowns") Irish factions skirmished regularly (and sometimes fatally) along the line, fueled by illicit whiskey consumption. On April 1, 1840, when it became apparent that the prevailing lower winter wage would be continued for the summer season, workers united in full-scale riot. After beating or driving off higher-paid masons, carpenters, and other skilled workers, as well as otherwise content laborers, a thousand Westchester workers—equal to the entire winter crew and a quarter of the usual summer force—commandeered work boats at the Harlem River and invaded upper Manhattan, scattering work crews there. "The Croton War"[30] ended bloodlessly several days later when the city militia, mustered for the first time in long memory, rode north to find the rioters had already dispersed. But spontaneous strikes and violence persisted along the entire line through the month, until it became clear that there were many unemployed laborers eager for work at depression wages.

National economic malaise did not significantly affect progress on the Croton because its bonds, though deflated, did continue to sell. By 1840 much of the work in Westchester was completed, or nearly so, and construction in Manhattan was moving forward.

The piers to support the fifteen arches of the High Bridge slowly rose from the bed of the Harlem River and the low Westchester plain. A pipe siphon on a low embankment would carry Croton water across the Harlem until the bridge's completion in 1848 [Figs. 22, 23].

From the Harlem River, the aqueduct ran underground (in brick conduit, cast iron pipe, and two tunnels) for four miles toward, and then down, the future line of Amsterdam Avenue and, after a gentle S-turn, to just west of today's Columbus Avenue, surfacing dramatically at what is now 102nd Street to cross the Clendening Valley.

Perceived today as merely a shallow dip in the Upper West Side, the valley was the Clendening family farm. By late 1840, a 50-foot-high, 1,900-foot-long arched stone wall supporting a section of brick conduit, lined with iron to prevent seepage damage to the stonework, rose from the bucolic valley [Fig. 24]. The nine arches—three which are 27-feet wide, and

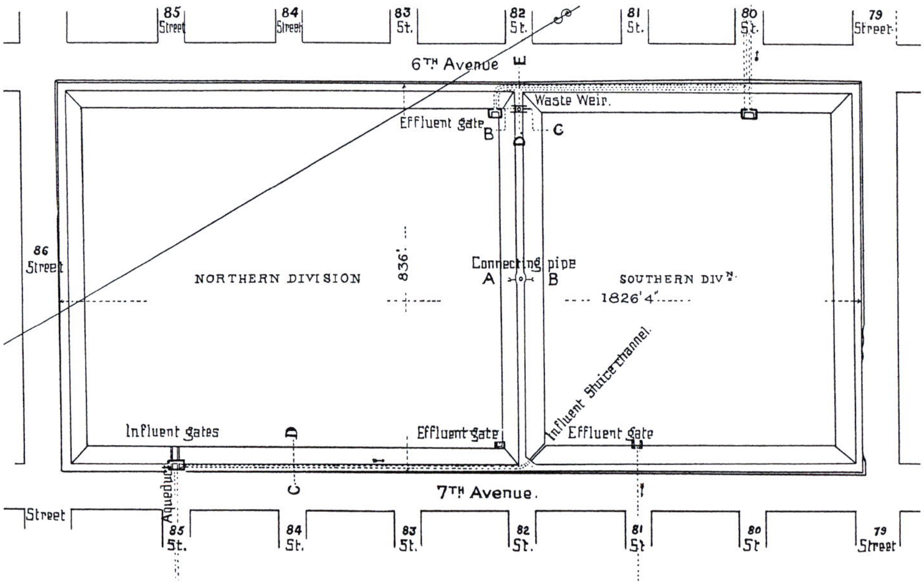

FIG. 25. *York Hill Receiving Reservoir. Plan with intended street layout (before implementation of Central Park plan). c. 1839.*

FIG. 26. *York Hill Receiving Reservoir. Section of effluent pipes and vaults for the north basin. c. 1839.*

flanked by 14-foot-wide arches—represent a bit of city governmental history. When the Whigs gained control of state government in 1840 (for two years), they impaneled a slate of cost-cutting water commissioners (led by old alderman Samuel Stevens) who won a vote by the politically divided Common Council to scrap the planned Clendening arches in favor of a cheaper solid wall. Democrat Isaac Varian responded with the first veto by a New York mayor. Varian explained that not only was the council order beyond the council's authority and a violation of existing contracts, but the unbroken wall it mandated would be a barrier to development. And so a Solomonic compromise was made: a solid wall would rise at the northern and southern ends, comprising half the valley crossing's length, and the partially constructed nine-arched midsection would be completed. Within twenty years, the Clendening farm was gone, and the streets and sidewalks of 98th, 99th, and 100th Streets passed beneath those arches. In the 1870s, development demanded more: the aqueduct section was shifted underground into a

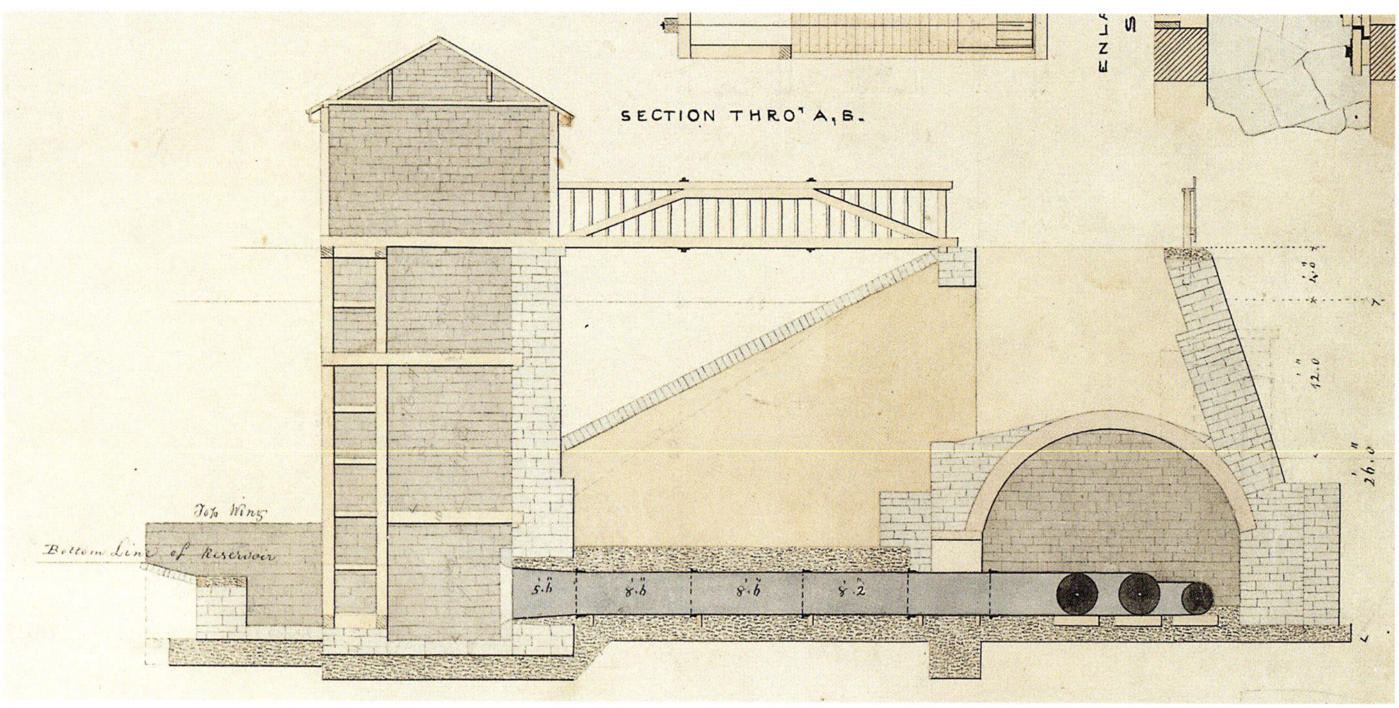

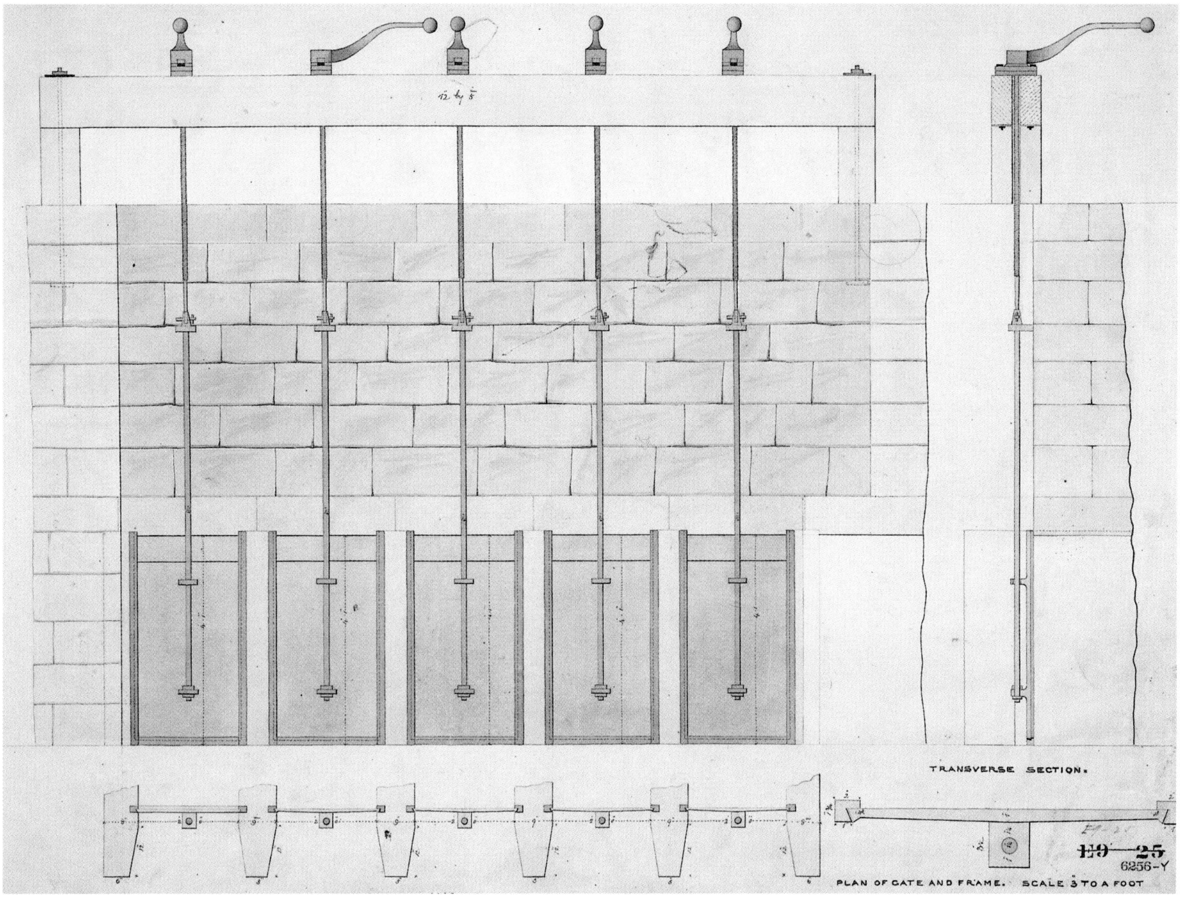

FIG. 27. *Old Croton Aqueduct. Sluice gates at termination of aqueduct at York Hill Receiving Reservoir. c. 1839.*

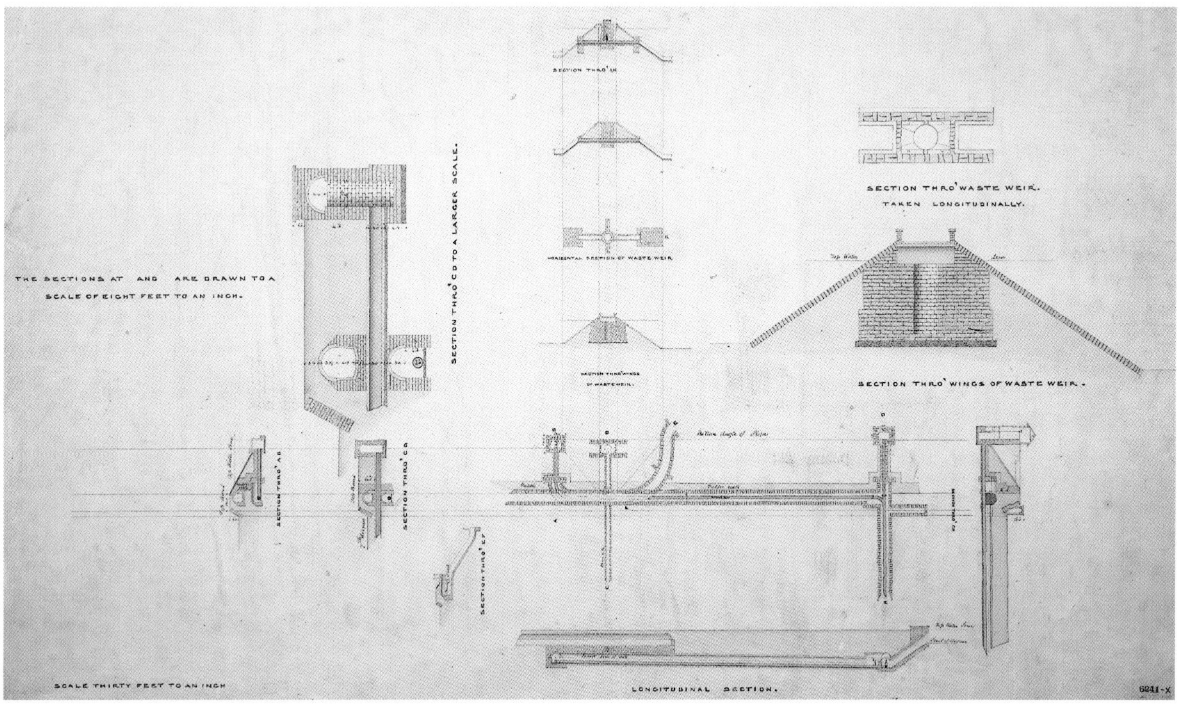

FIG. 28. *York Hill Receiving Reservoir. Sections through waste weir. c. 1839.*

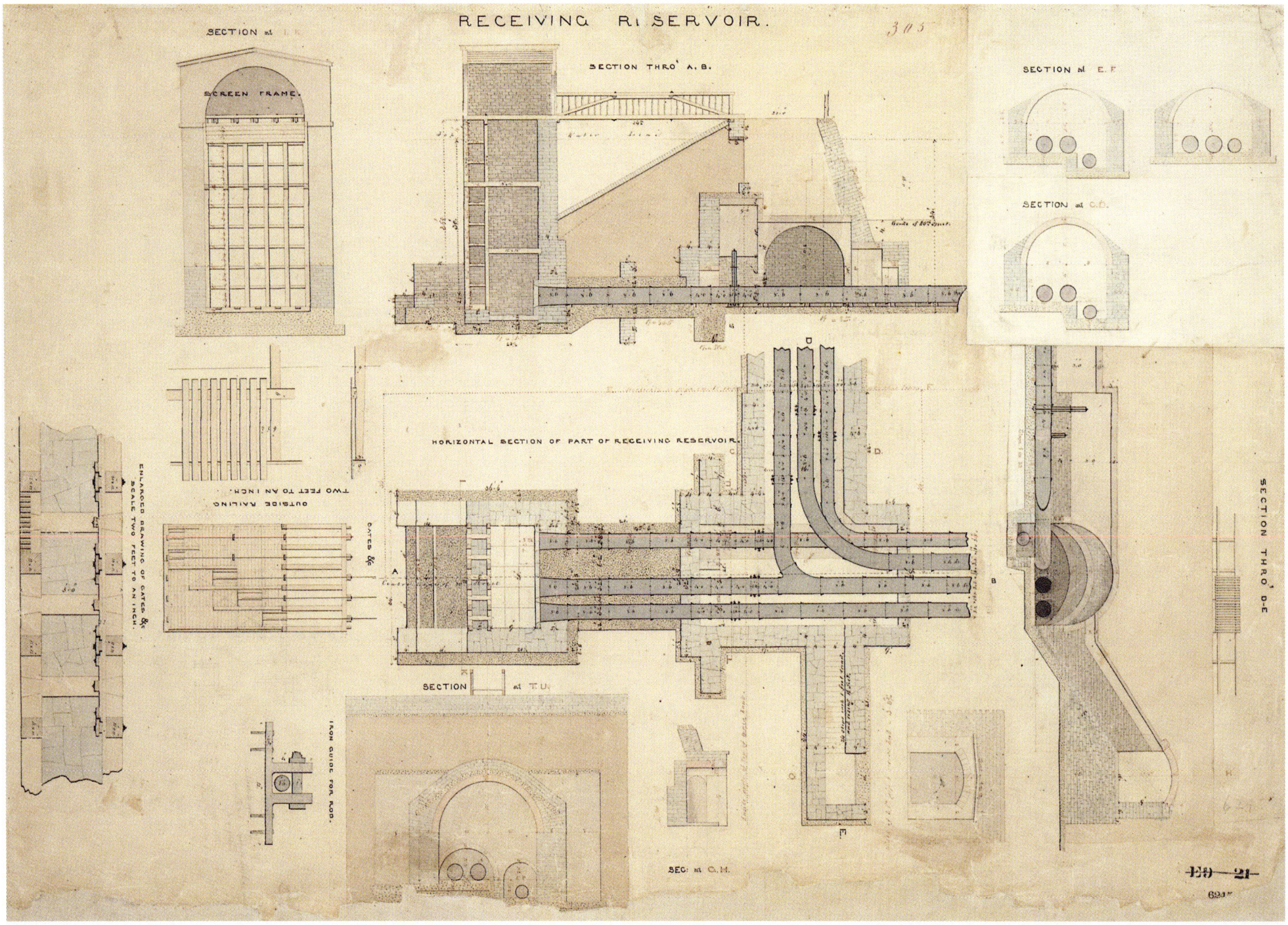

FIG. 29. *York Hill Receiving Reservoir. Details of gatehouse and pipe vaults. c. 1839.*

pipe siphon and the entire valley crossing—the arched central portion and the solid wall blocking 96th, 97th, and 101st Streets—was torn down. Some might say that the valley has never countenanced its development: after its degeneration into a dense region of shabby tenements, 1950s urban renewal delivered the current "banal . . . slab and balcony" high-rise apartments of Park West Village.[31]

From the looming tristesse of Clendening Valley, the brick conduit disappeared for its final underground mile to the York Hill receiving reservoir in rocky, open ground that the 1811 grid plan intended to eventually fill with fourteen blocks bounded by 79th and 86th Streets and Sixth and Seventh Avenues. Twenty years after it was completed, the reservoir was the centerpiece of Central Park—843 urban acres of re-created nature rescued from their fate of becoming part of the 153 blocks of grid (and inspired by Brooklyn's Green-Wood Cemetery, the pioneering exercise in urban pastoralism created in 1838 by displaced Croton chief David Douglass).

The York Hill receiving reservoir was comprised of two linked basins with unadorned sloping stone walls that rose in varying heights above the uneven terrain to maintain a maximum water depth of twenty-five feet [Fig. 25]. With a capacity of 150 million gallons, the receiving reservoir outpaced the storage of the Manhattan Company and the 13th Street Reservoir system three hundred times over [Figs. 26, 27, 28, 29, 30].

York Hill received its first water on the afternoon of June 27, 1842, to the cheers of twenty thousand citizens and the salute of thirty-eight guns. Fayette Tower, who during the past five years had been assigned increasing responsibility for major structures from the dam to the Clendening crossing, was present from early morning, when he took the hand of a distinguished visitor emerging from his exquisite carriage. "His head was bent forward beneath the weight of years and being introduced to one of the contractors for building the reservoir, he turned up one eye towards him and remarked 'I think you ought to make money here.' How characteristic of the man whose thoughts have turned to dollars."[32] Possibly the thoughts of America's richest man, seventy-three-year-old John Jacob Astor, were also turned to the costs of cholera and the approaching tenth anniversary of his daughter Magdelen's agonizing death.

From the receiving reservoir, twin three-foot-diameter iron mains led water south for two miles under the future line of Fifth Avenue to a 20 million gallon distribution reservoir at the provincial intersection of dirt roads, presently Fifth Avenue and 42nd Street. The receiving reservoir was primarily a functional creation; the distribution reservoir, on four acres of Murray Hill a mile north of the city limits, was a showpiece [Fig. 31]. Its massive sloping masonry walls were 45 feet high, towering over new lots, country cottages, and trees that thus far had survived clearing. The exterior was adorned with neo-Egyptian detailing. An iron railing ringed the flat top of the walls, creating a broad public promenade a third of a mile square [Figs. 32, 33]. To all who viewed it from a distance or viewed into the distance from its high promenade, the great reservoir augured the city to come, surging northward to meet, surround, and, by century's end, overwhelm it. When the reservoir was filled to its 36-foot depth, the level of the water was 115 feet above mean tide, and 51 feet below the water level at the Croton Dam 41 miles away.

With cannon booming at sunrise on Independence Day 1842, the water gates into the twin basins of the Murray Hill reservoir swung open. "At an hour when the morning guns had roused but few from their dreamy slumbers, and ere yet the rays of the sun had gilded the city's domes," Fayette Tower, perched on the reservoir parapet, "saw the first rush of the water as [it] entered the bottom and wandered about, as if each particle had

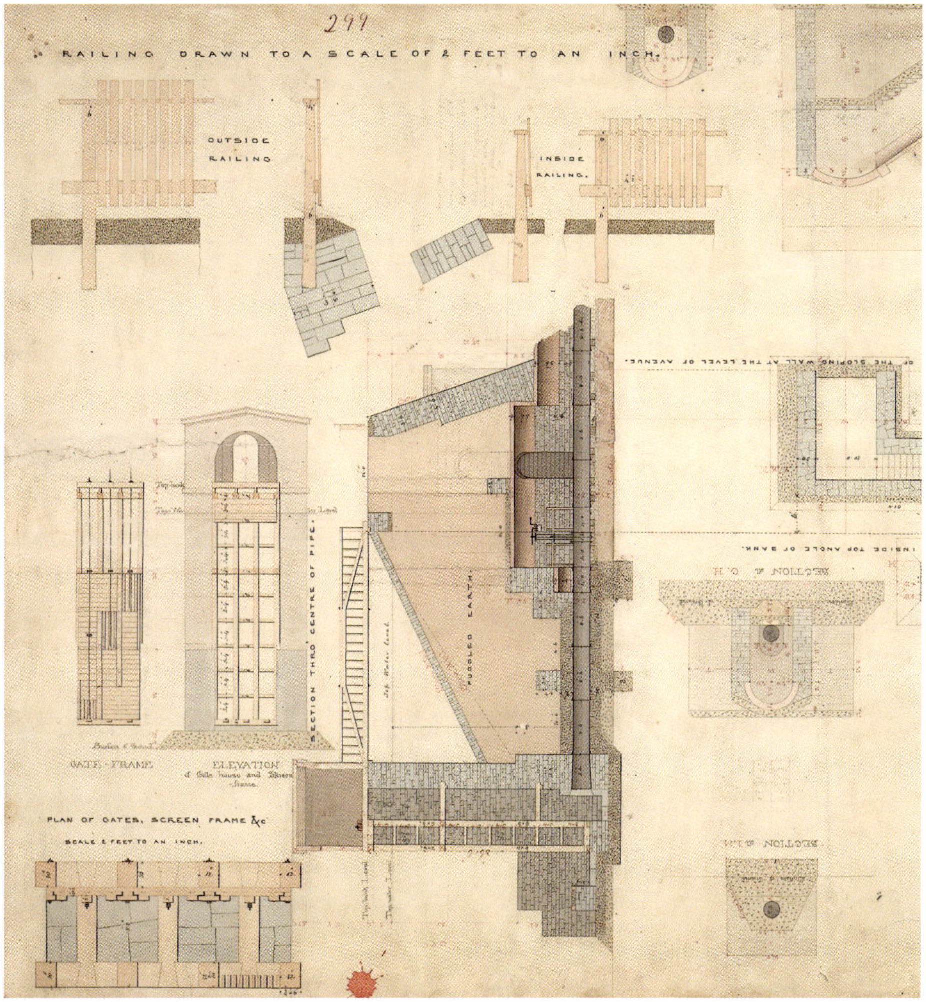

FIG. 30. *York Hill Receiving Reservoir. Cross section of gatehouse and perimeter embankment. c. 1839.*

FIG. 31. *Murray Hill Distributing Reservoir. 42nd Street and Fifth Avenue (future site of the New York Public Library).*

consciousness."[33] Later, the reservoir's bounty was served in cups to twenty-five thousand sweating citizens. By mid-October, hundreds of miles of distribution pipe had been laid, from the Battery to 21st Street, river to river, and spectacular fountains played in City Hall Park and Union Square. "Nothing is talked of or thought of in New York but Croton water; fountains, aqueducts, hydrants, and hose attract our attention and impede our progress through the streets," wrote Philip Hone in his famous diary; "Water! Water! is the universal note which is sounded through every part of the city, and infuses joy and exultation into the masses, even though they are somewhat out of spirits."[34] New Yorkers were out of spirits from six years of economic depression, which had something to do with the fact that few households had engaged a new breed—the Croton plumber—to hook up expensive service pipes. Only two weeks earlier the city officially announced that it would make free and unlimited water available at street hydrants and charge up to $12 a year for private service.

To promote the use of its new abundance, the city threw the "Croton Water Celebration" on October 14, 1842. Tens of thousands of marchers, in groups representing every layer of society, paraded before the population crowded along the winding citywide route festooned with flags and floral displays. Church bells pealed, bands played, artillery fired, and everyone endured a choral society rendition of the seven-stanza "Croton Ode," which opened with:

> Water leaps as if delighted,
> While her conquered foes retire!
> Pale contagion flied affrighted
> With the baffled demon Fire![35]

More lyrically, editor and best-selling writer Lydia Maria Child (chiefly remembered today for her Thanksgiving verse "Over the river and through the wood . . . ") reported to her newspaper readers in Boston: "Oh, who that has not been shut up in the great prison-cell of a city, and made to drink of its brackish springs, can estimate the blessings of the Croton Aqueduct? Clean, sweet, abundant water!"[36]

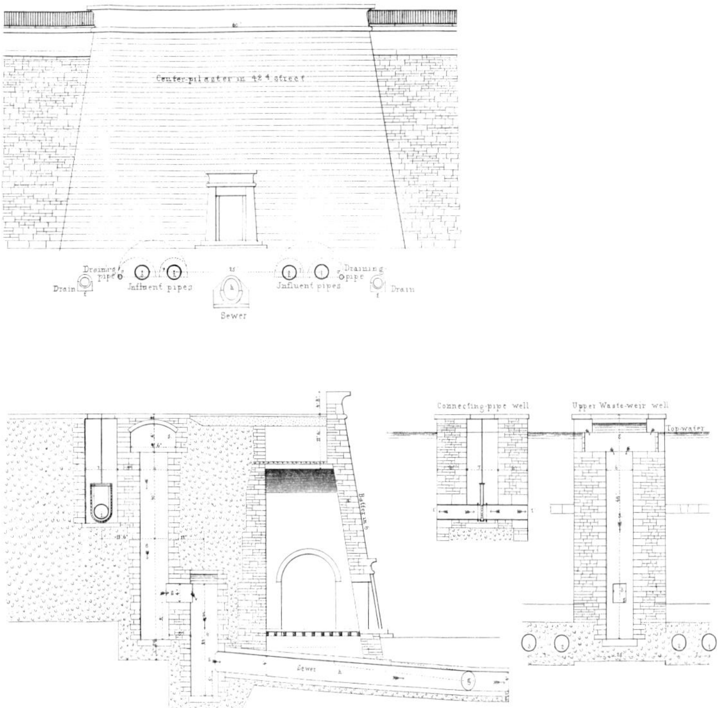

FIG. 32. *Murray Hill Distributing Reservoir. Elevation and cross section details at entry to pipe vaults.*

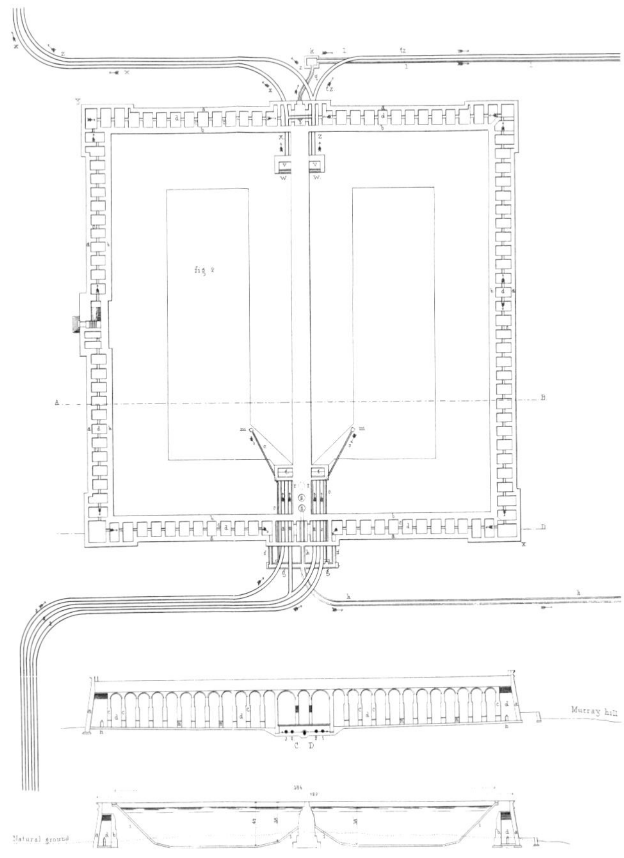

FIG. 33. *Murray Hill Distributing Reservoir. Plan and sections showing vaulted perimeter masonry walls.*

FIG. 34. *Old Croton Aqueduct. Section of aqueduct exposed during building excavation at West 105th Street, Manhattan. May 4, 1928.*

THE GEOLOGIST

Basically, my job as a geologist is to reach an understanding of the rock we're going through—to build a picture of what has happened over the last several million years and apply that knowledge to the work of construction. Once we know where we'd like to have the tunnel go, we drill core borings, test borings. To get these we drive a truck along the route, and set a drill up which goes down six, seven, eight hundred feet and brings back with it this piece of the puzzle. This is vital information for the contractors to know when they are making their bids. If you think about it, this is the only concrete, pardon the expression, information the General Contractor has to go on. They have to estimate how long it's going to take them to tunnel out, how they are going to do that, and how much it is all going to cost, without ever seeing anything besides this.

After the work begins, we geologists inspect all the rock as soon as it's revealed, looking for faults, cracks, loose shifting, settling sand, matter or material, which could lead to a cave-in or an unsafe condition. At the same time, we are trying to understand as much as we can about what has happened over the last millions of years. For example, the rock we're going through now in Red Hook Brooklyn is 450 million years old, from a time when the west coast of Africa, the Orkney Islands, and New York City were all a part of the same land mass. It's very humbling. I think it's good to be humbled in your job every once in a while.

Excerpt from *City Water Tunnel #3*,
written and performed by Marty Pottenger

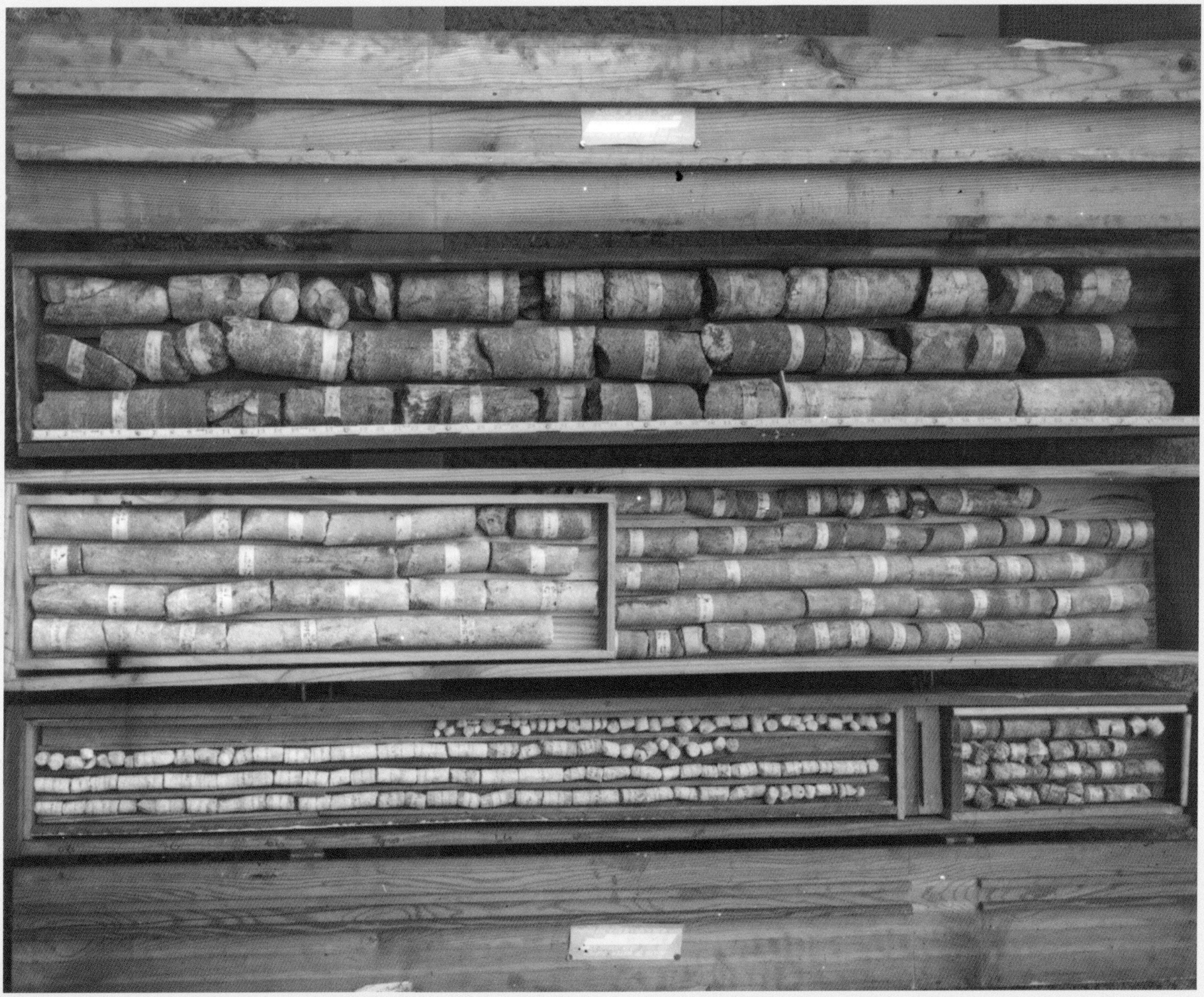

THE EXTENSION OF THE CROTON SYSTEM

KEVIN BONE

By 1849 New York City's first substantial public water supply system, the magnificent Croton waterworks, was complete. But the relief it provided from the onslaught of fire, disease, and epidemic illness was short-lived: no sooner had the water from the Croton River begun to flow through the taps of New York City than another water famine loomed on the horizon. The continued and unprecedented growth in population, the increasing use of the flush toilet, and excessive water waste added to consumption rates that far exceeded the projections of Jervis and his fellow Croton engineers. With demand soaring, the aqueduct was pushed to the limits of its capacity. This pattern of increased need, struggle, success, and shortfall would repeat itself many times over the next 125 years as New York City continued, almost without interruption, to seek new sources and build new structures for the capture, storage, and delivery of water.

With the completion of the High Bridge—the most visible symbol of its new water supply—the city abolished the Water Commission (appointed by the state in 1833) and the Croton Aqueduct Board (organized by the Common Council in 1842) and replaced them with the Croton Aqueduct Department. The department was in charge of structures stressed from overloading and in need of constant oversight and maintenance. The aqueduct itself, much of it located above grade on artificial berms, was exposed to erosion, differential settlement, and to the freeze/thaw cycles of the harsh New York winters. Soon after completion it became evident that the aqueduct would need constant upkeep. The collapse of retaining walls, the settlement of earth supports, and cracking in the brickwork because of thermal expansion and contraction all resulted in significant leakage. Large quantities of water were escaping from the aqueduct before reaching the thirsty city.

Cottages were built at intervals along the line so that superintendents could closely monitor the structure. A yearly inspection of the conduit was considered essential; even a small tree root snaking through the brick walls of the enclosure could start a leak that might lead to catastrophic failure. The annual maintenance review took at least three days to perform. A lead engineer, followed by work crews, would walk the entire length of the dewatered aqueduct, from the Croton Dam to the York Hill Receiving Reservoir, noting problems and directing repairs. Each year, the inspection became a time of concern for the city. Because the aqueduct functioned without backup, within the short period of seventy-two hours—the minimum time it took to shut the aqueduct down, drain the structure, and make a few key repairs—the amount of water in the Manhattan reservoirs would rapidly decrease. "During the November inspection of 1846 the aqueduct was closed for thirteen days and water levels dropped dramatically in both Manhattan reservoirs, and this was at a time when there were only 12,000 water-takers."[1] It was clear by 1846 that the city needed additional water storage.

In 1851 the Common Council directed the Croton Aqueduct Department to purchase ground for a new reservoir that could contain enough water to last the city for at least sixty days. The board voted to appropriate ninety-six acres of land between Fifth and Seventh Avenues and between 86th and 96th Streets to build a larger version of the rectangular receiving reservoir that was then in use. The site was a natural basin bounded largely by existing ridges and situated below the new grades that were being established for city streets. Embankments to complete the impounding area would only be required on the west and south boundaries.[2]

Condemnation proceedings and the logistical battle to acquire land from over one hundred different owners delayed the project

FIG. 1. *New Croton Dam. Spillway and downstream face. April 19, 1912.*

until a new law, "An Act to Facilitate the Acquisition of Lands for a New Reservoir," empowered the New York Supreme Court–appointed Commission of Appraisal to settle disputes. By 1856 the city had secured the land. But before work could begin on the new receiving reservoir, Central Park was conceived and the newly formed Park Commission proposed an exchange of territory. Instead of a rectangular reservoir, the commission suggested the new reservoir follow the contours of the topography, which would accord with the naturalist vision for the park [Fig. 2]. "Its curvilinear shape is largely due to the persuasiveness of the man who was then the engineer of the new park, Anton Viele. Viele argued that the new reservoir should not be rectangular like the old one but should have a more graceful and natural form."[3] This "new reservoir represented a change in design philosophy," where urban infrastructure would be "integrated into park landscaping, and enjoyed for scenic views and recreation."[4]

The new design not only augmented the landscape architecture; it was practical, too. Because it relied on the natural landforms, it required less excavation and fewer rock cuts and embankments to build. The Croton Aqueduct Department endorsed this change and construction began [Fig. 3]. During the initial work season of 1859, 1,200 men (along with 107 horses) were engaged in building the basin.[5] On a map of 1859, this new reservoir was called Manahatta Lake, and in the records of the Aqueduct Department it is called the Grand Reservoir. Popularly known as the New Reservoir, it was completed in 1862, covered ninety-six acres of the park, and could store up to one billion gallons of water (the amount needed for the projected sixty-day supply that had been called for) [Figs. 4, 5, 6, 7]. The old receiving reservoir, which held 150 million gallons of water, was kept in service until 1925:

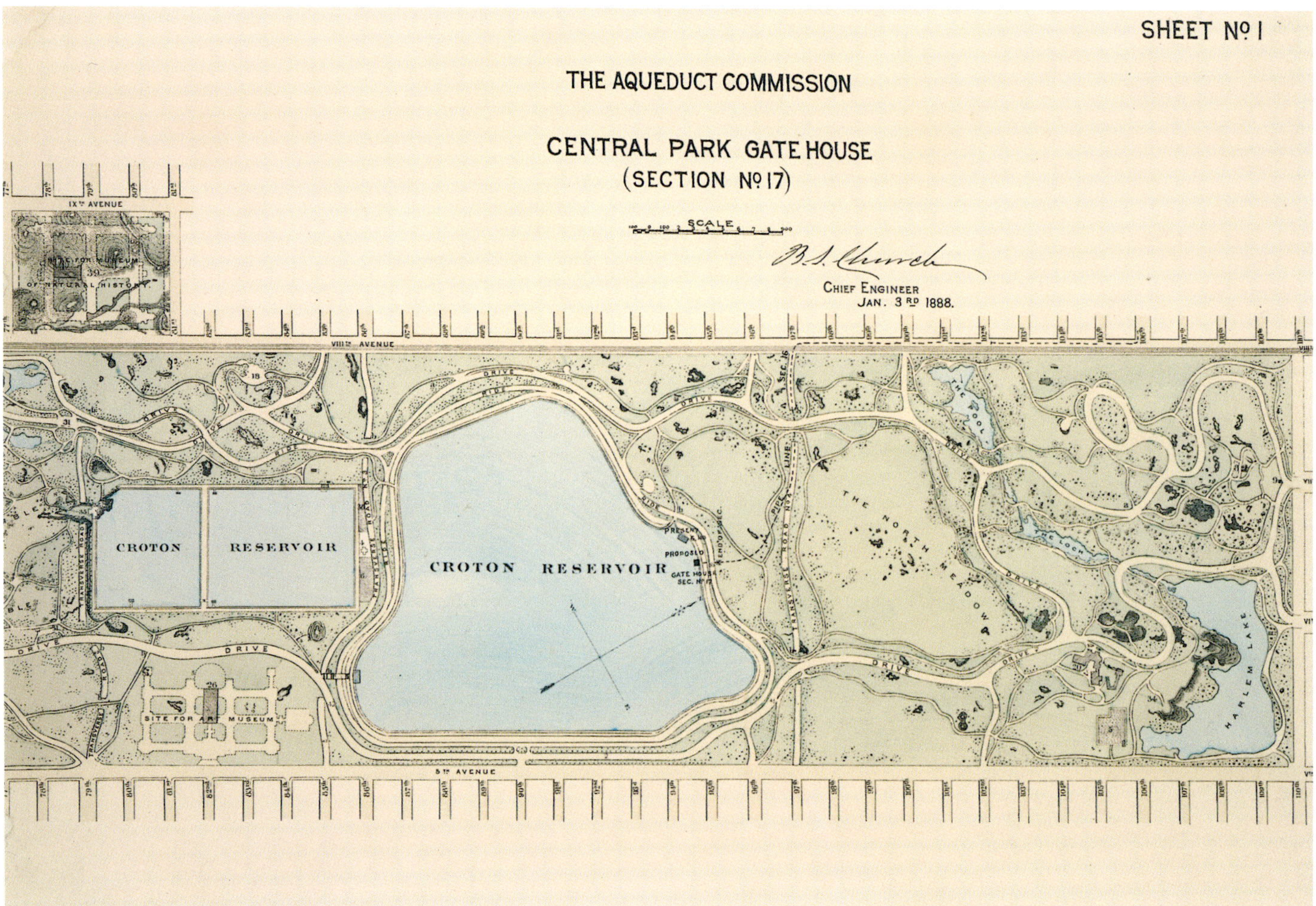

FIG. 2. *Plan showing the York Hill Receiving Reservoir (Croton Reservoir) and the New Receiving Reservoir in Central Park. January 3, 1888.*

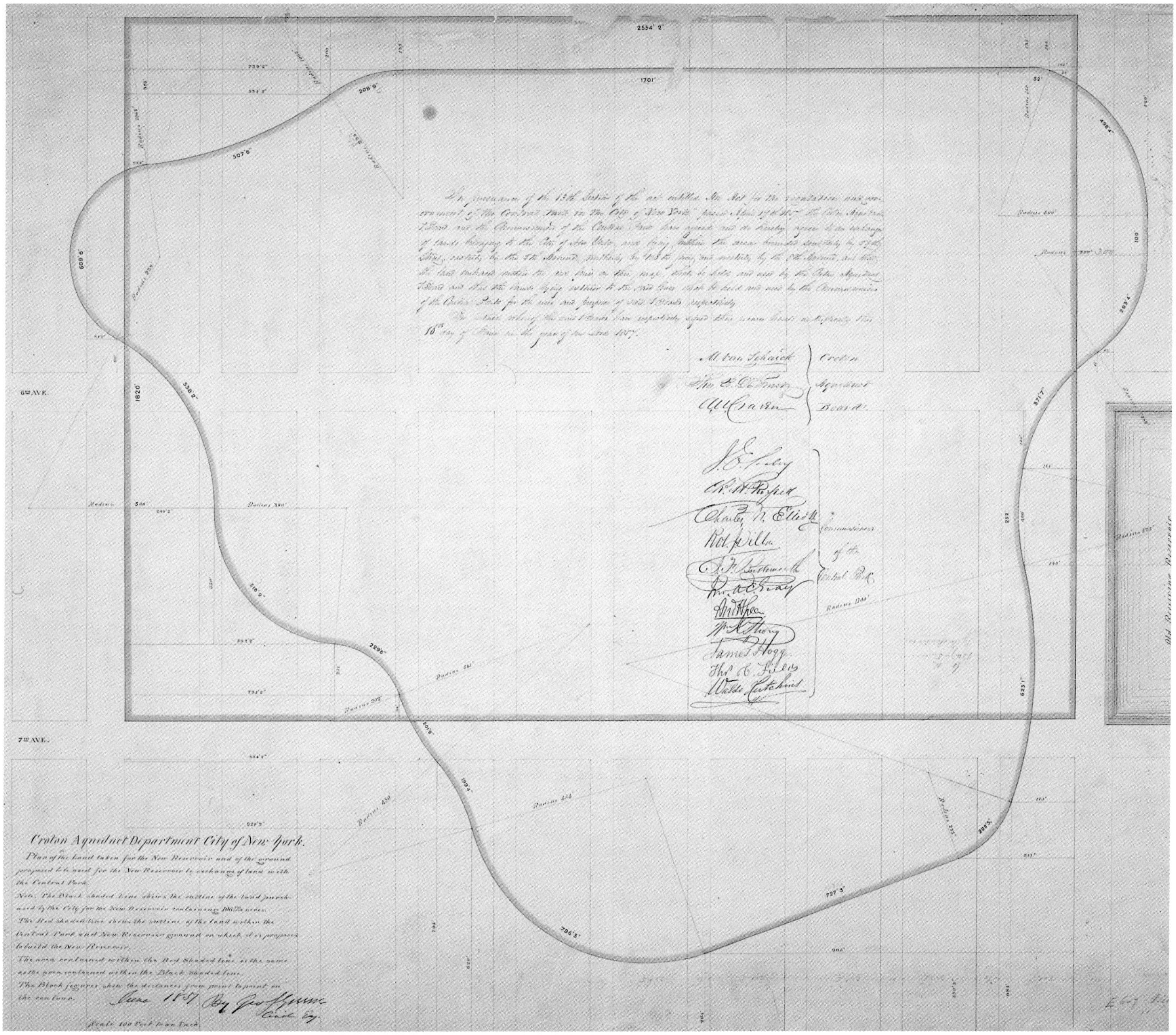

FIG. 3. *Plans for original and revised configurations of the New Receiving Reservoir in Central Park. June 16, 1857.*

When completed the reservoir will be enclosed by a compact earth embankment, connected everywhere with the rock bottom by the puddle bank, and lined on the whole of the inner face by a heavy stone slope wall, capable of resisting any waves to which it might be subjected, and impenetrable by any animals whatever. Every part of the work on this reservoir has been under the daily scrutiny of the engineer of the department, and the constant watch of the inspectors appointed to see that the specifications be strictly adhered to. The embankments, as compared to those of the existing receiving reservoir, present greater strength for the same and relative depth of water, which is the proper function for comparison of the resistance to be overcome. The receiving reservoir has stood sufficiently long to test its strength and it has shown no symptom of decay, nor indications of weakness.[6]

Before the New Reservoir could be completed, New York City would be reminded of its vulnerability twice in one year. For unspecified reasons, the aqueduct's annual inspection in 1860, typically

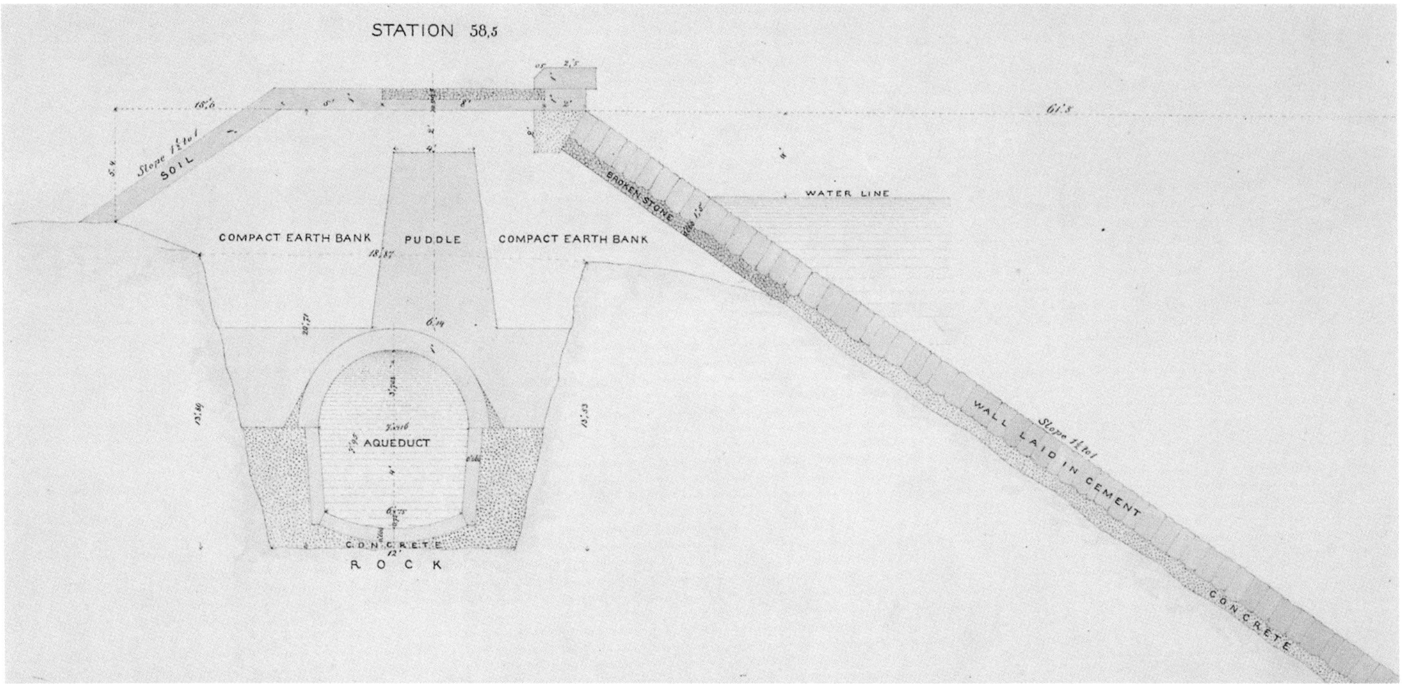

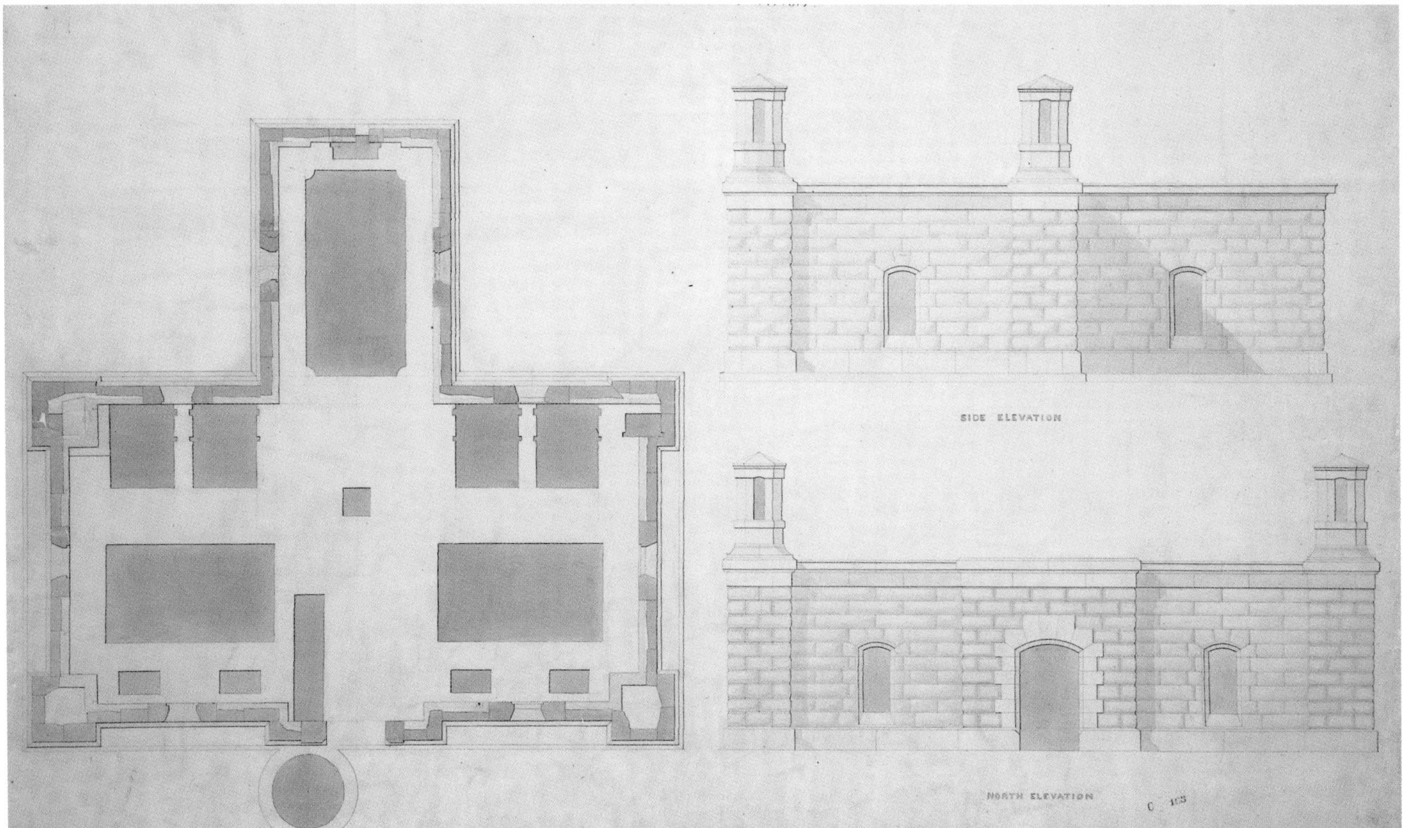

FIG. 4. *New Receiving Reservoir in Central Park. Cross section through the embankment. c. 1856.*

FIG. 5. *New Receiving Reservoir in Central Park. Plan and elevations of north gatehouse. c. 1857.*

scheduled for late October when consumption rates were lowest, was delayed until December 4. Usually the reservoirs within Manhattan would be filled to their maximum limit before the aqueduct was shut down, but on that day, they were below capacity. During tests made in the week prior to the shutoff, engineers concluded that more water was leaving the reservoirs than normal, in part because residents had already begun their winter practice of running water to prevent house pipes from freezing.

The department concluded that closing down the aqueduct at a time of low capacity and high use would risk depleting the small reserve and postponed the examination until spring. Word of the suspended inspection reached the Croton gatehouse keeper just hours before he was to begin the draining process by lowering the stopcocks and closing off the aqueduct [Fig. 8]. The timing was a remarkable moment of good luck. On the evening of December 5, both of the 36-inch conduit pipes on Fifth Avenue (the lines that connected the York Hill Receiving Reservoir to the Murray Hill Distributing Reservoir) collapsed into unstable ground as the result of a project to level the grade of Fifth Avenue. As the full pipes deflected downward into loose earth, the joints gave way. Water gushed out of the mains and quickly drained both reservoirs. Nearly the entire supply spilled over the adjacent fields. Crews were assembled, water was shut down wherever possible, and within fifty hours of the break, a new connection had been made. The accident was considered the most perilous moment in the history of the young Croton supply. The chief engineer telegraphed all police offices and ordered a fire watch. The next morning, orders were given to cease all manufacturing and all water supply to ships and ferries. The two smaller lines that remained in service from the receiving reservoir supplied only enough water to meet minimum domestic needs. Had the Croton Aqueduct been drained as scheduled, the city would have certainly been without water. This incident prompted the construction of a 48-inch line on Fourth Avenue as a backup to the Fifth Avenue mains.[7]

Several days later another failure affected the system. Two miles below the aqueduct's connection at Croton Lake, where the structure passes the valley of Bailey's Creek on a high embankment, water was found

FIG. 6. *New Receiving Reservoir in Central Park. Stone details for stairwell. c. 1857.*

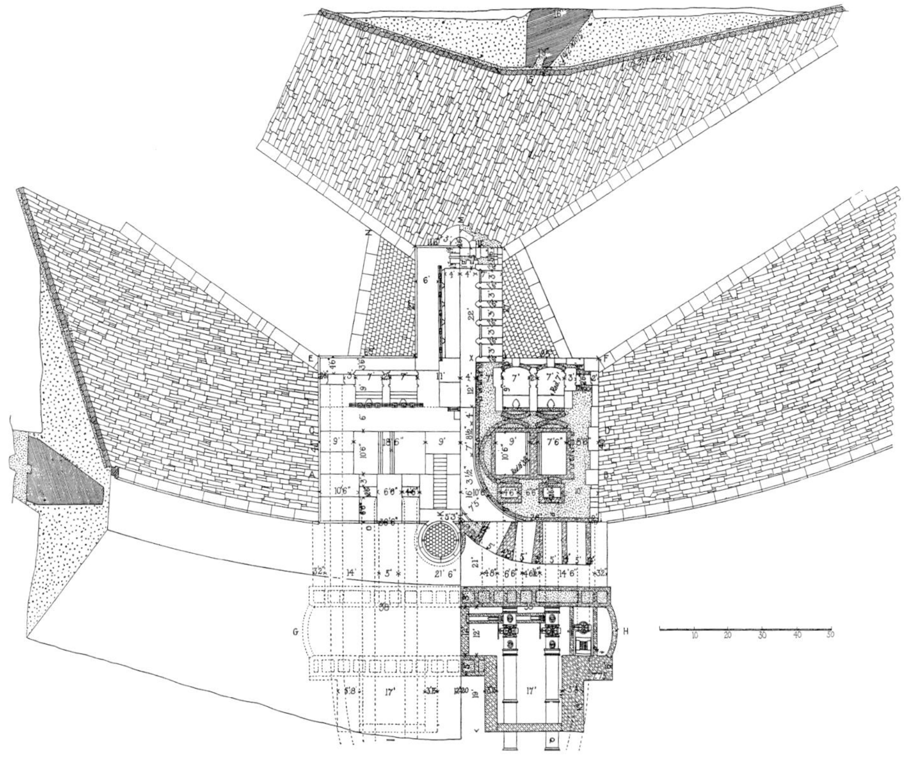

FIG. 7. *New Receiving Reservoir in Central Park. Plan of north gatehouse. c. 1857.*

pouring out of the side of the conduit. The chief engineer was telegraphed, but the damage was seen as so serious that the on-site superintendent shut down the main gates at the dam without waiting for orders. When the water had drained away, a crack one-quarter to one-inch wide and 425 feet in length was found on the invert arch at the bottom of the aqueduct. Although the emergency repair work was slowed by cold weather, the total time of the shutdown was limited to 38½ hours. These two incidents left the department with a clear understanding of the limitations of the new system. Even with restrictions placed on industry and manufacturing, the water stored in the two local city reservoirs could only last for several days. The department concluded that at no time could the aqueduct be out of service for more than three days. This severely limited the type of repairs that could be done: if permanent repairs could not be completed within that time, a temporary solution would have to be found. This was the case until two key improvements were realized: the completion of an expanded receiving reservoir within the city limits and an increase in the capacity of the pipes carried by the High Bridge. These two improvements marked the beginning of the Croton Waterworks Extension.[8]

As early as 1851, Croton engineers had identified the two 36-inch-diameter pipes crossing the High Bridge as a major constriction in the system. Because of their relatively small size, they could carry only a portion of the water available via the aqueduct. Moreover, the pipes were made of cast iron and prone to leakage; cast iron is a brittle material that cannot resist the bending and pushing of water under pressure. Plans were made in 1859 to lay a new 90.5-inch-diameter wrought iron pipe over the original pipes [Fig. 9]. This larger tube would be capable of carrying the maximum amount of water that the masonry portion of the aqueduct could deliver. Although wrought iron had been used as a building material for some time, industrialization combined with advancements in metallurgy contributed to the development of manufacturing techniques that made the production of larger sheets possible. Wrought iron has more tensile strength and can tolerate greater pressure than cast iron, allowing for a wider variety of construction applications; in this case, the largest pipes yet employed in New York. The 90.5-inch-diameter pipe was set on cast iron stanchions placed in between the two 36-inch lines, which were left in place in case the new pipe failed. Rollers were installed on top of the stanchions to accommodate expansion and contraction of the pipe and reduce the stress on the connections between its sections [Figs. 10, 11].

The top of the bridge was covered with a masonry arch to protect the pipes and establish a roadway surface. The stone parapets of the original bridge were raised about six feet, and the starkly simple original design was modified with the addition of an ornamental cornice [Fig. 12]. The bridge remained a key civic attraction, and the Bronx River Park on the Westchester side was terraced, landscaped, and integrated into the architectural assemblage of the High Bridge area.

Several years later, other waterworks were built at the site of the High Bridge. Croton water, because of the elevation of the upstate reservoir and the hydraulics of the aqueduct, had a certain head and under its own pressure could rise to a height about 125 feet above mean high tide. As New York City expanded into the higher ground of upper Manhattan, where elevations are substantially above that level, the system could summon little pressure. As a remedy, a facility was built that mechanically pumped water from a small reservoir into an elevated storage tank perched atop a masonry tower. Local water was drawn down from this pressure tank. This so-called high-service was first discussed in 1851, but it was not initiated

FACING PAGE
FIG. 8. *Six-inch stopcock for a pipeline. Plan, sections, and elevations. January 3, 1888.*

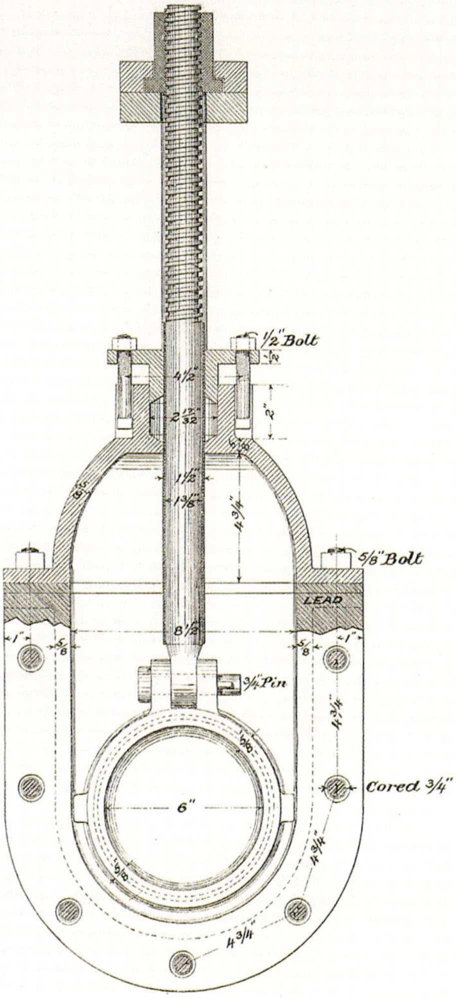

Section aa

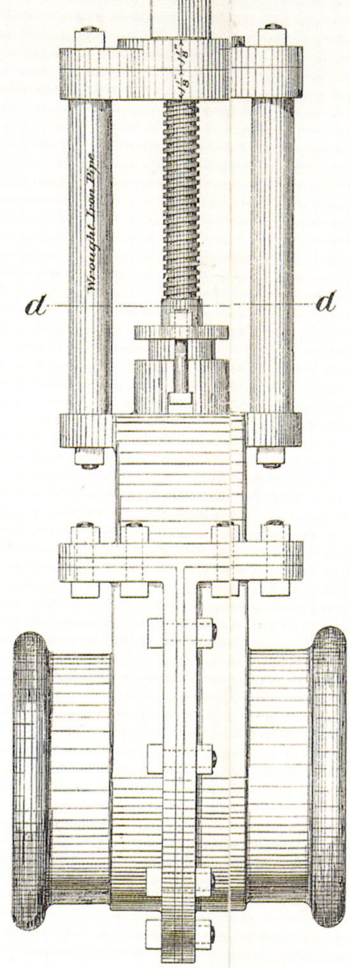

Side Elevation

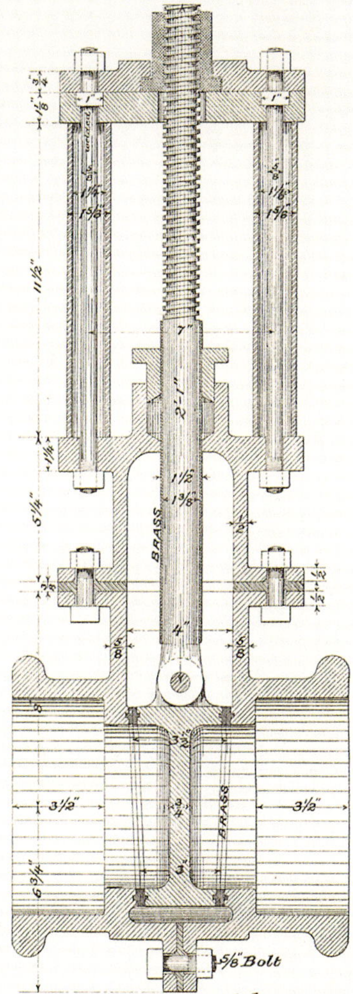

Section bb

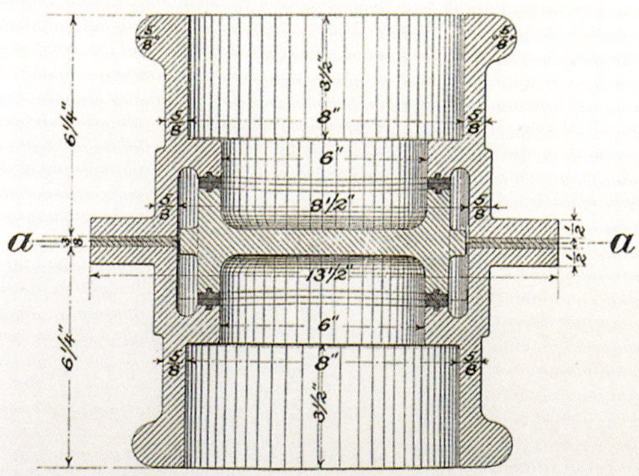

Section cc

SCALE 4" = 1'

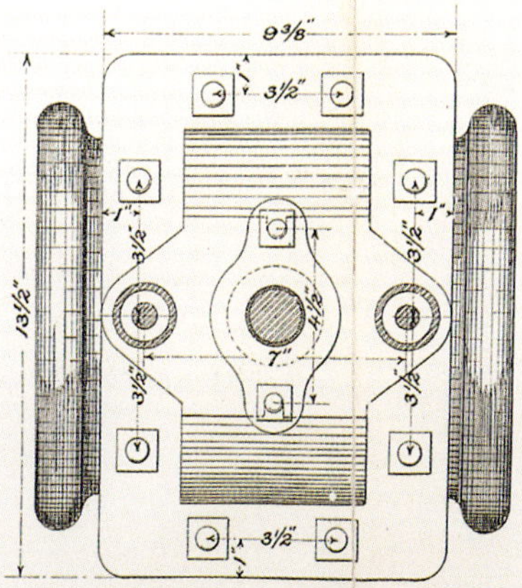

Section dd

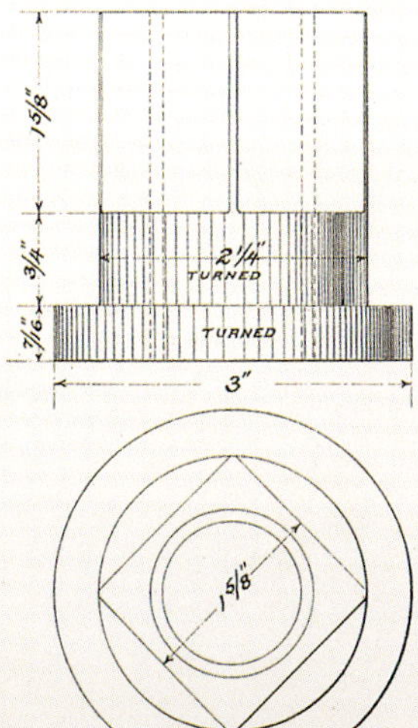

Brass Nut
(Full Size)

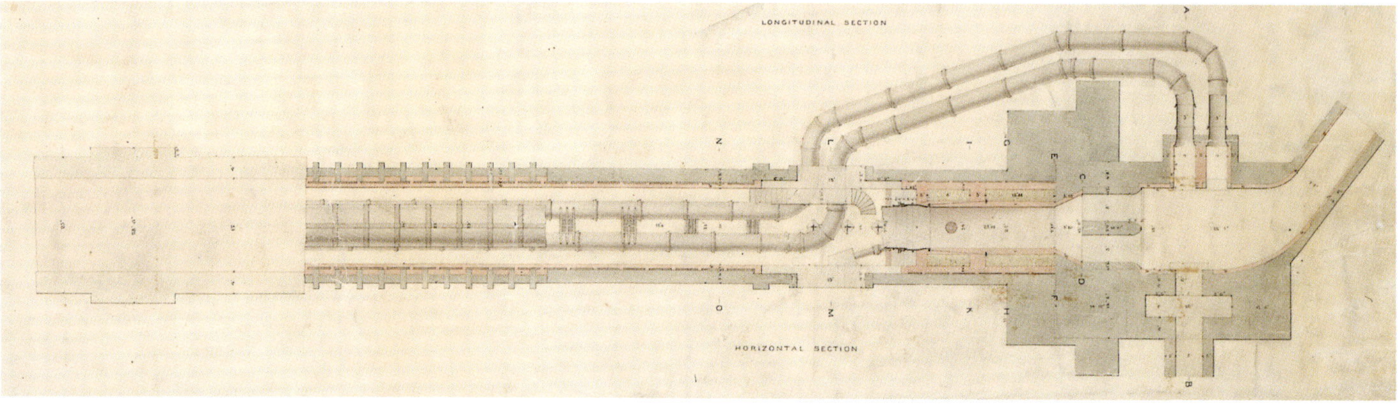

FIG. 9. *High Bridge. Plans for the addition of the 90.5-inch diameter line. c. 1859.*

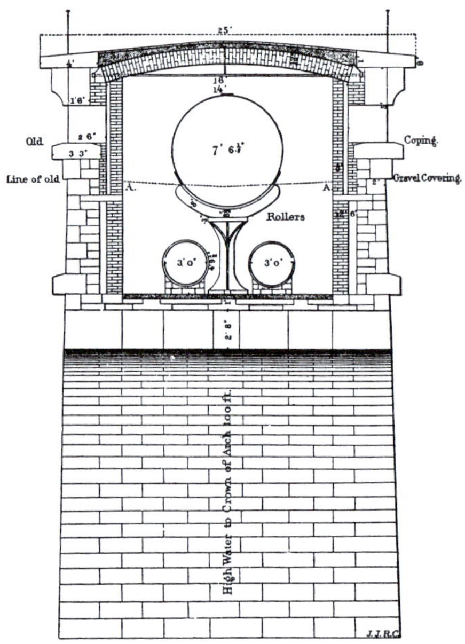

FIG. 10. *High Bridge. Cross section showing the 1860–61 addition of the 90.5-inch diameter line and modifications to the brick enclosure.*

FIG. 11. *High Bridge. Improvements underway. 1860–61.*

until 1863, when authority was given to build the High-Service Tower and Reservoir at High Bridge. The high-service works consisted of a rectangular reservoir about seven acres in size, a boiler and pump house with a grand chimney, a coal house and wharf for receiving shipments of coal, and a stone water tower 170 feet high, with a 47,000 gallon spherical wrought iron pressure tank at the top [Figs. 13, 14, 15, 16]. The water within the tank was 324 feet above high tide—57 feet higher than the highest point in Manhattan. The tower remains one of the defining landmarks of the Croton-era waterworks. (The reservoir was demolished, and the site is now a municipal swimming pool.)

Despite the construction of the Croton waterworks, as late as 1859 the city still had 140 working pumps and wells, though these vestiges of the previous era's water supply were slowly being reevaluated and abandoned. That year, four wells were completely filled and fifty, all located below 28th Street, were covered over with a timber roof deck. Although wells were becoming increasingly irrelevant, the board was committed to maintaining the better quality wells for emergencies.

As construction of the New Receiving Reservoir in Central Park and the High Bridge improvements moved forward, other extensions to the Croton waterworks were planned. In 1857, the Croton Aqueduct Department began to survey the Croton watershed to establish suitable locations for additional storage reservoirs and collect more data on its hydrological potential.

The surveys carried out by the department covered 352 square miles of watershed, identified thirty-one natural lakes and ponds, and laid the foundation for the expansion of the Croton system. Fifteen sites for reservoirs were studied along with potential water storage capacities and estimated costs of construction. Planning continued during the Civil War, but actual construction was severely limited by both human resources and funds. The inadequacies of the Croton water supply were exacerbated by droughts in 1864 and 1865. These prolonged dry spells depleted the watershed, and no water spilled over the Croton Dam for weeks; instead, all of the water from the Croton River was diverted south down the aqueduct for use by New Yorkers. In non-drought years, unused water spilled over the reservoir during times of high flow. These waters needed to be captured for use in times of drought, and new reservoirs would have to be built to accomplish this. The Croton Aqueduct Department's annual reports from this period stress the need for additional reservoirs.

The first of these was sited up the West Branch of the Croton River in Putnam County, near the small settlement of Boyd's Corners. Plans were prepared for a masonry dam 670 feet long and 57 feet high. Boyd's Corners Dam is simple and beautiful, with rusticated stone ashlars facing both the vertical upstream face and the battered downstream side of the dam. The heart, or the "hearting," of the dam was composed of large undressed stones set in concrete. A small stone gatehouse was located at the base of the downstream face, and the overflow was designed to spill down a wasteway cut from the natural bedrock. Work began in 1866 and was completed in 1873. Although somewhat

FIG. 12. *High Bridge and gatehouse. Sections and elevations. c. 1859.*

FACING PAGE
FIG. 13. *High Bridge across the Harlem River with the High-Service Tower at its Manhattan terminus. May 12, 1916.*

FIG. 14. *High Bridge and High-Service Tower. View north up the Harlem River. October 20, 1916.*

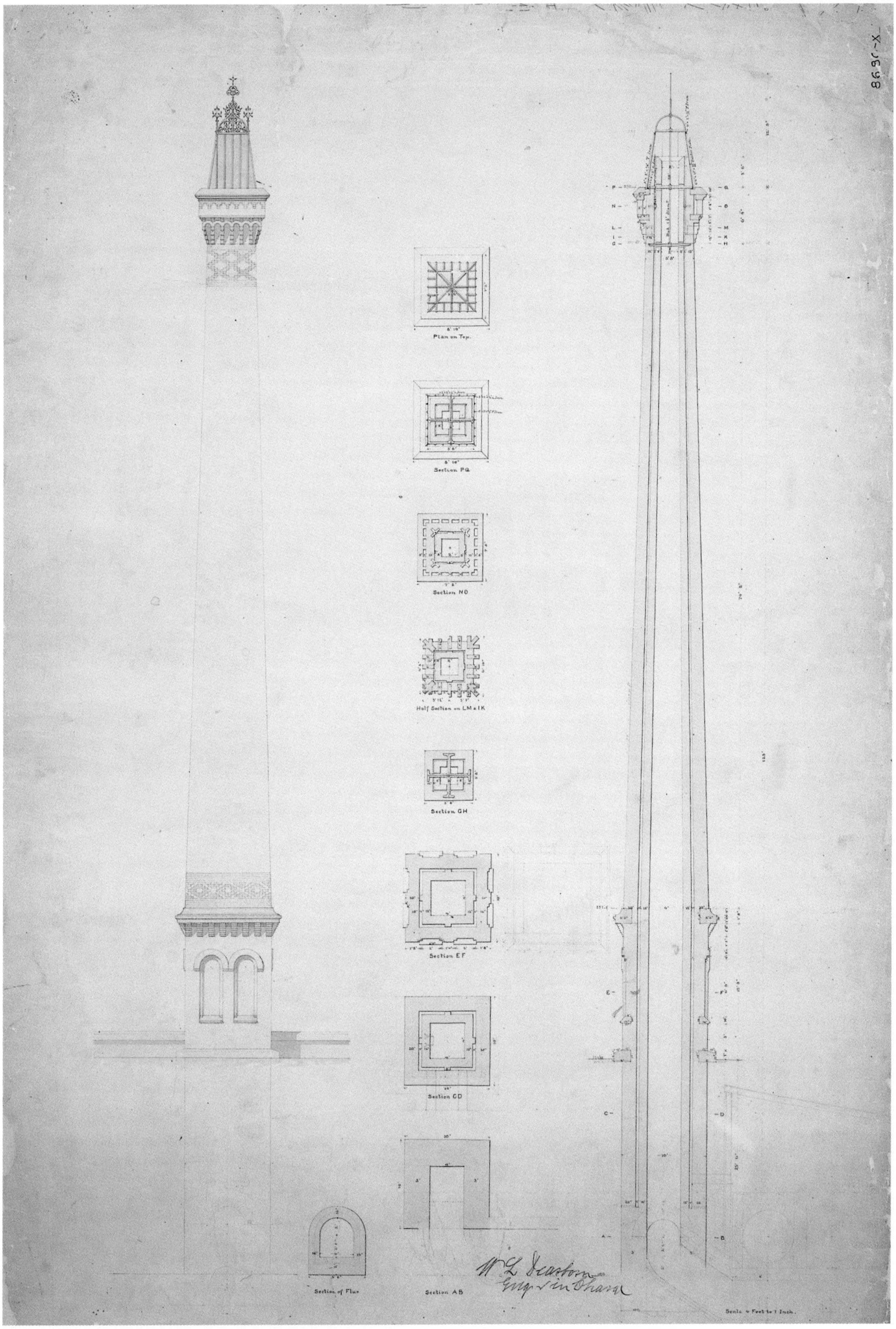

FIG. 15. *High-Service Tower at High Bridge. Boiler house chimney. Sections and elevations. c. 1859.*

FIG. 16. *High-Service Tower at High Bridge. Engine and boiler house. Elevations. c. 1859.*

modified from its original design, the dam remains in service today [Fig. 17].

Boyd's Corners Dam was built by the Croton Aqueduct Department, but legislation in 1870 reorganized governmental power and created the Department of Public Works. William Tweed, an astute politician who had risen to power as the grand sachem of a craftsman's political society known as Tammany Hall, managed to appoint himself the first commissioner of the new agency and became responsible for an incredible range of infrastructure initiatives.[9] The Boss Tweed era became a time best known for governmental excess and the illegal diversion of public funds. It was also a time during which the important work of improving the water supply infrastructure was compromised by corruption. Immediately after the completion of Boyd's Corners Dam, officials from the new Department of Public Works questioned the strength of its masonry wall and the overall soundness of the design. The new department ordered an earthen embankment to be built against the upstream side, ostensibly to strengthen the structure by resisting some of the lateral pressure of the impounded water. But it seems the huge and unnecessary embankment project was just another opportunity for graft, and the work, like much of the construction during the Tweed era, was poorly done with little regard for the requirements or specifications of the project. The engineer originally responsible for the construction of the dam observed that the earth embankment was badly built of improperly compacted and porous materials. It was soon fully saturated, and probably did more to increase pressure on the structure than to reduce it.[10]

What this and many other Tammany Hall–era projects did was to change the attitude of the public:

> The corrupt administration of the Croton waterworks under Tweed fundamentally shook the faith of the public . . . It ended an era of innocence during which there had been absolute confidence in the motives and honesty of the administrators and engineers of the system, who were entrusted with the decisions concerning the vast works and

huge outlays of public money that are inherent to the supply of water to great cities. It also began a schism between the engineers of the former Croton Aqueduct Department and the new masters, whose primary interest was in controlling the flow of money rather than water.[11]

Tweed's control of public works began to decline in 1871 when the *New York Times* charged that he had defrauded the city of millions of dollars. Boss Tweed was convicted on 204 counts of corruption in the fall of 1873 and died in prison five years later.[12]

With Tweed out of power, serious execution of new waterworks projects gained momentum. For the next expansion project, the engineers designed a reservoir to be built on the Middle Branch of the Croton River. A site was found about a mile upstream from the confluence of the East Branch and West Branch of the Croton River. Surveys were conducted in 1872 and 1873, and by 1874, construction had begun on the masonry core wall of the 515-foot-long, 94-foot-high earthen dam. Completed in 1878, the dam created a reservoir that covered 452 acres with a maximum storage capacity of about three billion gallons. It had been twenty years since the plan for the fifteen Croton storage reservoirs had been developed, yet only two had been constructed. Certainly, this

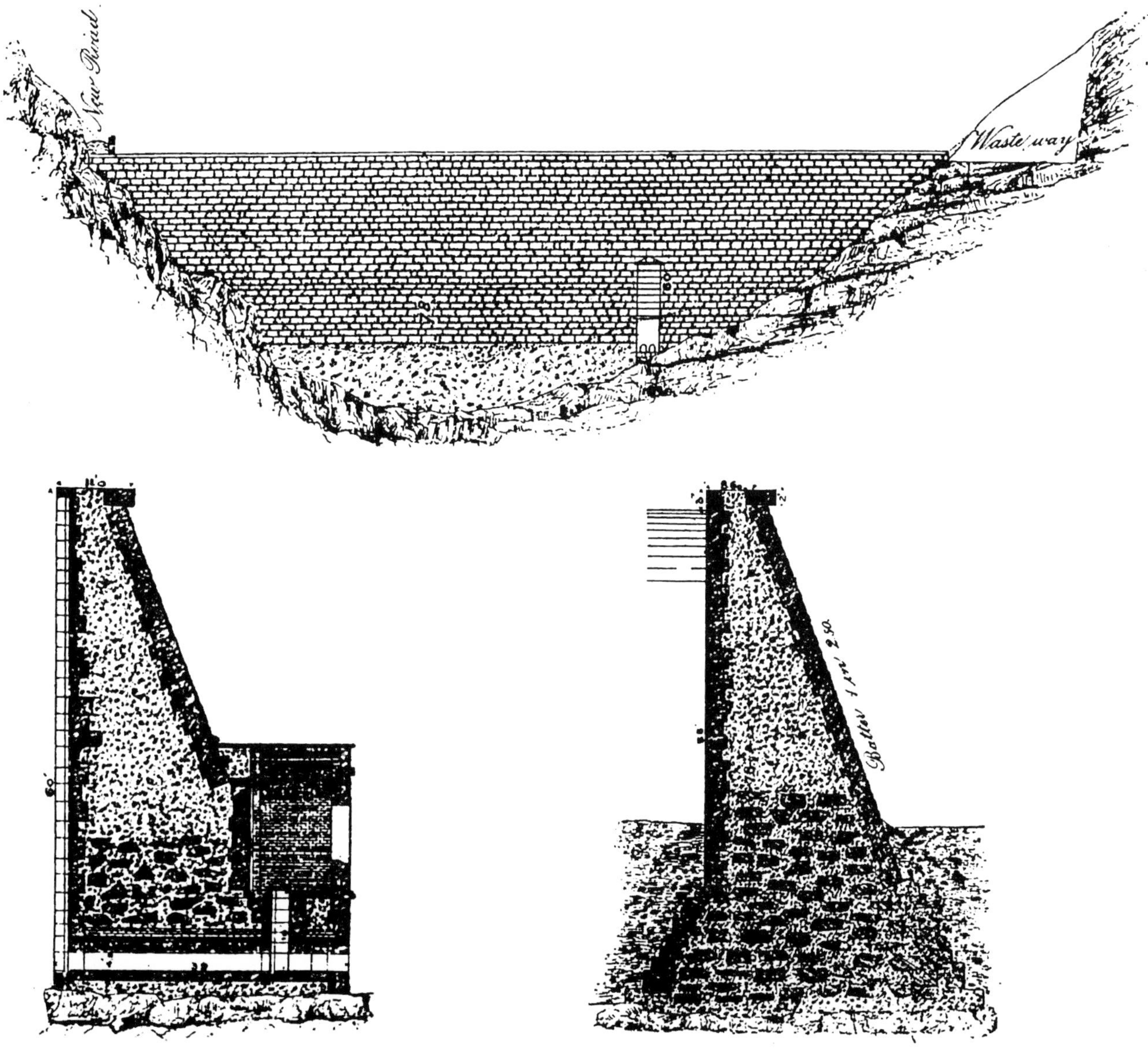

FIG. 17. *Boyd's Corners Dam. Downstream elevation and sections. 1866–72.*

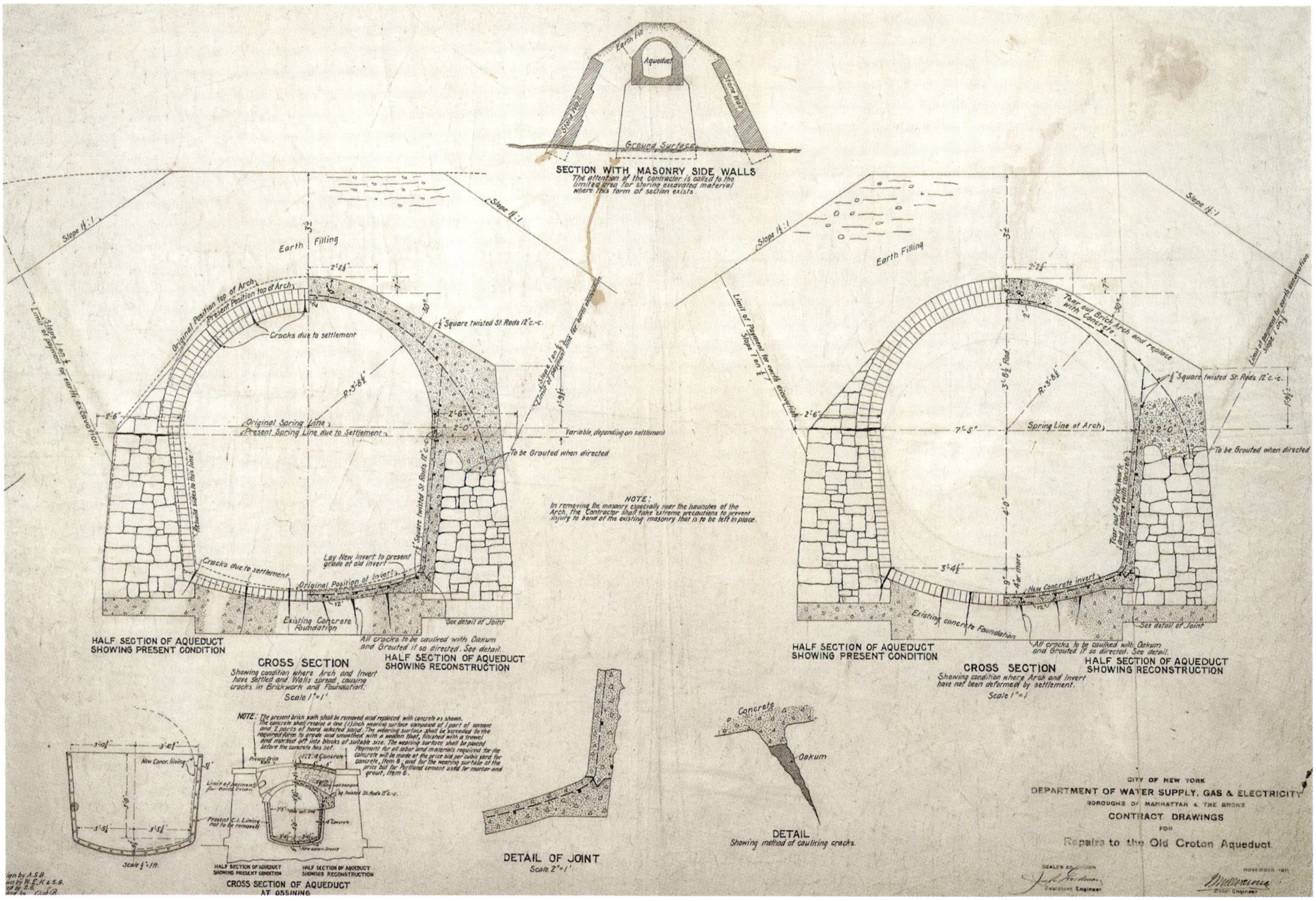

FIG. 18. *Old Croton Aqueduct. Cross section showing methods of repair used in 1911.*

was a consequence of the years lost to the Civil War, but it was also due to an administration that was unfit to govern and that cared little for real civic advancement.

By 1875 New York's water consumption was again pushing the limit of its supply. The brick aqueduct, now over thirty years old, was called upon to carry ever-larger quantities of water to meet the rising demand, which steadily increased stress upon the entire chain of structures. Aqueduct failures and associated maintenance were gobbling up the resources dedicated to water supply. The aqueduct, originally designed to carry a maximum of 72 million gallons of water per day (mgd), carried more than 100 million gallons—which meant that water in the conduit rose well above the flow line from which the enclosing roof arch sprang. Water above this point put outward pressure on the roof enclosure, a part of the aqueduct that was designed and built to resist only the downward pressure of the earthen cover. A brick arch, very strong in compression and good at resisting this downward load, was ineffective in resisting the outward push of water from the overfilled conduit. The result was "like violently boiling water lifting the lid off a kettle; the arched roof of the aqueduct was constructed to keep the water clean, not to withstand pressure."[13] This problem was complicated by structural settling, which caused the aqueduct to drop thirteen inches out of its correct position in several places, bringing the water level to the top of the arch and putting entire sections under pressure. Work was needed to reduce the stress from the higher loads that were being imposed upon the structure.

The prospect of a major water supply interruption due to failure along the only supply line clearly put the city at risk. In response, the department developed plans to strengthen the aqueduct in order to withstand overloading. To enable the masonry conduit to resist the extra strain to which it was subjected, an additional layer of brick was added [Fig. 18]. After this modification, which commenced in

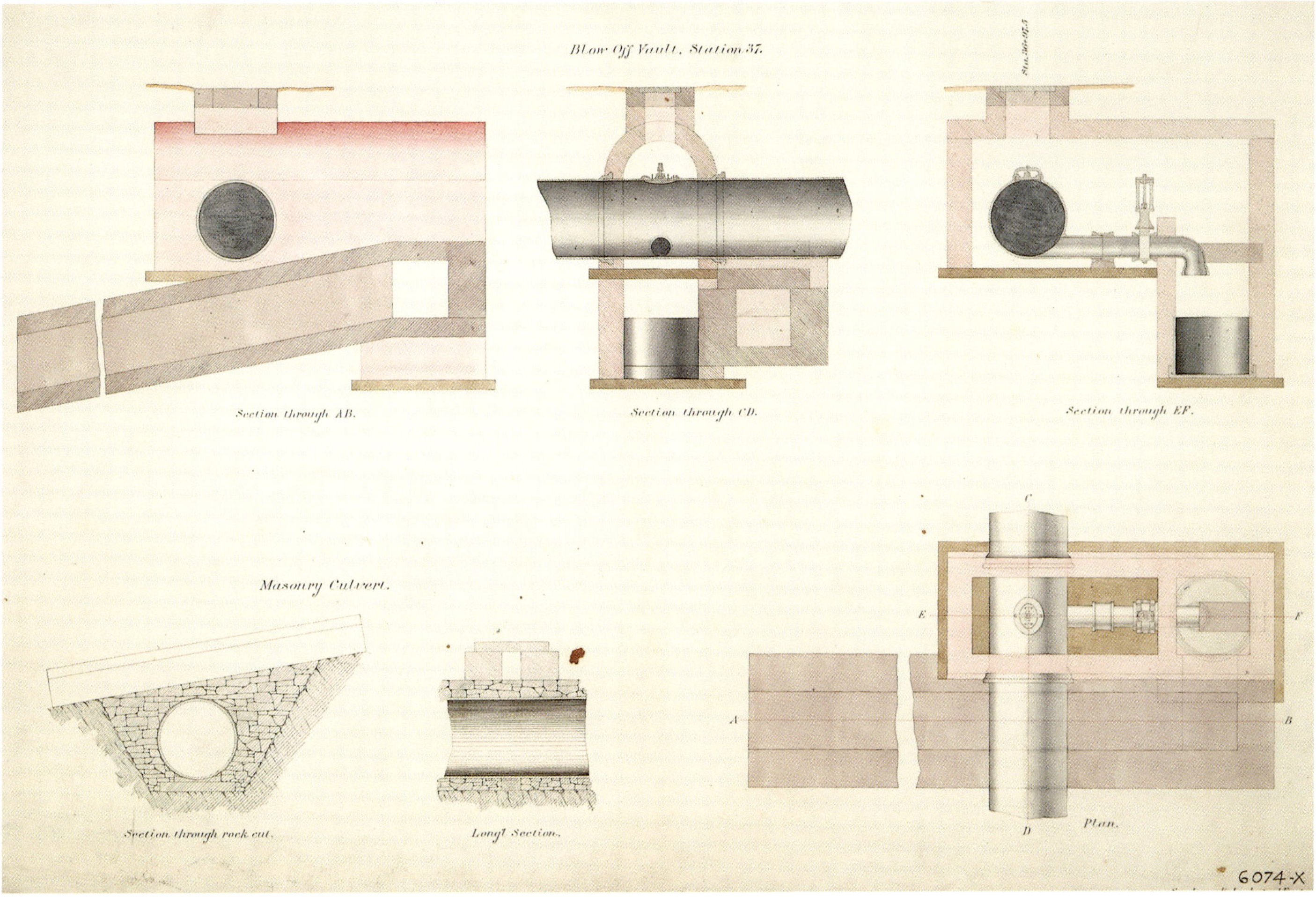

FIG. 19. *Aqueduct from old Kensico Reservoir to Williamsbridge Receiving Reservoir (the 48-inch line). Details of blow-off vaults. c. 1884.*

FIG. 20. *Williamsbridge Reservoir. Details of blow-off valves for the 36-inch lines. c. 1884.*

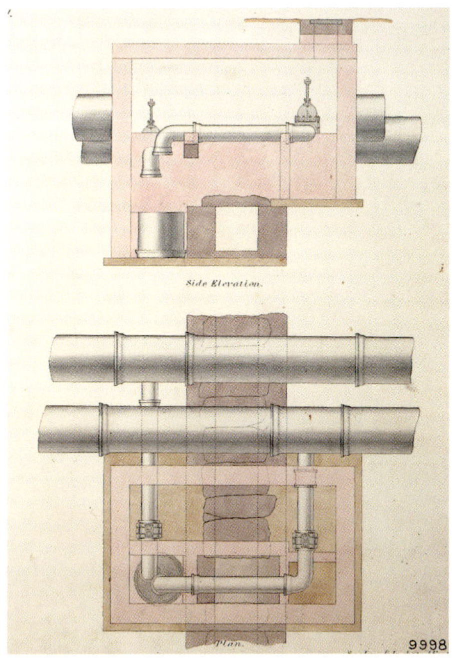

1875, the Croton Aqueduct was able to deliver a maximum discharge of 95 mgd (23 mgd above the original design capacity), but the population continued to grow. An ever-greater number of residents clamored for indoor plumbing, and manufacturing and industry were fast becoming important economic activities of the harbor city. Some of these businesses consumed huge quantities of water (for example, sugar refineries used up to 45 million gallons a year) and industry complained constantly about water shortages.[14] And yet, while New York slipped into another period of chronic water scarcity, there were times when millions of gallons were running to waste over the top of the Croton Dam. Though the Croton watershed yielded at minimum 250 mgd, according to department surveys from 1857, the aqueduct could not transport it all, in spite of the previous rounds of upgrading. Clearly, a second tunnel was needed to meet demand.

Preliminary surveys for a new aqueduct to carry the surplus water from the Croton had been made as early as 1873, but paradoxically a drought would delay that construction. The years 1876 and 1877 were very dry, and in an effort to locate a source that could be readily put to use, Department of Public Works officials reconsidered the Bronx River. First studied by Dr. Joseph Browne in 1798 and again

FIG. 21. *Williamsbridge Reservoir. c. 1890.*

by Canvass White in 1820, the river that was determined to be inadequate fifty-five years earlier now seemed an easily available supplemental supply that could alleviate the need for water in the expanding Bronx section of the city. Commissioner of Public Works Allan Campbell agreed that the best way to gain additional supply would be the construction of a New Croton Aqueduct. But given the depleted state of city finances (a legacy of the Tweed era), he thought it inadvisable to pursue such an expensive undertaking. Instead, Campbell outlined a plan that would utilize the waters of the Bronx and Byram Rivers in lower Westchester County. His strategy included a small 15-foot-high dam to be built at the outlet of the Little Rye Pond in order to create a 280-acre reservoir; a 45-foot-high dam across the Bronx River near the Kensico station of the Harlem Railroad to make a 230-acre reservoir; and a third dam, 15 feet high across the Byram River, that would produce a 75-acre reservoir. The water from these three modest reservoirs would be delivered through a 48-inch cast iron pipe running from the Kensico Reservoir to a new 120 million-gallon receiving reservoir [Figs. 19, 20, 21, 22] to be built at Williamsbridge (in what is now the Bronx).[15]

Work on this assembly of dams, the largest of which was the First Kensico Dam, began in 1880 and was completed by 1885. The impounded waters of the three dams collectively formed the fourth reservoir in the city system. The new 15.2-mile-long cast iron pipe, eventually known as "the 48-inch line," became the second of New York City's aqueducts. It delivered a rather modest 19 mgd.

But the new Bronx and Byram projects were tainted by residual Tammany Hall influence, since the Department of Public Works still employed Tweed-era appointees. As a result of past experience, civic leaders questioned the wisdom and soundness of the project. An 1882 report by the Union League (formed in 1863 to support the Union cause, and by 1873 the largest club in New York) stated that if a new aqueduct was required, "one from the Croton would doubtless have been selected in lieu of one from the Bronx. The selection of the latter must now be regarded as [an] official mistake not likely

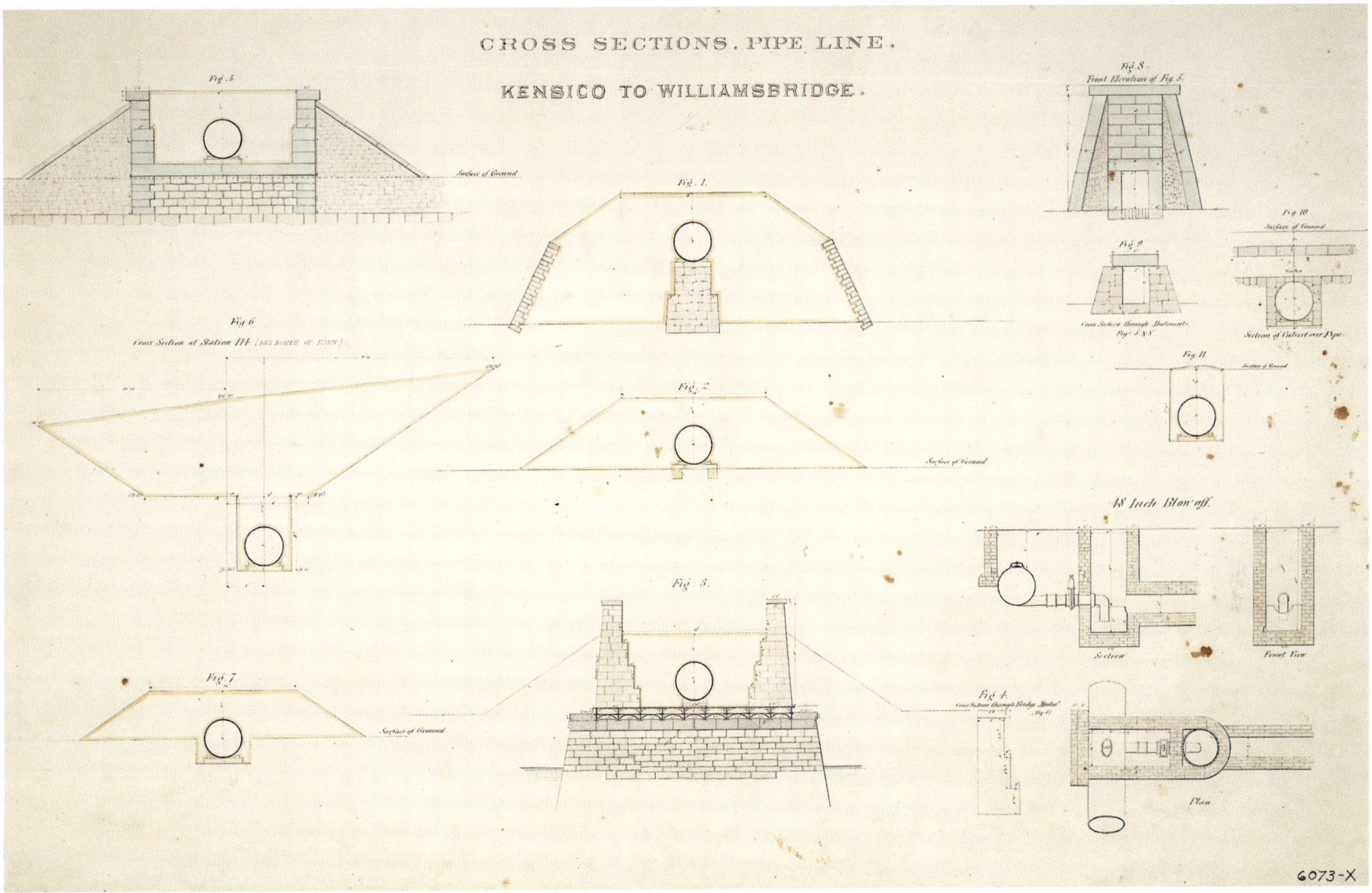

FIG. 22. *Aqueduct from the old Kensico Reservoir to Williamsbridge Receiving Reservoir (the 48-inch line). Cross sections. c. 1879.*

to have been made by a body of intelligent commissioners." More importantly, the report advocated a change in the approach to waterworks: "The Club Favors a Comprehensive System, and the Construction by a Commission to be Appointed by the Governor and Senate, of such Additional Works as may be needed."[16]

As work progressed on the controversial Bronx and Byram River systems, changes to the water authority hierarchy were implemented, allowing for more effective improvements to proceed. Eighteen months after the Union League made its case for a commission, the state legislature passed "an Act to provide new reservoirs, dams and a new aqueduct with the appurtenances thereto, for the purpose of supplying the City of New York with an increased supply of pure and wholesome water."[17] The law, which established the Aqueduct Commission and made it responsible for the building of the system, was clearly an effort to shift responsibility to a semi-independent body, where state appointees would balance the politics of City Hall. The commission included three such appointees, as well as the city's mayor, comptroller, and commissioner of the Department of Public Works. The Croton Aqueduct Department was dissolved, but the Department of Public Works remained. It retained responsibility for developing plans and for the operation of the system, once structures were completed.

At its first meeting, the new Aqueduct Commission directed the Department of Public Works to:

> secure all the water that can be obtained from the Croton River and its tributaries with a proper and suitable aqueduct and reservoirs . . . to develop a plan for the immediate construction of a work on the East Branch, the Sodom Reservoir . . . and to prepare plans for two routes for an aqueduct, one to the [planned] Quaker Bridge Dam site and one to the [then existing] Croton Dam, at the terminus of the present Aqueduct.[18]

In 1881, and again in 1883, Isaac Newton, chief engineer of the Croton Aqueduct, presented materials that outlined his plans for the future of the Croton system. Newton did a careful analysis on storage and consumption and demonstrated

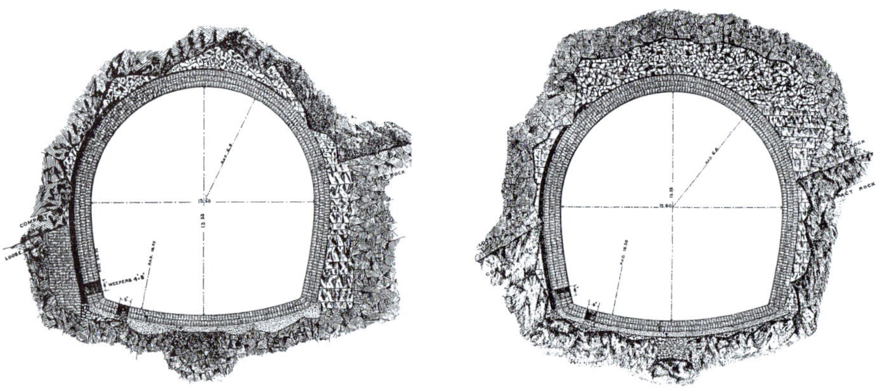

FIG. 23. *New Croton Aqueduct. Finished shaft and headhouse details. December 31, 1886.*

FIG. 24. *New Croton Aqueduct. Cross sections of aqueduct in compact and loose rock. 1886.*

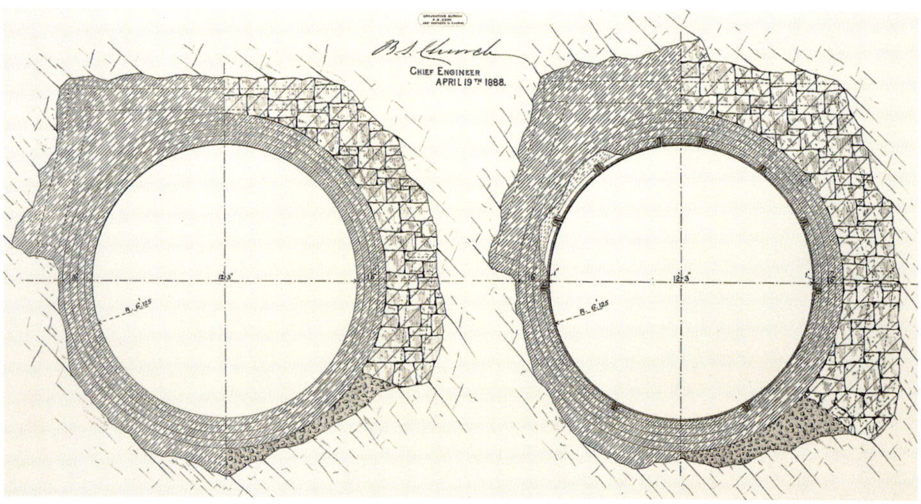

FIG. 25. *New Croton Aqueduct. Cross sections of pressure tunnel construction. Iron-lined pressure tunnel assembly at right. April 19, 1888.*

his argument for a large Croton River reservoir as far downstream in the watershed as possible [Fig. 54]. His studies showed that earlier plans for multiple reservoirs had miscalculated watershed areas and had thereby overstated the potential yield. Newton also showed that the fifteen-reservoir plan would fail to collect water from over 115 square miles of the lower Croton watershed, an area whose water could only be captured by a larger dam somewhere well below the site of the first Croton dam: "The location of the dam, which supplies the aqueduct with water, is the pivot on which turns the whole question of the availability of Croton water shed for a large additional supply for the city of New York."[19]

Aqueduct Commissioner Hubert Thompson and his associates submitted materials that included the original Quaker Bridge plans as well as schematic aqueduct plans drawn up by Isaac Newton. Thompson favored Newton's plan to build one large storage reservoir rather than multiple smaller impounding areas. To bring the water from the new reservoir, Newton preferred a grade tunnel (in which water flows on the hydraulic gradient, not under pressure), circular in cross section and capable of carrying 250 mgd. Thompson and his commission supported Newton's view that the aqueduct should be a tunnel, which would be relatively straight, instead of a grade structure that would have to follow the contours of the land. By being well below the surface of private property, a tunnel would reduce the costs of damage and legal claims. The structural problems with the Croton aqueduct may have also influenced the decision to build a tunnel, as the original surface structure, with its elevated embankments and shallow cover, had proved to be a demanding maintenance problem. The commissioners' plan proposed a change in the route that followed the Hudson riverbank, more or less along the line of the Old Croton Aqueduct that Newton had developed. Thompson moved the line east, straightening and aligning it so the aqueduct could draw water directly from the old Croton Lake

FIG. 26. *New Croton Aqueduct. Branch aqueduct for Jerome Park Reservoir under construction. September 15, 1899.*

until the final decision on a future reservoir was made. The plans also called for a terminal reservoir to be built at Jerome Park, which would supplement the amount of storage within the city limits and provide a reserve supply during system shutdowns [Figs. 27, 28].

The New Croton Aqueduct was designed for a capacity 3.3 times that of the original Croton Aqueduct. Construction started in 1884, and though it was not substantially completed until 1893, the structure was put into use as early as 1890. Revisions, driven by economics, resulted in the adoption of a horseshoe-shaped cross section (similar to the first Croton Aqueduct) wherever possible [Figs. 23, 24]. Circular tunnels were used in certain sections where the aqueduct was under pressure [Fig. 25]. The new aqueduct, like its predecessor, was built with brick and enclosed with a continuous brick arch. This time the walls of the structure were built to one foot thickness. Because the structure was built primarily within tunnels excavated out of the bedrock, the stone of the surrounding tunnel reinforced the brickwork. This configuration was very solid and resulted in a structure able to withstand the pressures and stresses of 250 million gallons of rushing water per day.

The tunnel ranges from 20 to 390 feet below the surface. There is approximately 8 million cubic feet of masonry wall in the aqueduct, making the structure among the largest brick constructions ever made

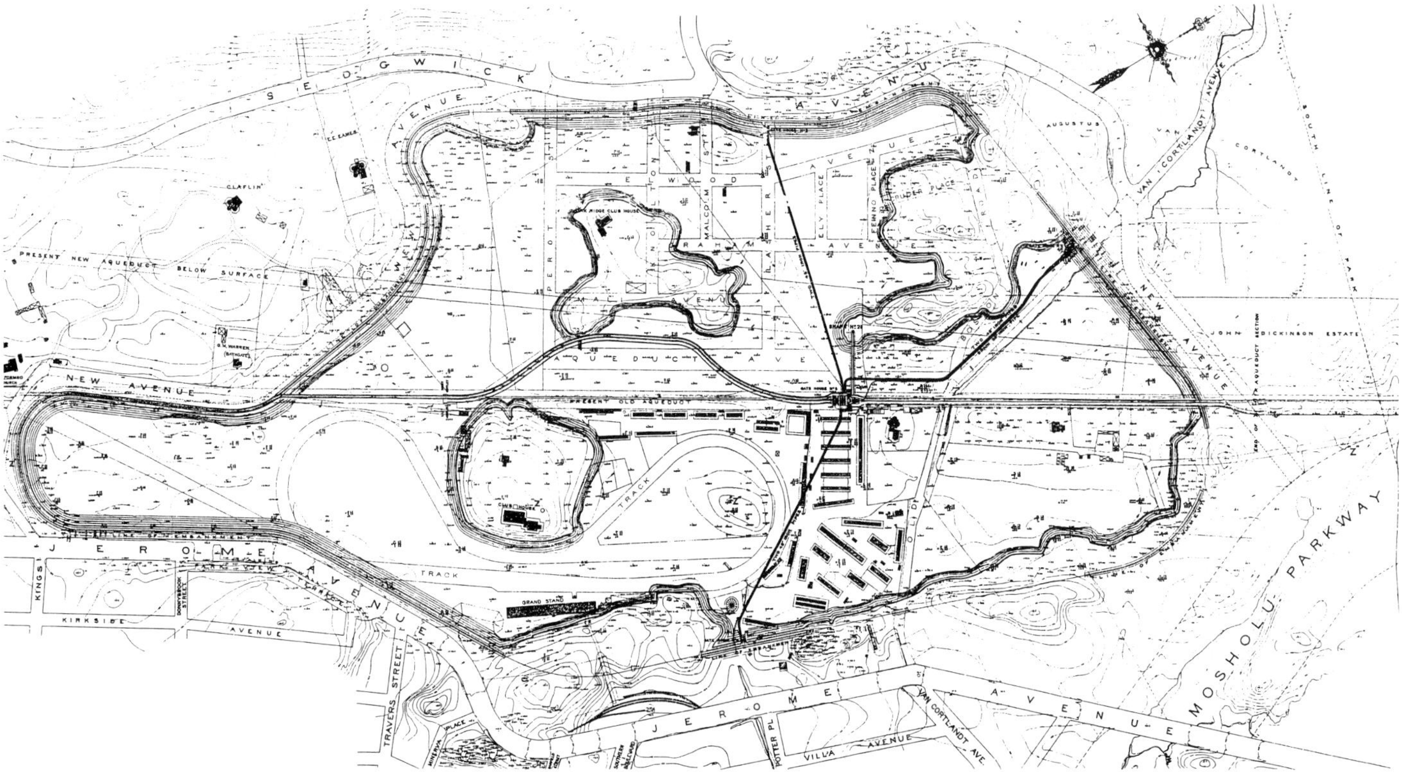

FIG. 27. *Jerome Park Reservoir. Early plan showing proposed islands and peninsula (later eliminated to increase capacity). 1895.*

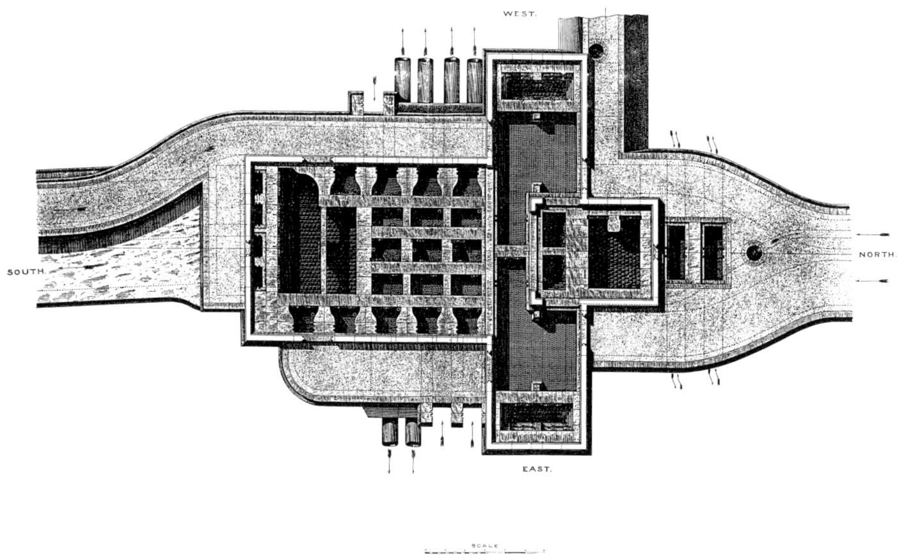

FIG. 28. *Jerome Park Reservoir. Plan of gatehouse. 1895.*

[Fig. 26]. It was, like the Croton Aqueduct before it, an unparalleled engineering and construction accomplishment. At the time, the tunnel from the Croton Lake to the Jerome Park Reservoir was the longest tunnel ever built. The 7½-mile inverted siphon that connected the Jerome Park Reservoir to the 135th Street Gatehouse and by which the New Croton Aqueduct passes under the Harlem River into upper Manhattan was far more complex than any siphon ever attempted. It was the first siphon designed for dewatering. Additionally, the diameter of the siphon decreased at the deepest sections under the river, thereby increasing the velocity of water flow and acting to flush out accumulated silt. The siphon, beginning at the point where the free-flowing tunnel ends, steps down from a 12 foot 3 inch diameter brick conduit to a 10 foot 6 inch tunnel below the river. From its starting point at the New Croton gatehouse, the New Croton Aqueduct travels forty-two miles to

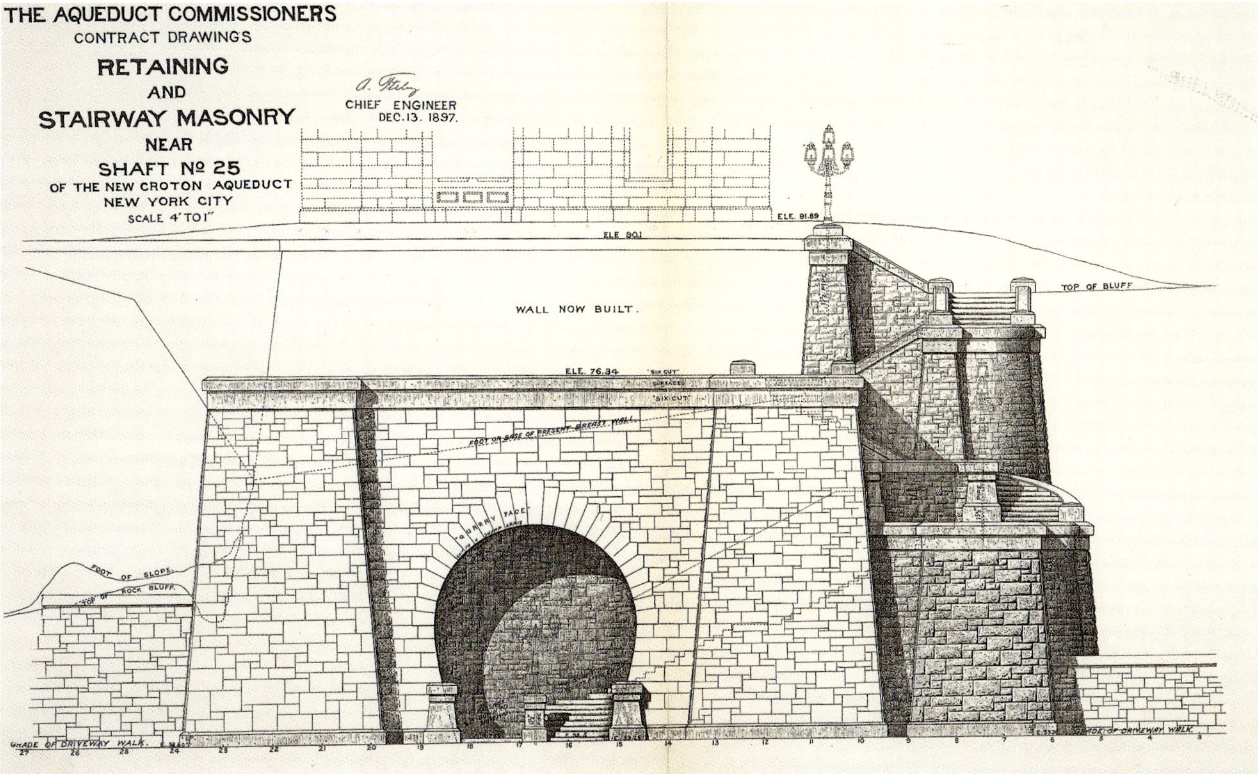

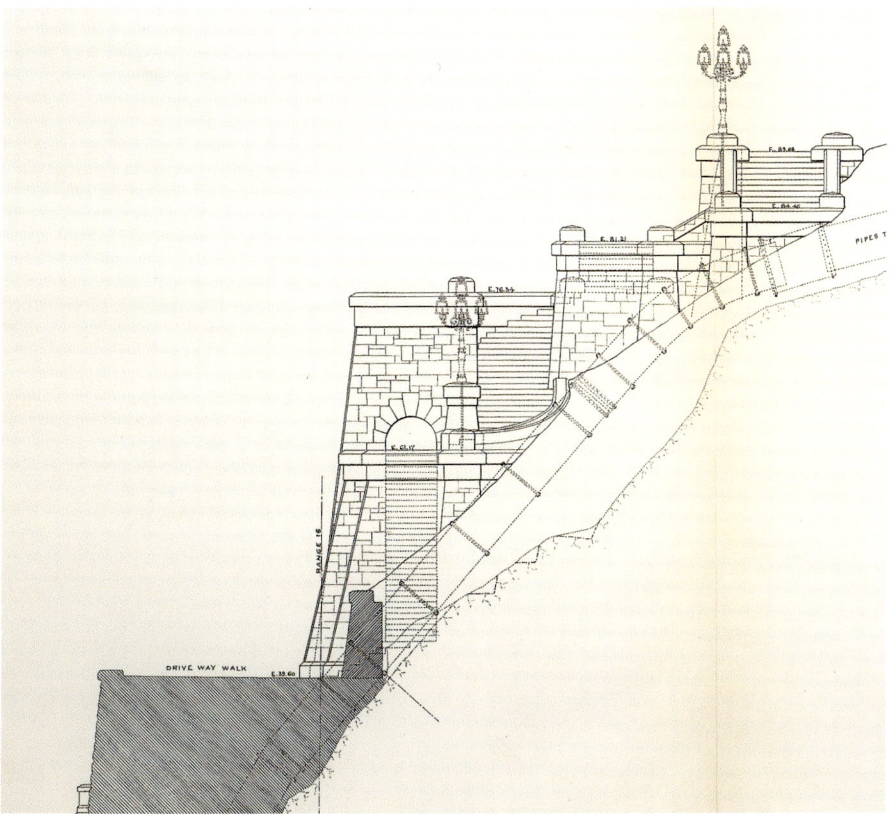

FIGS. 29 AND 30. *New Croton Aqueduct. Retaining walls and stairway built on bluff above the Harlem River at Shaft 25. December 13, 1897.*

FIG. 31. *New Croton Aqueduct. Gatehouse at Central Park Reservoir.*

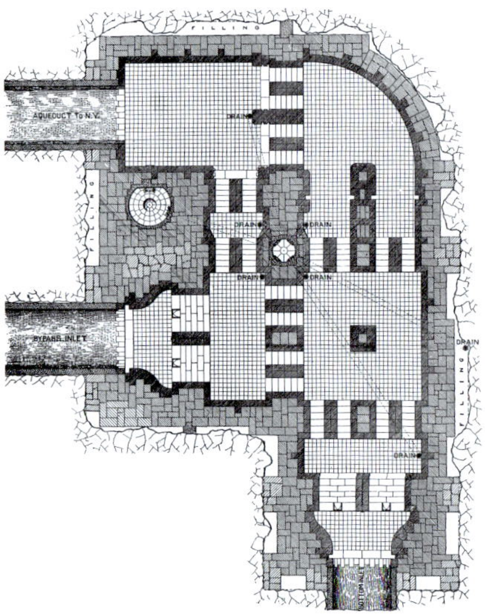

FIG. 32. *New Croton Aqueduct. New gatehouse at Old Croton Dam. c. 1886.*

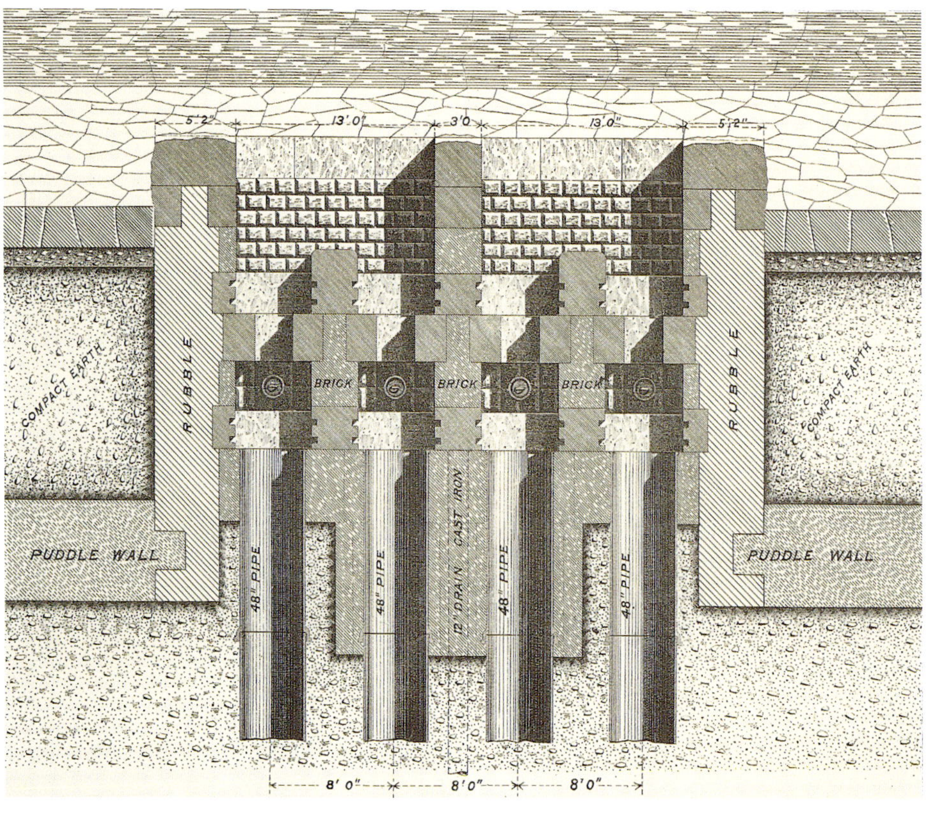

FIG. 33. *New Croton Aqueduct. Gatehouse at Central Park Reservoir. Plan showing four 48-inch incoming water mains. January 3, 1888.*

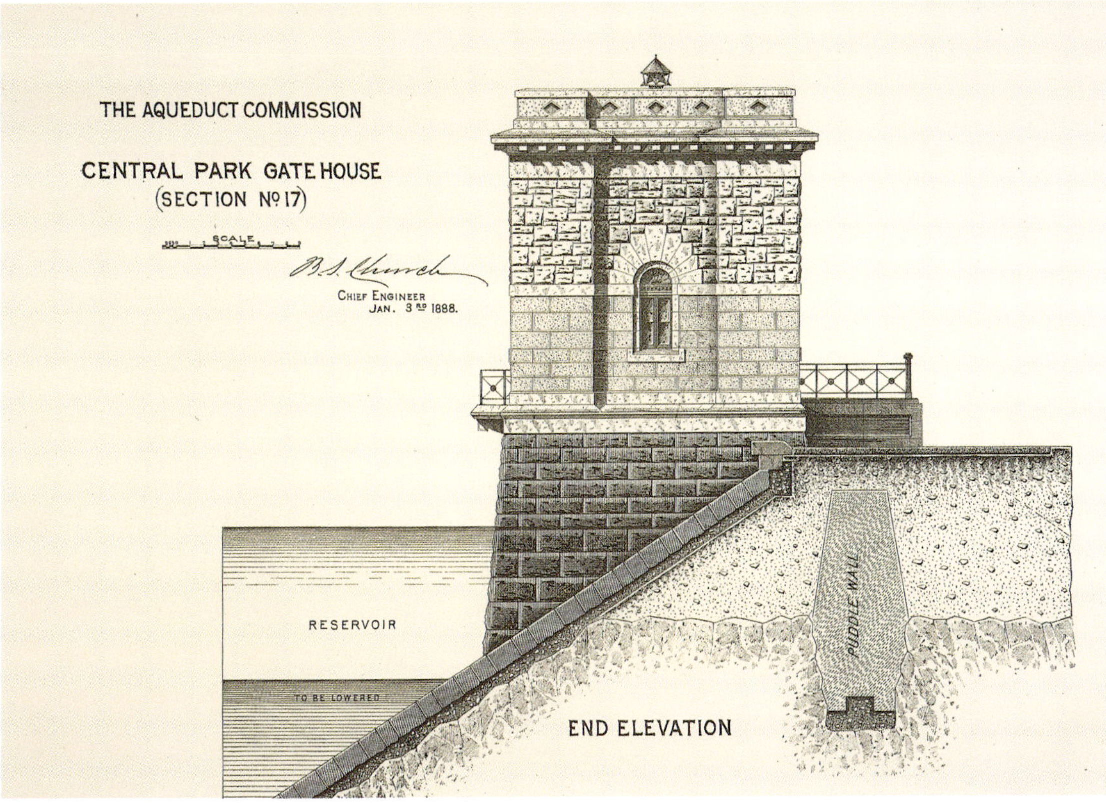

FIG. 34. *New Croton Aqueduct. Gatehouse at Central Park Reservoir. Side elevation and section through embankment. January 3, 1888.*

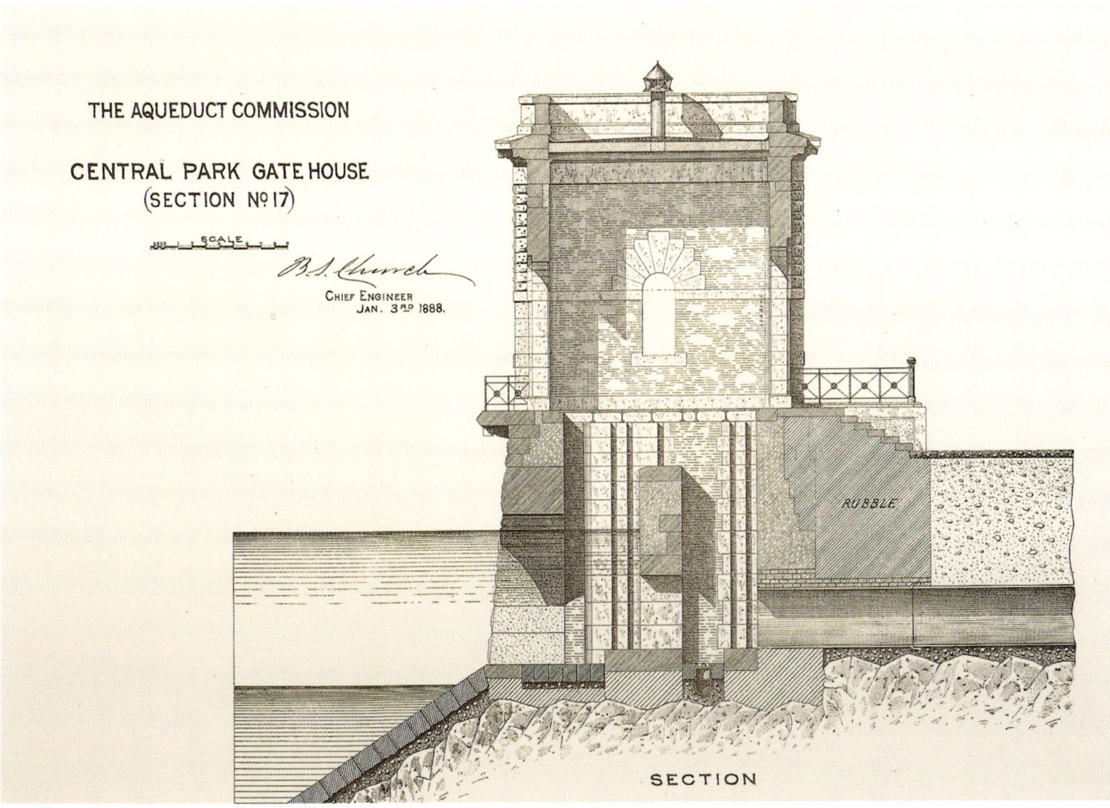

FIG. 35. *New Croton Aqueduct. Gatehouse at Central Park Reservoir. Section through sluice gates and 48-inch main. January 3, 1888.*

its termination at the 135th Street Gatehouse. Twelve lines of 49-inch-diameter cast iron pipes composed the final link from the 135th Street gatehouse to the Central Park Receiving Reservoir [Figs. 31, 32, 33, 34, 35].

Progress on the aqueduct was, however, impeded by scandal. From the beginning, engineers had trouble getting the quality of work that was required:

> Few contractors seemed motivated by the idea of doing good work. The very nature of the work on the tunnel, extending for miles and divided among many contractors, invited fraud and deception, the direct result of which was work of poor quality. Empty spaces were found behind the brick lining of the tunnel, spaces filled with loose rubble, and false partitions created to hide cavernous holes between the lining and the excavation—conditions that would render the whole conduit useless under the pressure of water.[20]

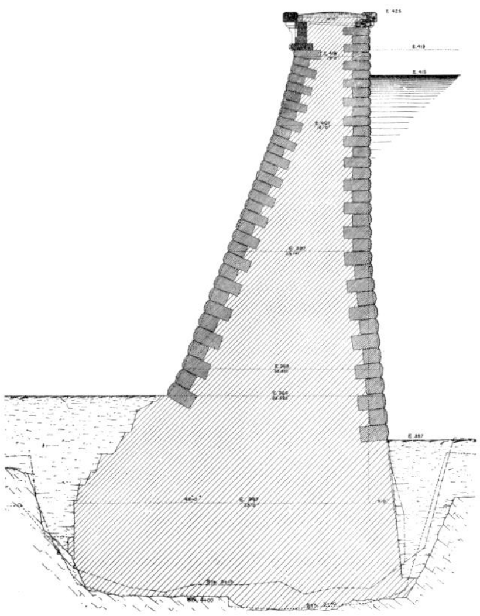

FIG. 36. *Sodom Dam, East Branch Reservoir. Cross section. 1888–93.*

By 1888, word of the inferior work had filtered back to officials and an investigation was ordered. A committee of two engineers and five masons inspected the entire line of aqueduct and documented the deficient work. Large areas were ordered removed and rebuilt, and repairs were made by injecting grout behind the completed brickwork to fill the smaller voids. As many as 384,000 barrels of cement were used for grouting voids, an indication of the magnitude of the problem. The repairs delayed the completion of the aqueduct by thirty months.

By 1890, five and a half years after construction started, work was sufficiently advanced to allow water to flow through the aqueduct. The New Croton Aqueduct had been in service less than a year when, in 1891, consumption jumped from 110 to 165 mgd. By 1895 it would reach 183 mgd—an increase of 73 mgd in four years. Shortages were once again a predominant concern.

Once the New Croton Aqueduct was under construction, the commission resumed its focus on increasing water storage in the Croton drainage. The public was skeptical about the plans for an unprecedentedly high dam. Even with popular support, a project of such grand scale would be years away at best, so the commission returned to Isaac Newton's earlier strategy of building numerous smaller dams higher in the watershed. In 1888, work began to control the East Branch of the Croton River. The water was impounded by three different structures that became the fifth, sixth, and seventh dams to be built by New York City in the Croton watershed. These dams backup water into two drainages, the East Branch and the Bog Brook, fill two separate valleys, and create the third and fourth reservoirs of the Croton system: the East Branch (first called the Sodom) Reservoir and the Bog Brook Reservoir. The Sodom Dam, an impressive wall of solid masonry with a near-vertical face upstream and a gently battered face downstream, rose to a height of almost 75 feet above grade and extended down into the bedrock about 30 feet [Figs. 36, 37]. The dam is flanked by a 600-foot-long earthen embankment with a masonry core wall that closes a low area in the ridge to the south. The main body of the dam is built completely of rubble masonry—rough-cut stones stacked like bricks and set in mortar—and faced on both sides with dressed granite blocks [Figs. 38, 39]. It was the only dam built in the New York City system that used this traditional method of masonry block work. Though the dam employed the most traditional construction assemblies, its methods

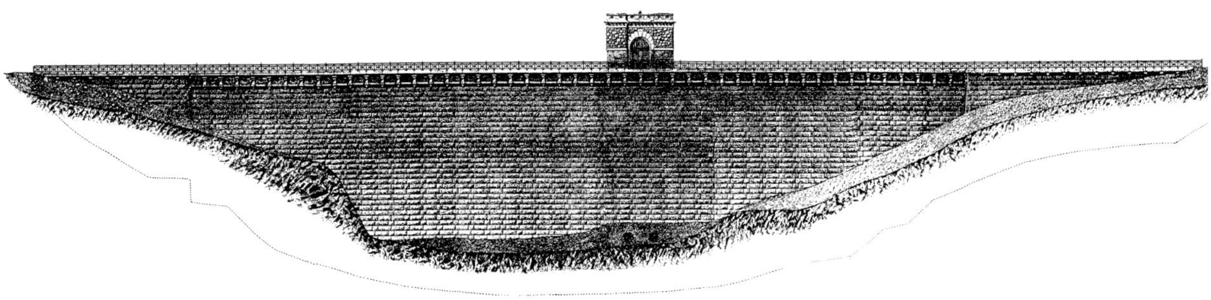

FIG. 37. *Sodom Dam, East Branch Reservoir. Downstream elevation. 1888–93.*

FIG. 38. *Sodom Dam under construction showing use of solid stone blocks throughout. 1888–93.*

FIG. 39. *Sodom Dam, East Branch Reservoir. Downstream face. c. 1895.*

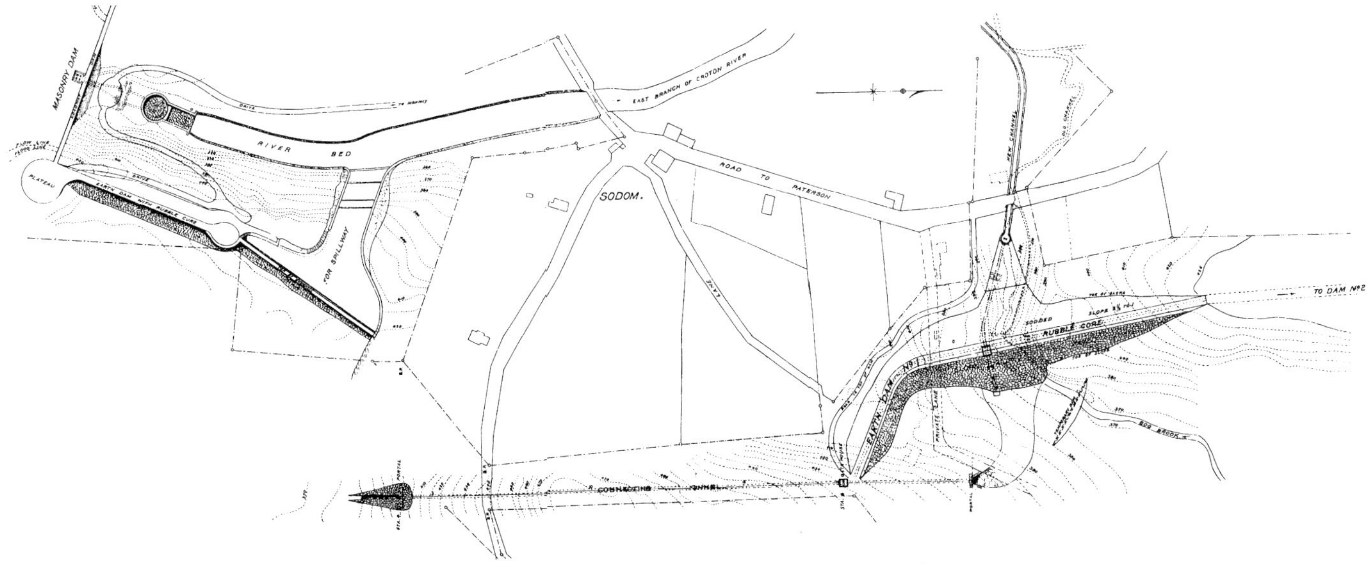

FIG. 40. *East Branch Reservoir. Sodom Dam on left and Bog Brook Dam on right. 1889–93.*

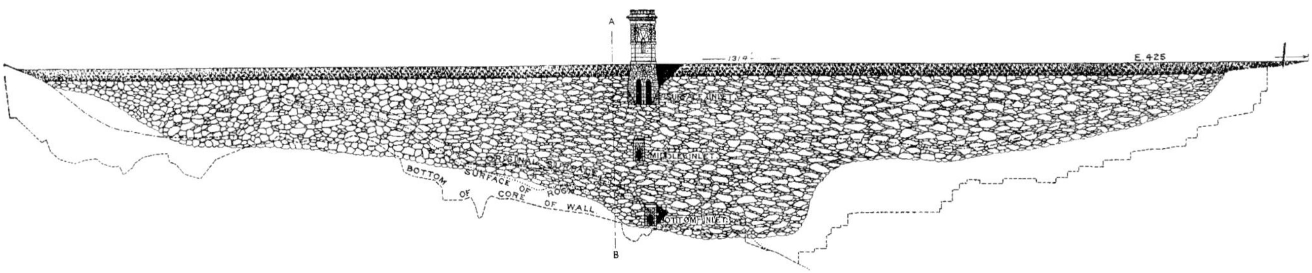

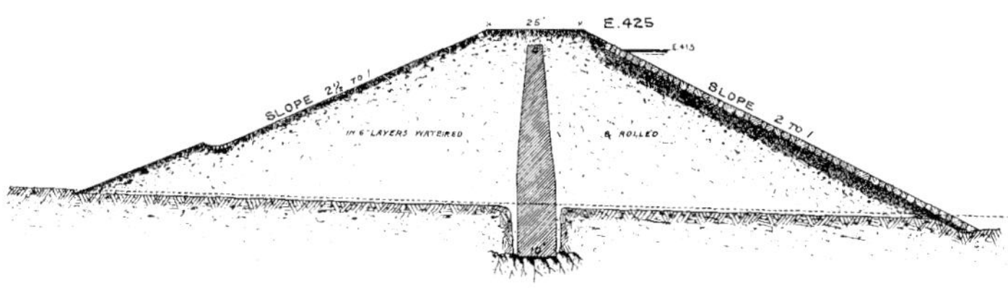

FIG. 41. *Bog Brook Dam, East Branch Reservoir. Upstream elevation and cross section. 1889–93.*

FIG. 42. *Titicus Dam. Upstream face and gatehouse prior to impounding of water. c. 1895.*

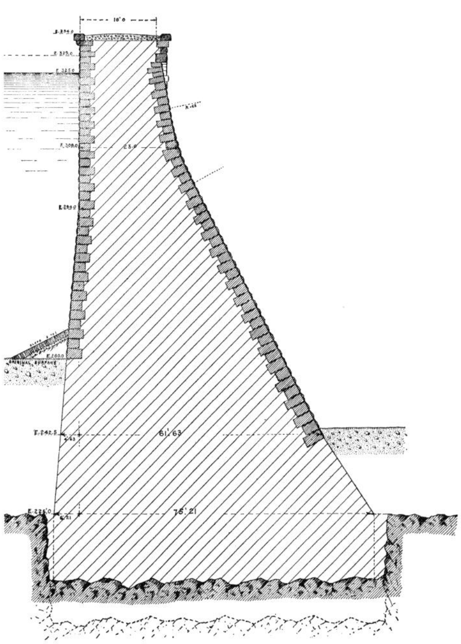

FIG. 43. *Titicus Dam. Cross section and corbelled coping detail. 1890–96.*

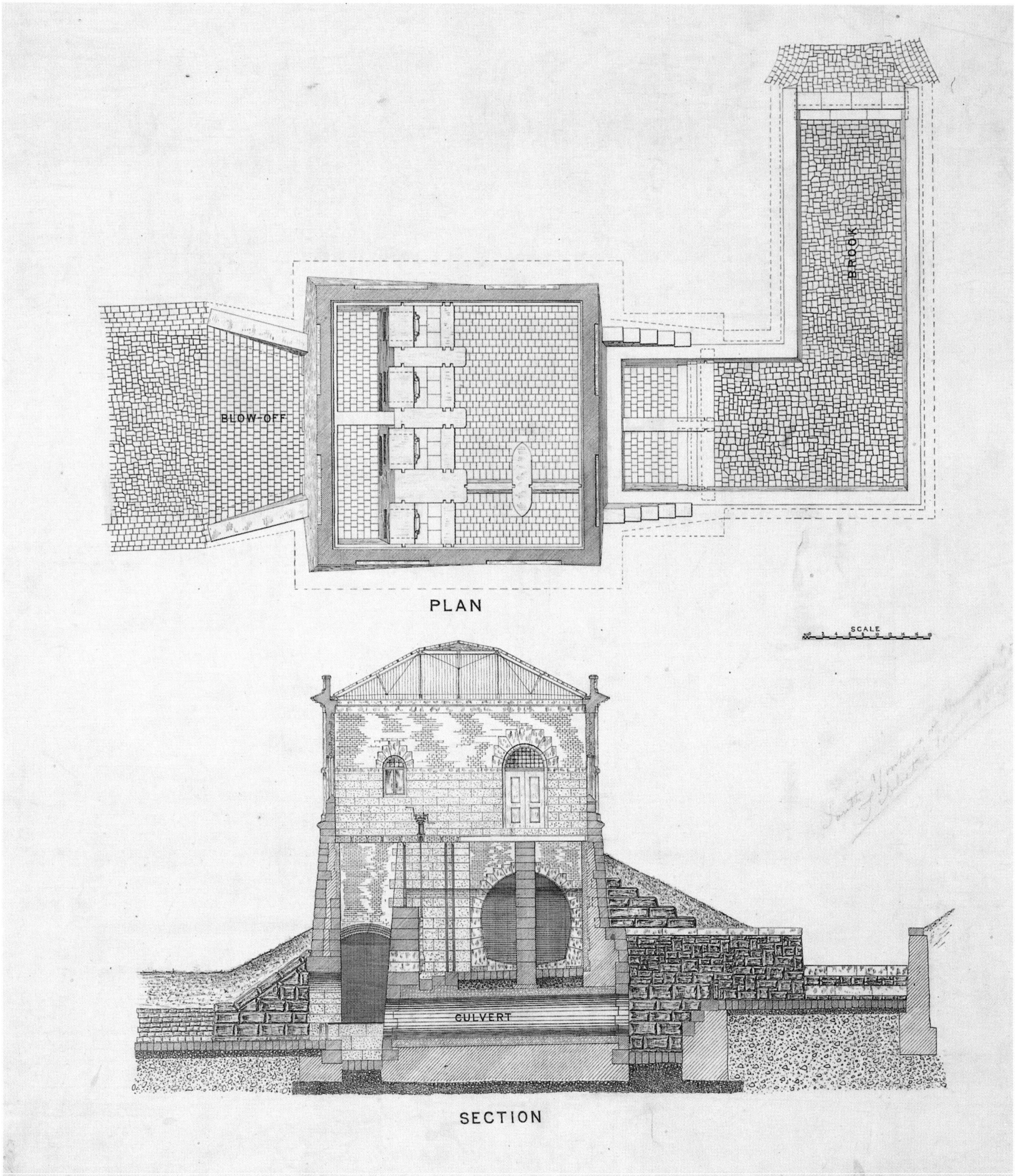
BROOK
BLOW-OFF
PLAN
SCALE
CULVERT
SECTION

of construction reflected evolving advances in building technology. The Sodom Dam was constructed using a two-inch-diameter steel cable stretched across the valley between two timber towers anchored to the bedrock. An overhead hoist, with runners and pulleys, ran along these steel cables to move the stones into place. As noted by Charles Weidner around 1960, "The overhead cableway has been for over half a century universally accepted as essential to the construction of high dams . . . this instance of its use for that purpose may well be the first."[21]

One year after work started on the Sodom, contracts were let on the Bog Brook Dams. These two dams would be built as earthen embankments with masonry core walls embedded in solid rock [Figs. 40, 41]. The dams were modest in height but spanned long, wide valleys. Dam #1 is 1,340 feet long and 60 feet high; dam #2 is 1,956 feet long and 24 feet high. Both are covered in sod on the downstream bank and their inner slopes are paved with dry rubble masonry. A circular tunnel 10 feet in diameter connects the two reservoirs. The tunnel, 1,778 feet long, is lined in brick and was excavated entirely from solid bedrock. The Bog Brook Reservoir came into service in April 1892.

During 1890, two more storage reservoirs were started. The city built an earth and masonry dam at a site originally designated "M," near Purdy Station on the Titicus River (a tributary downstream from the confluence of the three main branches). For several years the Titicus Dam would be the largest structure with the largest reservoir in the system. The masonry portion of the dam is particularly beautiful, with subtle rusticated stonework and a simple geometric cornice [Fig. 43]. A gatehouse rises like a tower from within the reservoir [Fig. 42]. The Titicus Dam is 1,519 long. It has a central masonry section 534 feet long with a 200-foot long, 109-foot-high overflow weir [Figs. 46, 47]. The earthen wings were built using a cross section similar to the Bog Brook: an earthen embankment with a heavy masonry core wall [Figs. 49, 50, 51]. Alphonse Fteley, chief engineer for the

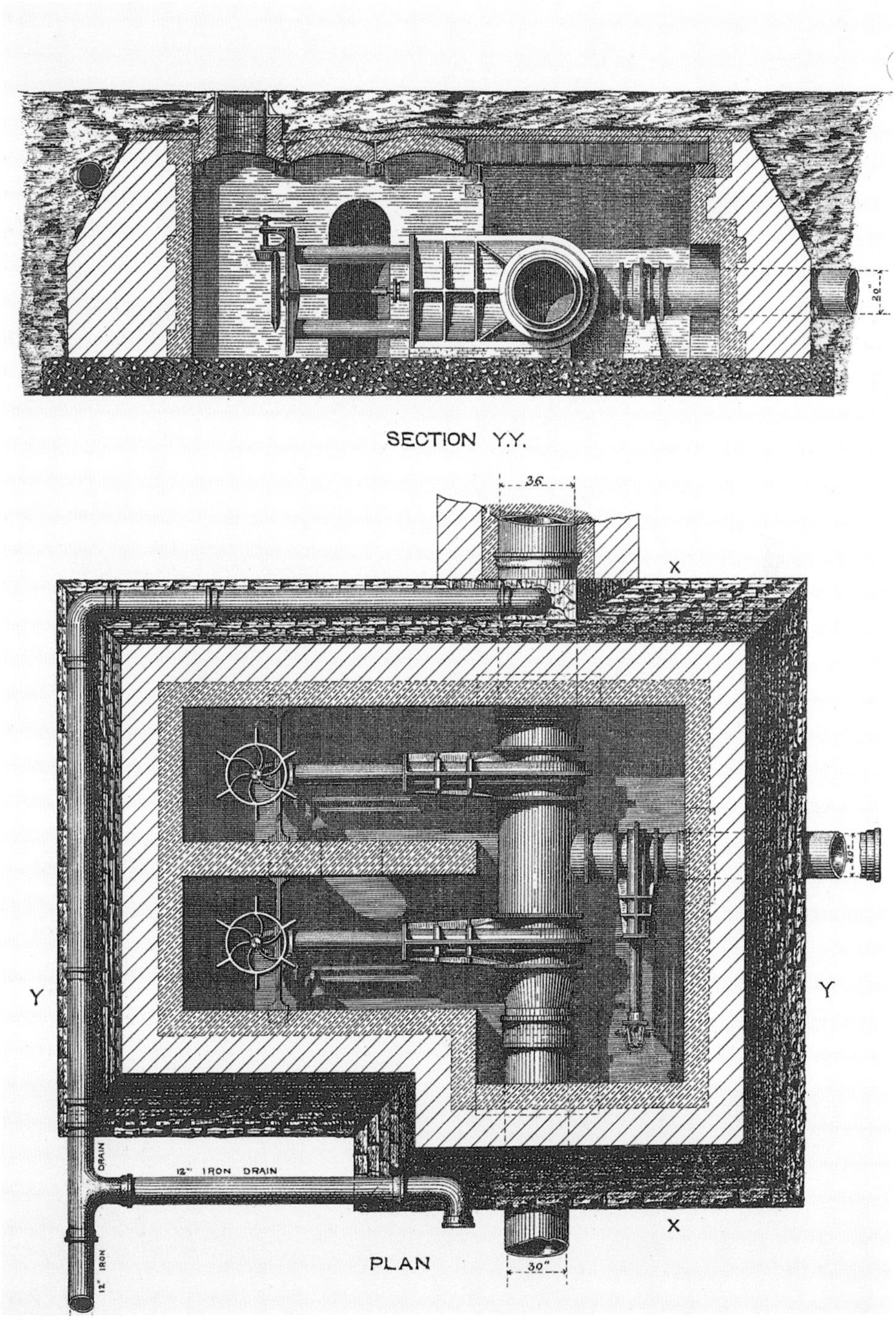

FACING PAGE
FIG. 44. *New Croton Aqueduct. Blow-off at Tibbets Brook, Yonkers, New York. Plan and section. c. 1886.*

FIG. 45. *New Croton Aqueduct. Blow-off at Shaft 24. Plan and section. February 17, 1890.*

FIG. 46. *Titicus Dam. Downstream face and stepped spillway. c. 1895.*

FIG. 47. *Titicus Dam. Downstream face with water cresting over spillway. c. 1896.*

Aqueduct Commission since 1888, drew up the plans. The gates of the dam were closed and water began to accumulate in the Titicus Reservoir in January 1895.

The Department of Public Works, despite its official role as planner and operator of the system, began its own building project in 1889 on the Muscoot River, a tributary on the western side of the Croton River watershed. The 1,280-foot-long earthen embankment, similar in construction to the West Branch Dam, was named after the local town of Amawalk. The Amawalk Dam created a reservoir about three miles long covering close to 600 acres that holds a maximum of 7 billion gallons. Completed in 1897, the Amawalk Reservoir remains in service.

While the Department of Public Works pushed forward with its Amawalk Dam, the Aqueduct Commission was building a pair of dams that together would form the 998-acre, 9-billion-gallon-capacity West Branch Reservoir. The reservoir fills to within a quarter mile of the Bog Brook Dam, which all but eliminates any natural flow of the West Branch waterway. The main dam is a 1,800-foot-long earthen structure with a masonry core wall and a solid stone-lined spillway [Figs. 52, 53]. The auxiliary dam, built of earthen construction to close a gap in the hills to the east, is 750 feet long. The West Branch Reservoir later became an integral part of the modern water system when it was incorporated into the Delaware System as a receiving and balancing reservoir for the Delaware Aqueduct.

The plans for the New Croton Aqueduct had, from the beginning, included the eventual construction of a large storage reservoir to replace the modest and aging Old Croton Dam. Early plans sited the proposed structure at the Quaker Bridge, in a deep, narrow part of the Croton River ravine, 4½ miles downstream from the first Croton Dam and about three miles from the Hudson River. The reservoir would flood 3,900 acres and have a maximum capacity of 32 billion gallons. The dam would reach nearly 270 feet in height, an unprecedented 100 feet higher than any dam previously built. The structure was referred to as the Quaker Bridge Dam, named after the local Croton River crossing. Though Newton's plans were

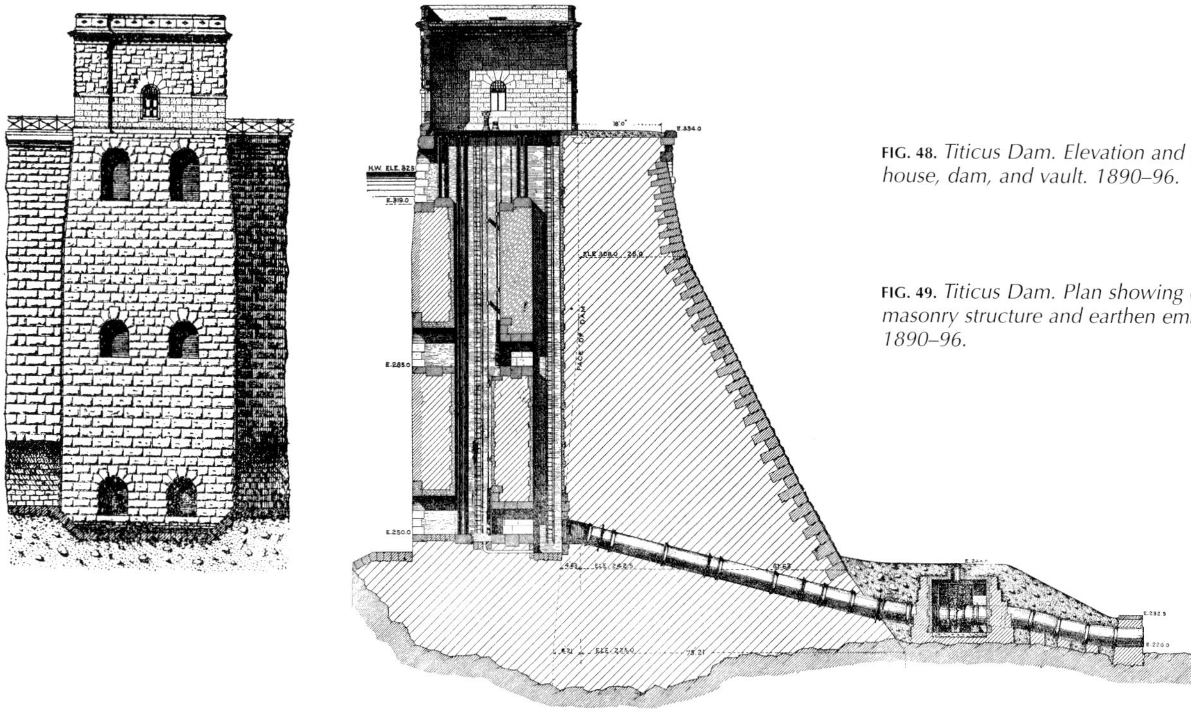

FIG. 48. *Titicus Dam. Elevation and section of gatehouse, dam, and vault. 1890–96.*

FIG. 49. *Titicus Dam. Plan showing central masonry structure and earthen embankments. 1890–96.*

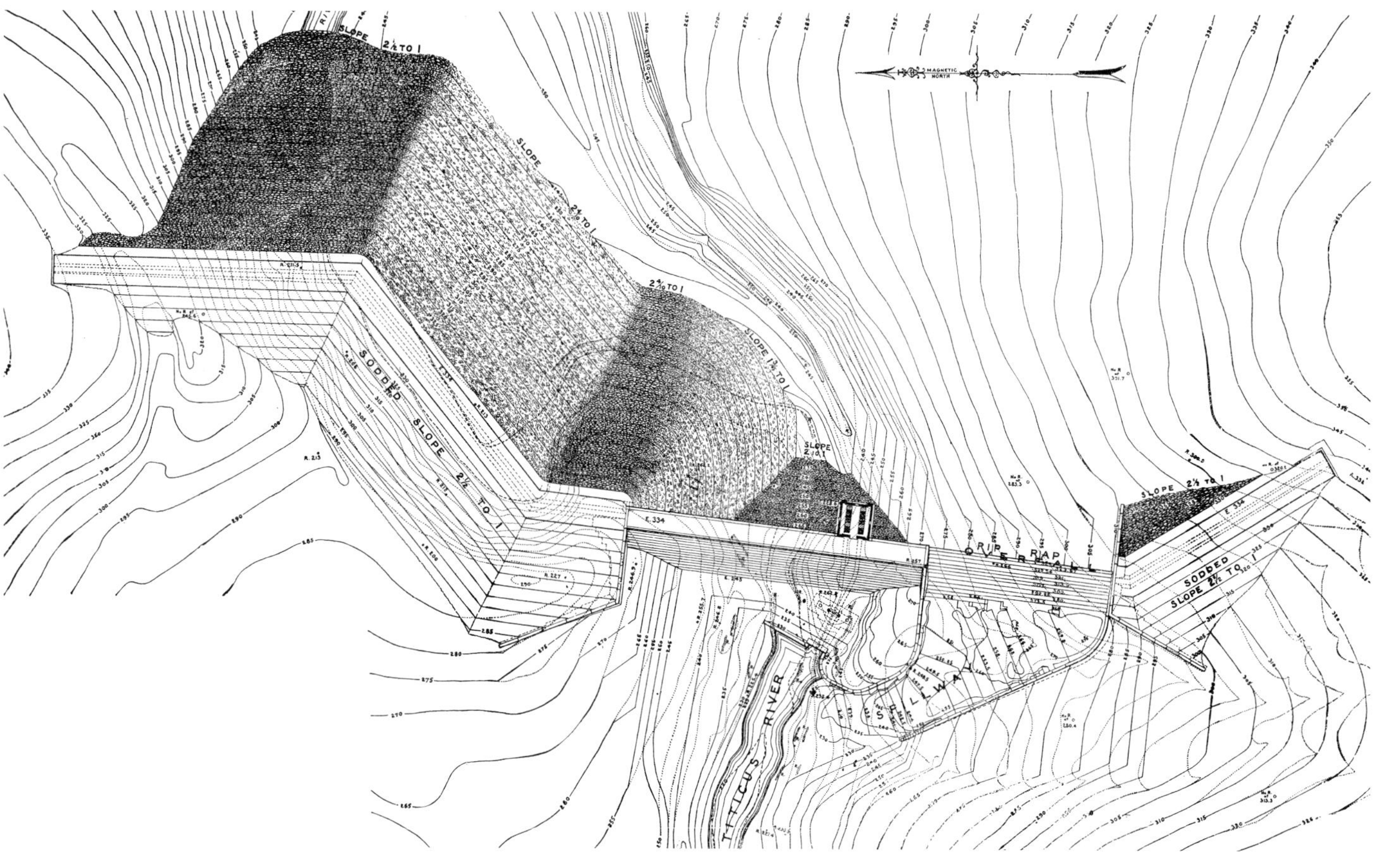

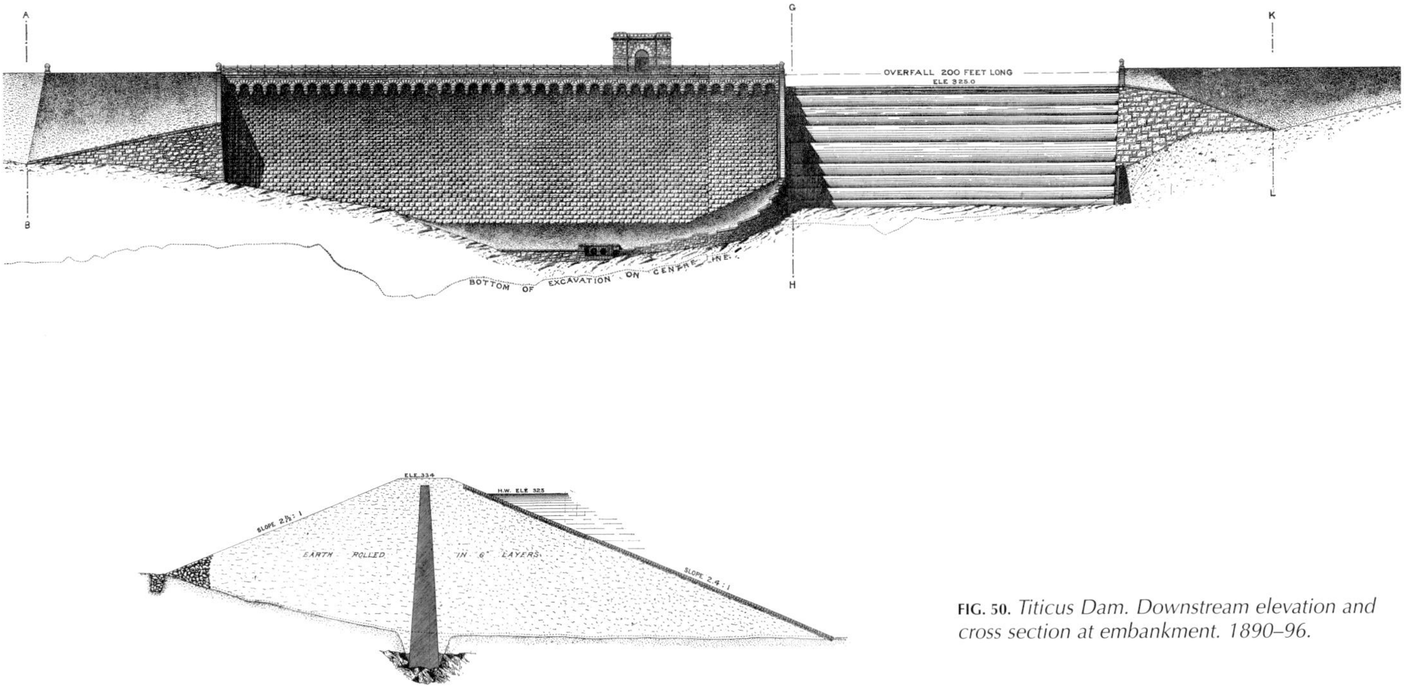

FIG. 50. *Titicus Dam. Downstream elevation and cross section at embankment. 1890–96.*

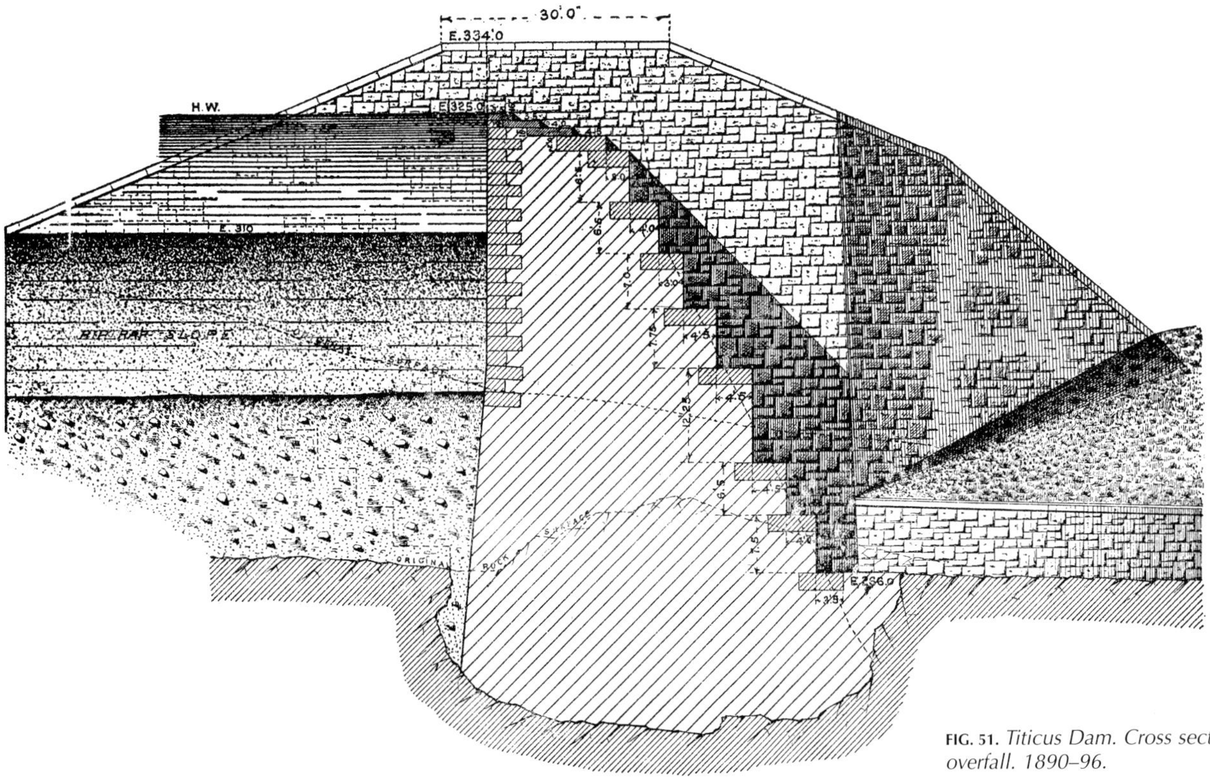

FIG. 51. *Titicus Dam. Cross section of spillway overfall. 1890–96.*

prepared for the Department of Public Works, implementation of the design and administration of construction remained the responsibility of the Aqueduct Commission [Fig. 59]. Because the design, impressive as it was, exceeded all previous attempts at dam construction, the Aqueduct commissioners ordered that the proposed structure, as well as large masonry dams in general, be further researched by their own engineers. B. S. Church, chief engineer of the Aqueduct Commission and Alphonse Fteley, Church's deputy at the time, led the study. Mathematical investigations were assigned to Edward Wegmann, the engineer who contributed his expertise to a number of important Croton structures and later wrote the first definitive engineering history of the early New York City water supply system. The Aqueduct Commission engineers studied the math, established safe limits for pressures within the materials, refined the cross sectional profiles and, in general, reviewed the principals, reliability, and conclusions of the engineering.

The Fteley-Wegmann team recommended a straight (rather than curved) dam in plan but modified the original overall dimensions and refined the shape. Practical adjustments were made by combining some sloping sections with vertical sections. The team produced a cross-sectional profile based entirely upon the load of the horizontal thrust of the retained water: vertical forces were considered negligible and omitted from the calculations. Nor was wind pressure considered. Detailed calculations allowed a specific shape to be designed and tested using the emerging science of statistics. The revised profile and the one developed by the Department of Public Works engineers were tested against the formulas, and the pressures were found to be in the range of 30,800 pounds per square foot (at the downstream face of the dam) to 33,200 pounds per square foot (at the base of the upstream site). These numbers represented the extreme stresses on the structure, and the studies concluded that the first design would not safely meet all the requirements of a dam of this dimension. The detailed calculations, backed by experience with material strengths gained in earlier New York waterworks projects, allowed the commission engineers to present a convincing case for their huge dam's structural stability.

Despite the fact that the Aqueduct Commission engineers were generally supportive of the Olive Bridge plans of Newton and the Department of Public Works, the fact that they raised any questions at all was interpreted as a challenge to the overall strategy of a high dam. The commissioners appointed a Board of Experts, which consisted of respected civil engineers from outside the government, to consider the revised Quaker Bridge Dam proposal and make recommendations as to its construction and the appropriate

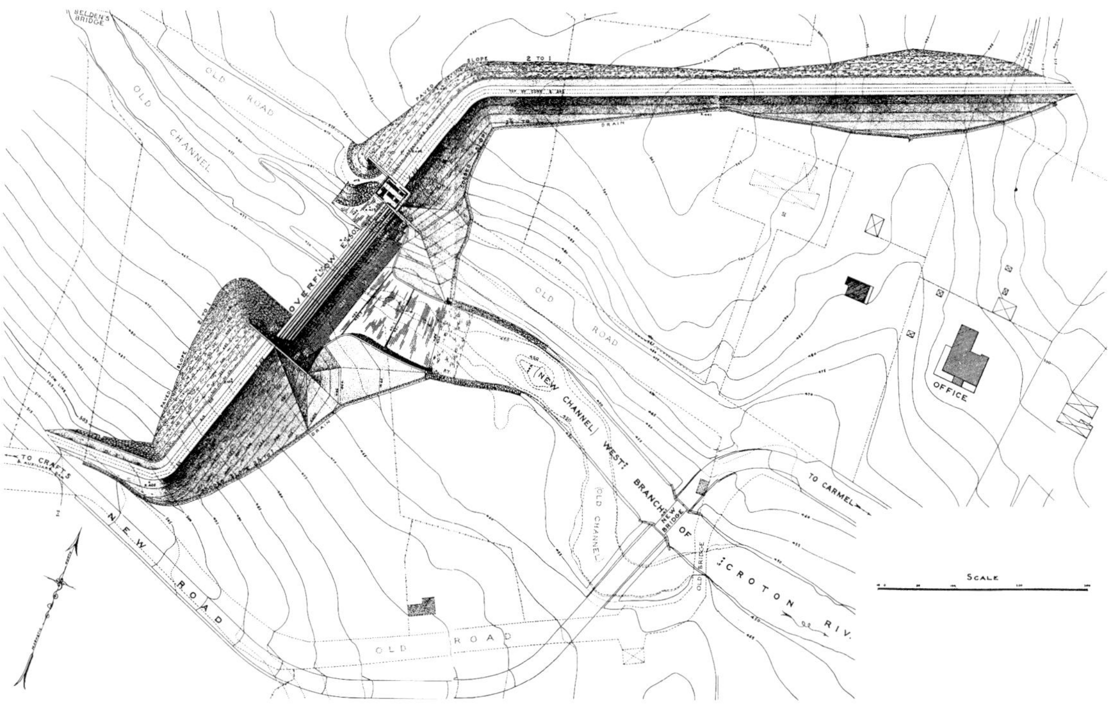

FIG. 52. *Carmel Dam, West Branch Reservoir. Plan showing earthen embankments and masonry overflow. 1890–96.*

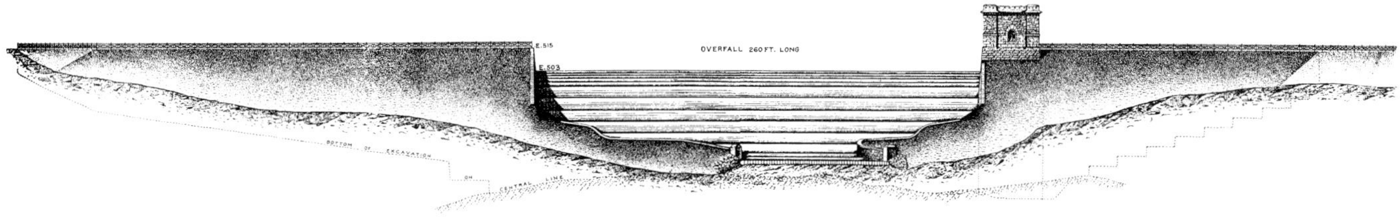

FIG. 53. *Carmel Dam, West Branch Reservoir, Downstream elevation. 1890–96.*

materials to be used. In a report presented to the commissioners, the engineers stated:

> The proposed dam is at a point on the Croton River . . . where steep sides of the valley approach to form a ravine. The ravine is about 1,300 feet wide at an elevation 230 feet above tide level, 300 feet wide at the level of the riverbed, 35 feet above tide and has a rock bottom 87 feet below the stream level, or 52 feet below the tide level of the Hudson River. It is proposed to close this ravine with masonry, which will impound the waters of the stream and raise the water to a level of 200 feet above tide. The greatest height of the dam from foundation to the top of the road parapet will be, therefore, from 265 feet to 270 feet . . . It will be 100 feet higher than any dam yet built . . . It is to impound upwards of 5,000,000,000 x 7.48 gallons [a conversion equation from cubic feet to gallons] in an artificial Lake, 16 miles long and 165 deep at the lower end. The length of the spillways will be about 1,300 feet. A depth of about 2.25 feet on the crest would pass the largest recorded flood in the valley . . . The greatest recorded flood of the river, measured at the Croton Dam, is 1,070,000,000 x 7.48 gallons in 24 hours . . . The waste-way will be constructed in rock cuts . . . so that the overflowing water will not touch the main dam . . . In our search for information on the expansion of ice in place, caused by increase of temperature, we found little of value recorded . . . but personal interviews supplemented by experimental data . . . has led us to the opinion that the dam should be proportioned to resist a thrust at the highest ice line of 43,000 pounds per line foot.[22]

The report describes other forces that might be brought to bear on the structure: the weight of the masonry, the pressure produced by the impounded waters, and forces produced by floating masses of debris or other material. It also outlined the impact of an extensive landslide or earthquake shocks.

The experts evaluated the shape of the dam in plan. Prior to the Fteley-Wegmann recommendations, there had been some consideration of a curved rather than a straight dam across the valley, since a curved wall could act as an arch to resist the lateral thrust of the restrained water. The engineers examined a very early design that employed a relatively tight 300-foot-radius curve, but the width of the ravine seemed to exclude such a radius. The board concluded that a project of this magnitude and importance should create a pleasing architectural effect, a criteria that they regarded as second only to structural stability, and the engineers believed that a curved dam face would contribute to that end. Their final recommendations called for a curved plan with a radius of about 1,200 feet, the exact location to be based upon further subsurface geologic investigations. The report from the Board of Experts concluded that the profile developed by the Department of Public Works engineers and the Aqueduct Commission engineers would not meet all the requirements of a dam of this dimension safely. They prepared a revised profile that added about 40,000 cubic yards of masonry to the upper 100 feet of the dam [Fig. 60]. By proposing substantial changes to the original plans, the Board of Experts inadvertently questioned the reliability of the engineering itself, and the whole matter of building a high dam was thrown into confusion.

The controversy left both the engineers and commissioners unwilling to push the project foward without civic support. The matter was taken to public hearings in January 1889, and the response was unfavorable. As a result, officials postponed further discussions indefinitely. It would take two more years before the decision to proceed would be made.

During this time, several more locations between the Hudson and the Old Croton Dam were studied. Fteley, who succeeded Church in 1888 as chief engineer of the Aqueduct Commission, conducted extensive borings to establish the geologic profiles throughout the lower Croton River valley [Fig. 55]. Fteley reversed his former position after this latest round of investigation and advocated the construction of a smaller dam near the Old Croton Dam at a site identified as "Location #2." The commission voted to accept the location near the Old Croton Dam but added that a "high" dam should be built there. The Aqueduct Commission requested that the Department of Public Works develop plans and property maps for a dam at that location, but one week later, on January 22, 1891, the commission reconsidered. In an unexplained move, members voted to build three miles downriver and about 1⅛ miles upstream from the Quaker Bridge, at a site they called Cornell (after the property owner) [Fig. 62]. Subsequent plans referred to the structure as the Cornell Dam, but it would soon become known as the New Croton Dam. The proposed New Croton Dam retained the improved profile that the

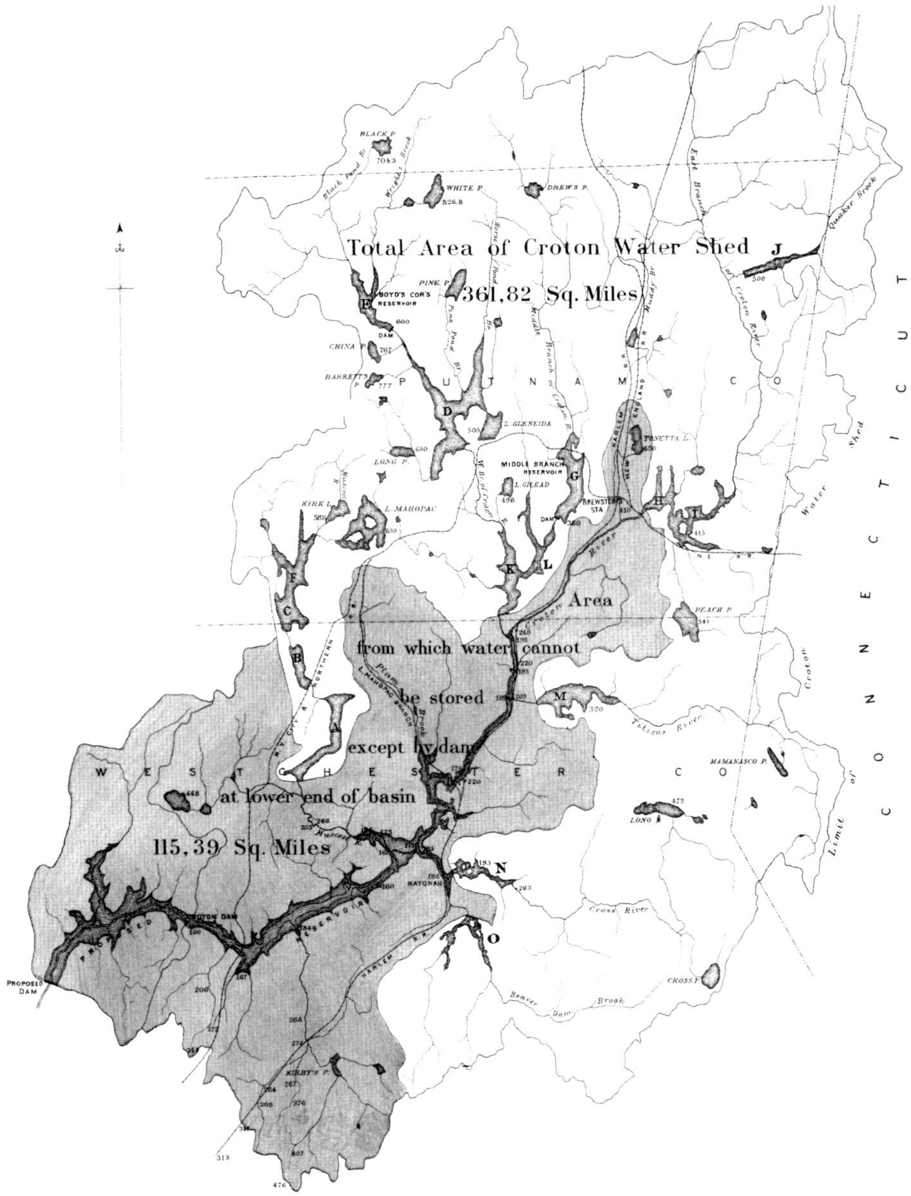

FIG. 54. *Map of Croton watershed from Isaac Newton's Report on Storage Reservoirs showing proposed location of New Croton Dam and Reservoir. February 21, 1883.*

Aqueduct Commission's engineers had developed for the Quaker Bridge site, with some further refinements. The original plans called for a central masonry dam 600 feet long, a masonry waste weir about 1,000 feet long, and an earth dam with a masonry core wall about 600 feet long on the south side of the valley [Fig. 56]. These were all to be founded on rock and form a continuous wall across the valley.

Construction commenced on the spectacularly large 2,168-foot-long and 270-foot-high dam in September 1892. The geologic studies had concluded that at the center of the ravine there was solid bedrock at least 65 feet below the surface of the valley floor. The excavation project was enormous. The base of the dam was over 200 feet wide, and a trench of this width had to be cut down into the bedrock before any building could begin [Fig. 57]. The construction of a 65-foot-deep, 200-foot-wide trench across the bottom of a valley, with a river flowing at the rate of about 500 mgd, was, for the times, an almost unimaginable logistical problem. Four years were spent rerouting the river in order to excavate and prepare the foundation trench [Fig. 58]. A new 1,100-foot-long and 125-foot-wide channel was blasted out of the cliffs on the north side of the ravine. Two temporary dams, each roughly the size of the original Croton Dam and totaling almost 1,600 feet in length, were built along the sides of the new channel to protect the six-story excavation from flooding. In the center of the valley, sand, gravel, and loose rock were excavated to about 75 feet below the valley floor before bedrock was encountered. The rock varied dramatically in hardness and structure, and in some places more than 50 feet of rock had to be cut away to get to a sound substrata. Water was a constant problem as the excavation went deeper. The dredges and steam shovels that excavated the pit had to float on platforms, and pumps were operated around the clock to keep water levels down. Over 2 million cubic yards of material were removed to make the foundation trench. The point between bedrock and the first layer of stonework is critical to the

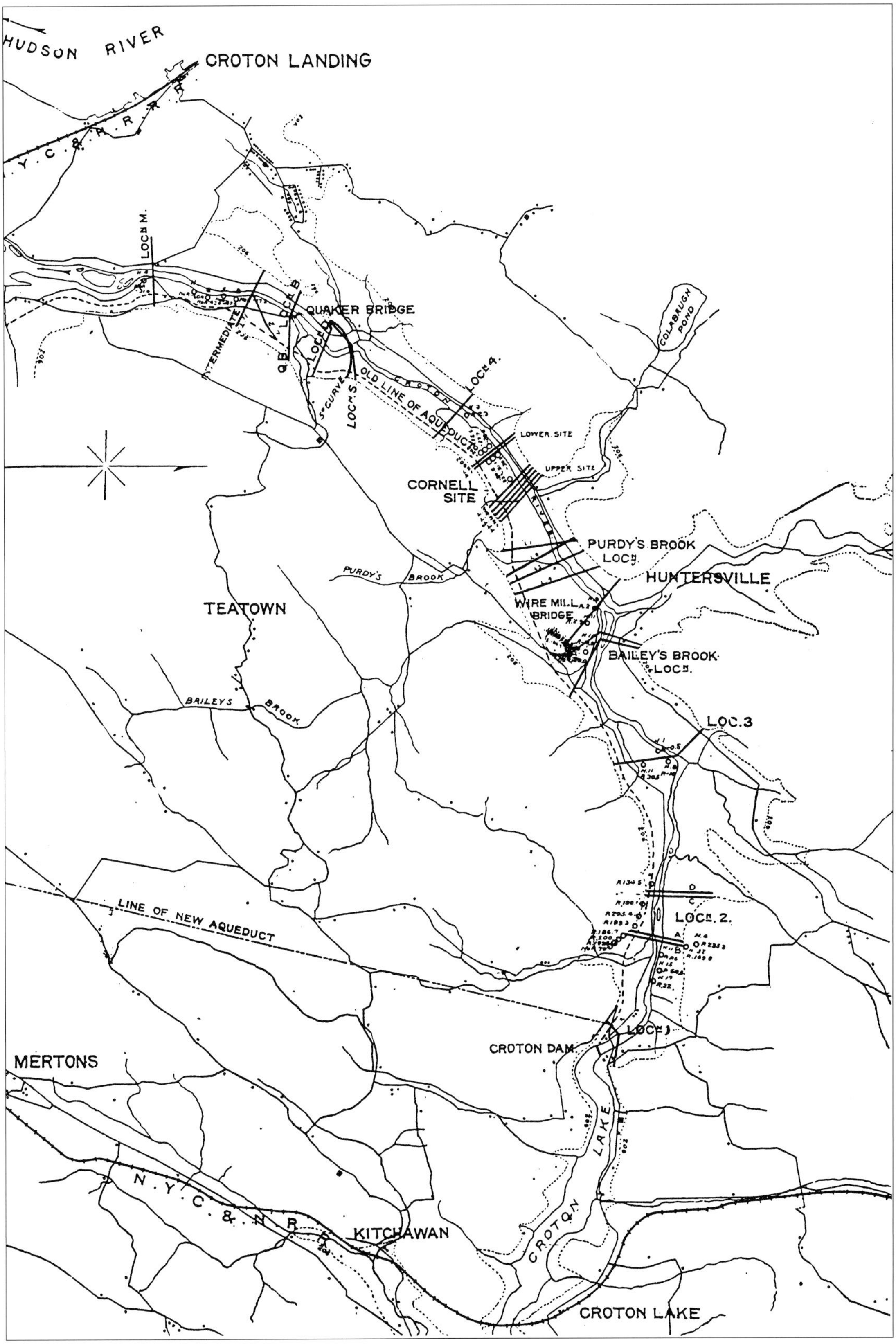

FIG. 55. *New Croton Dam. Map of sites investigated for the dam. 1888–91.*

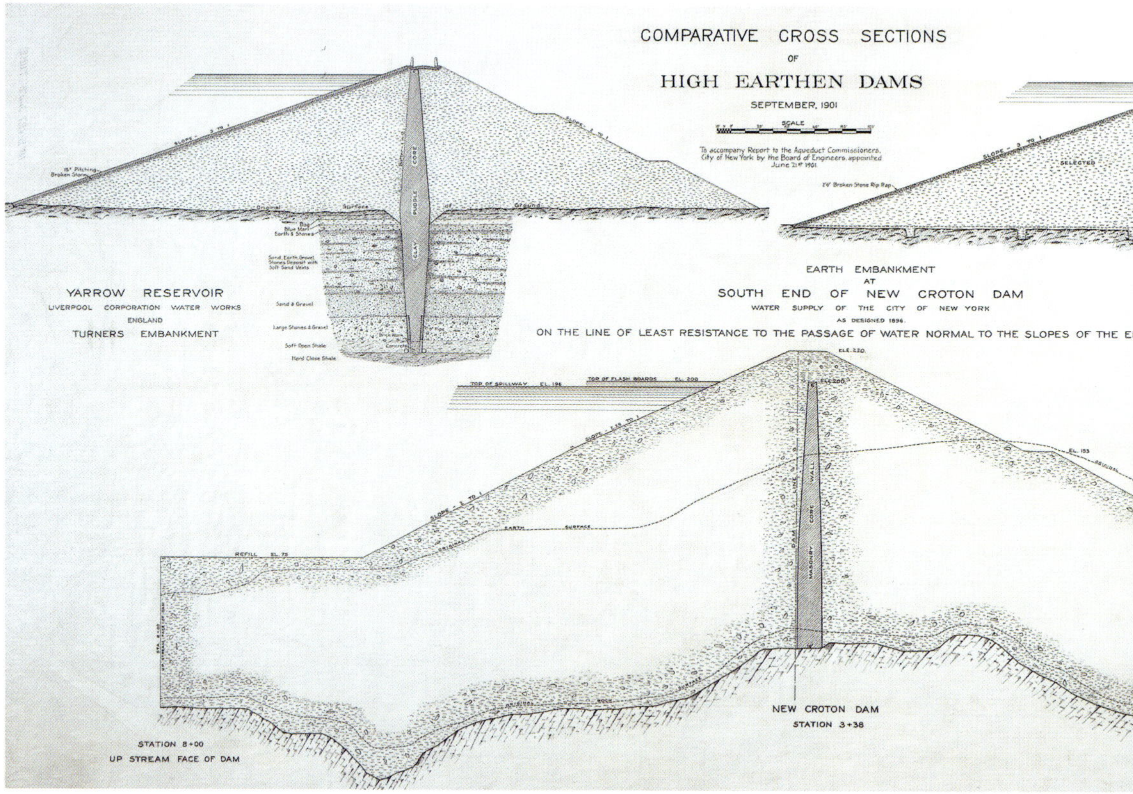

FIG. 56. *Comparative cross sections of high earthen dams, including the original section of the earthen portion of the New Croton Dam. September 1901.*

FIG. 57. *New Croton Dam. Profiles showing progress of foundation excavation and progress of placing masonry. September 16, 1896.*

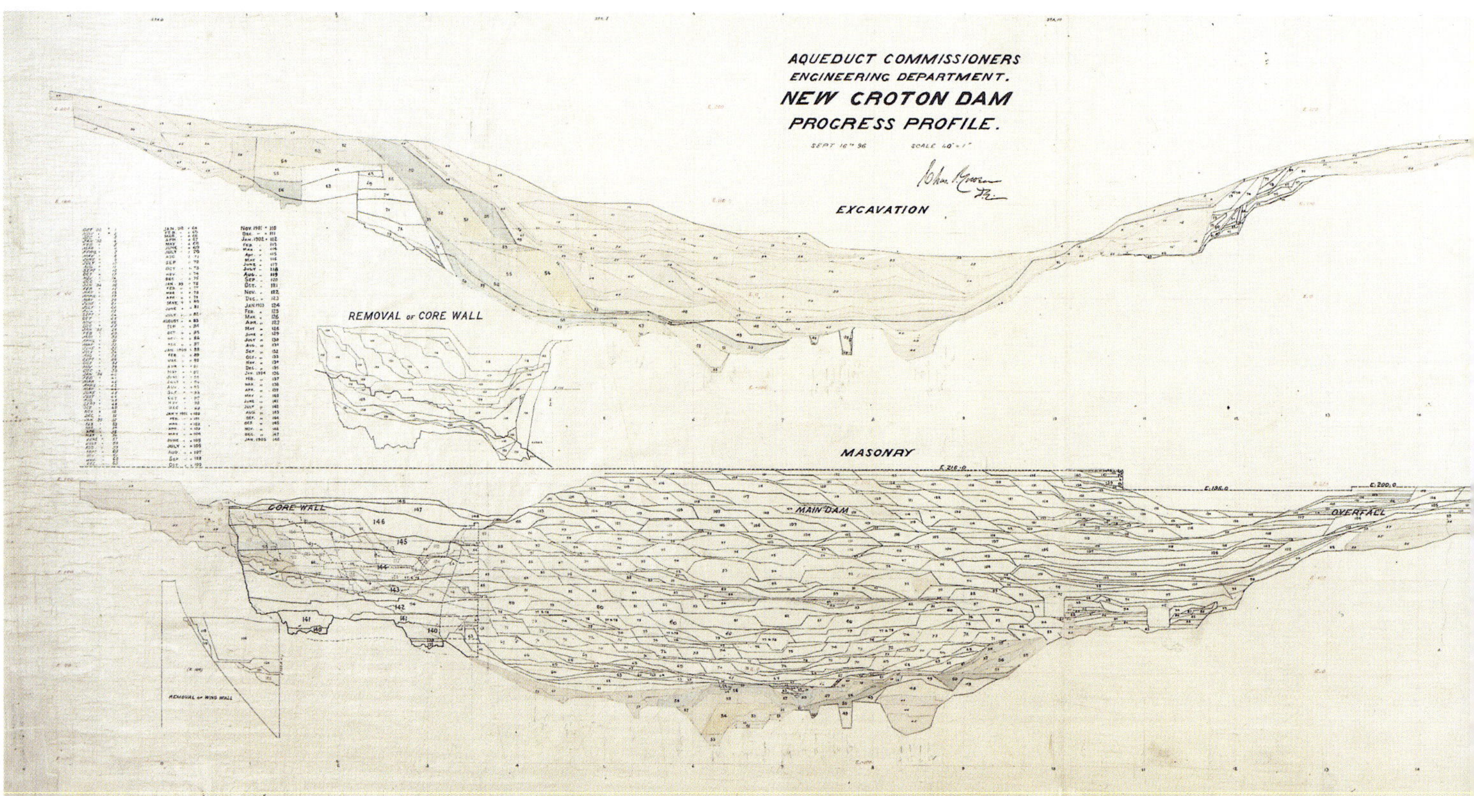

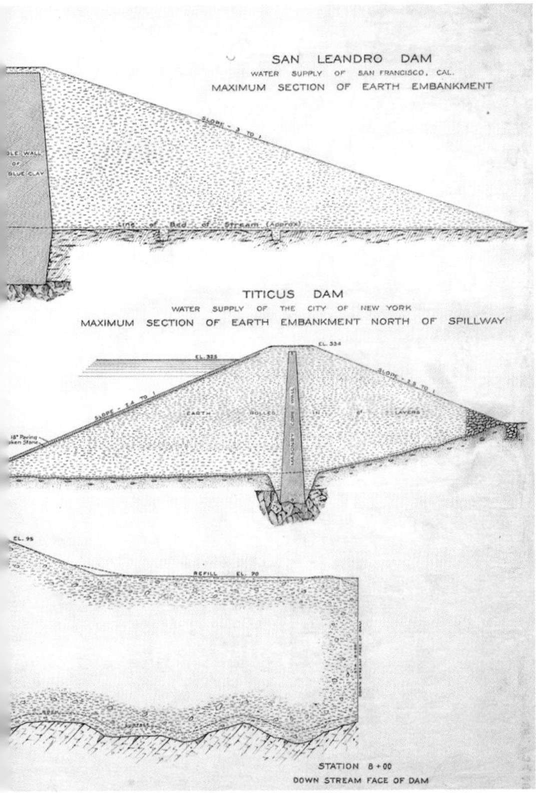

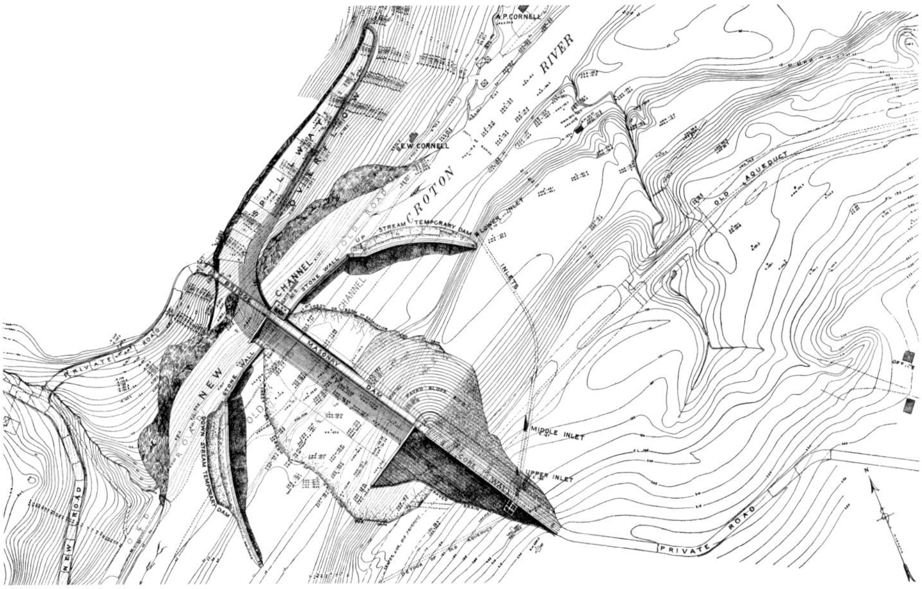

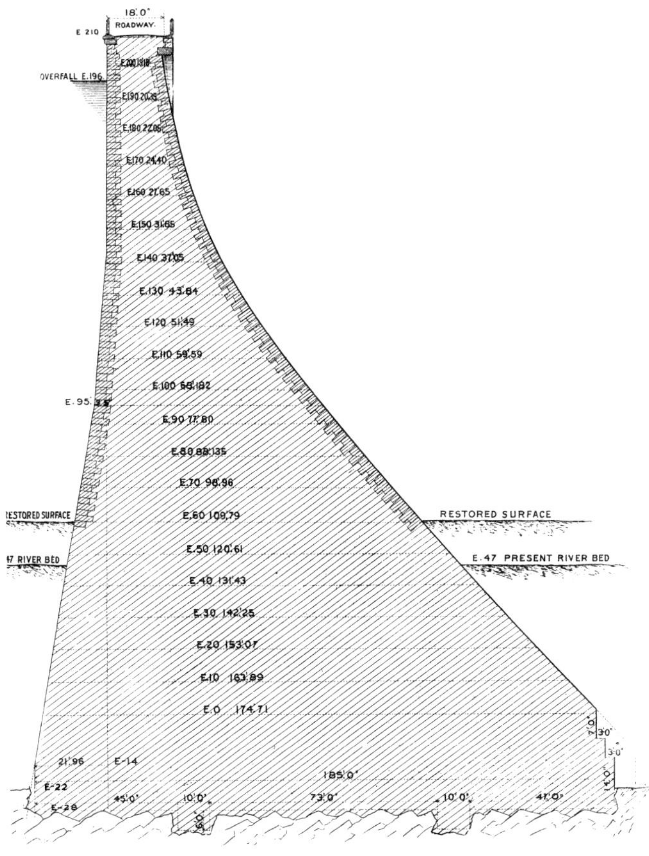

FIG. 58. *New Croton Dam at Cornell site. Contour plan of the dam showing temporary diverting channel and original configuration with earthen embankment (later eliminated). c. 1899.*

FIG. 59. *New Croton Dam. Cross section. 1892–1906.*

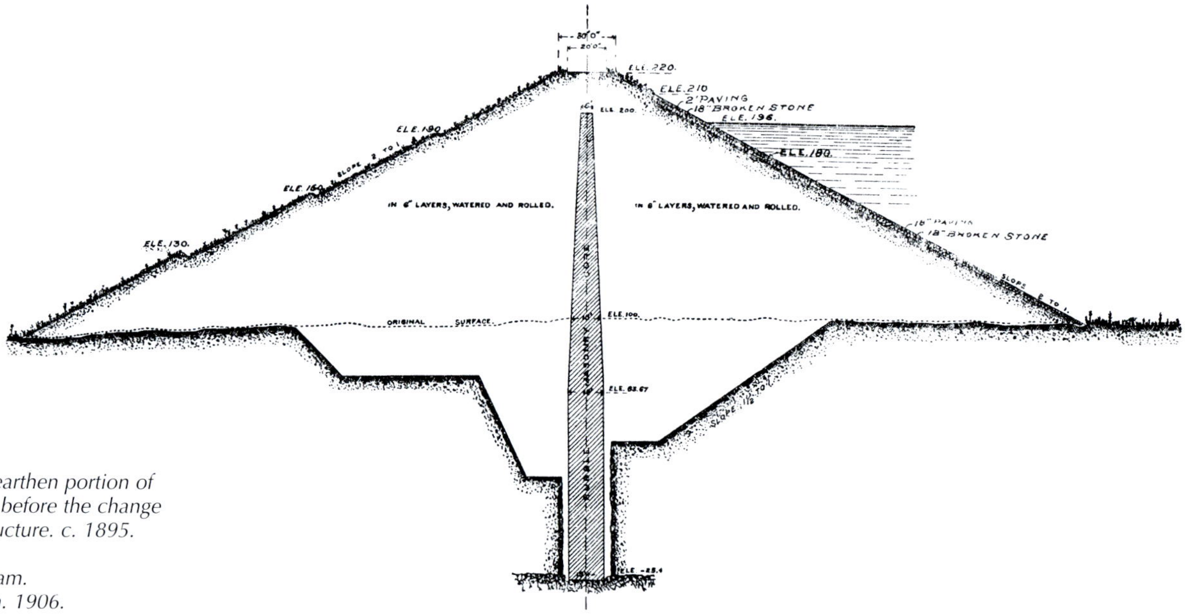

FIG. 60. *Section of the earthen portion of the New Croton Dam before the change to a solid masonry structure. c. 1895.*

FIG. 61. *New Croton Dam. Downstream elevation. 1906.*

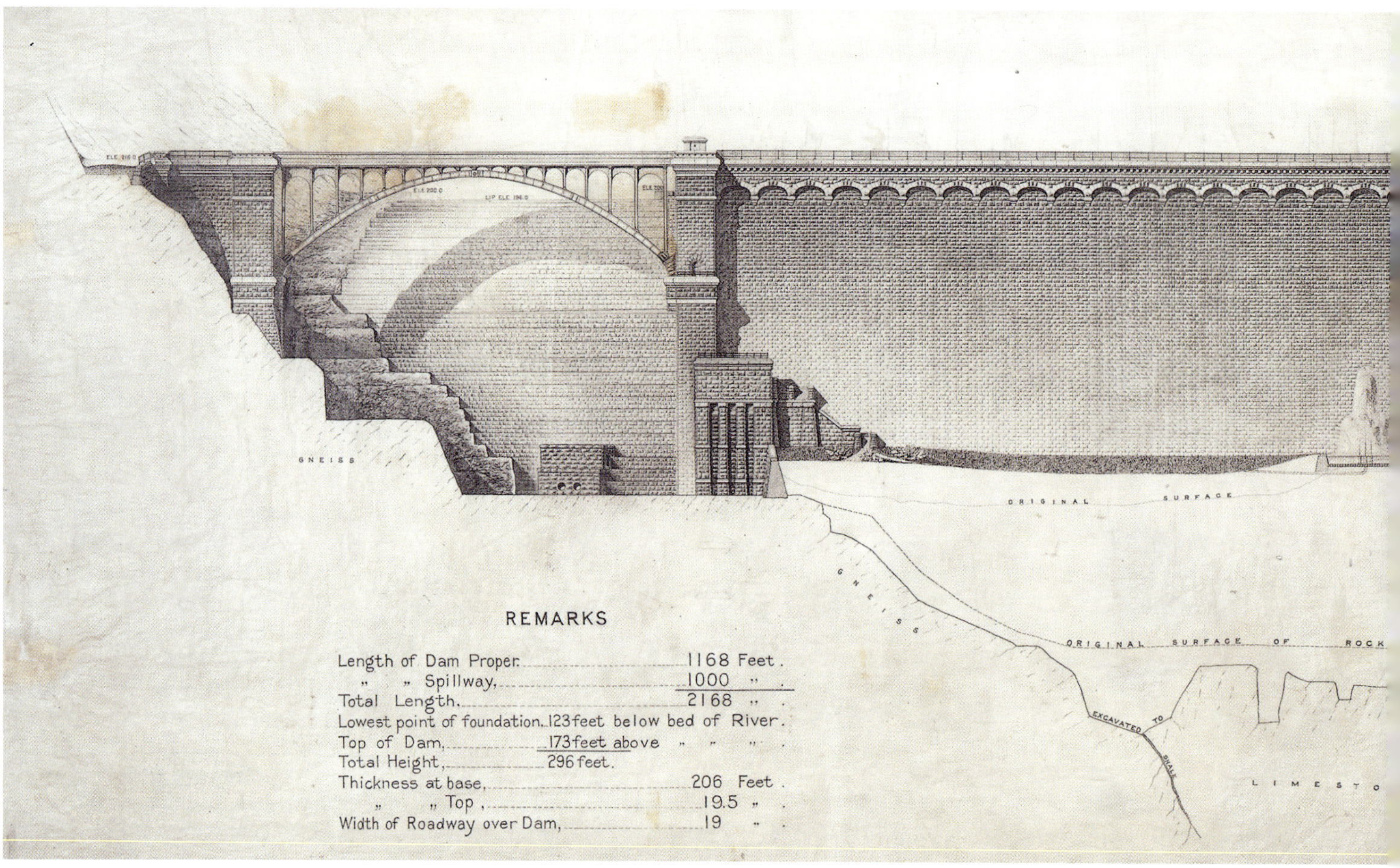

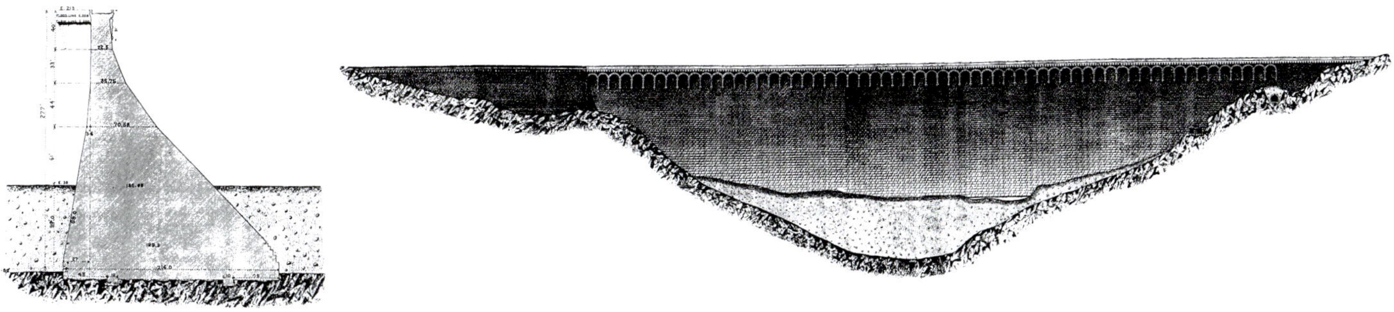

FIG. 62. *Quaker Bridge Dam. Cross section and elevation of Croton Dam as first proposed at the Quaker Bridge site. December 31, 1886.*

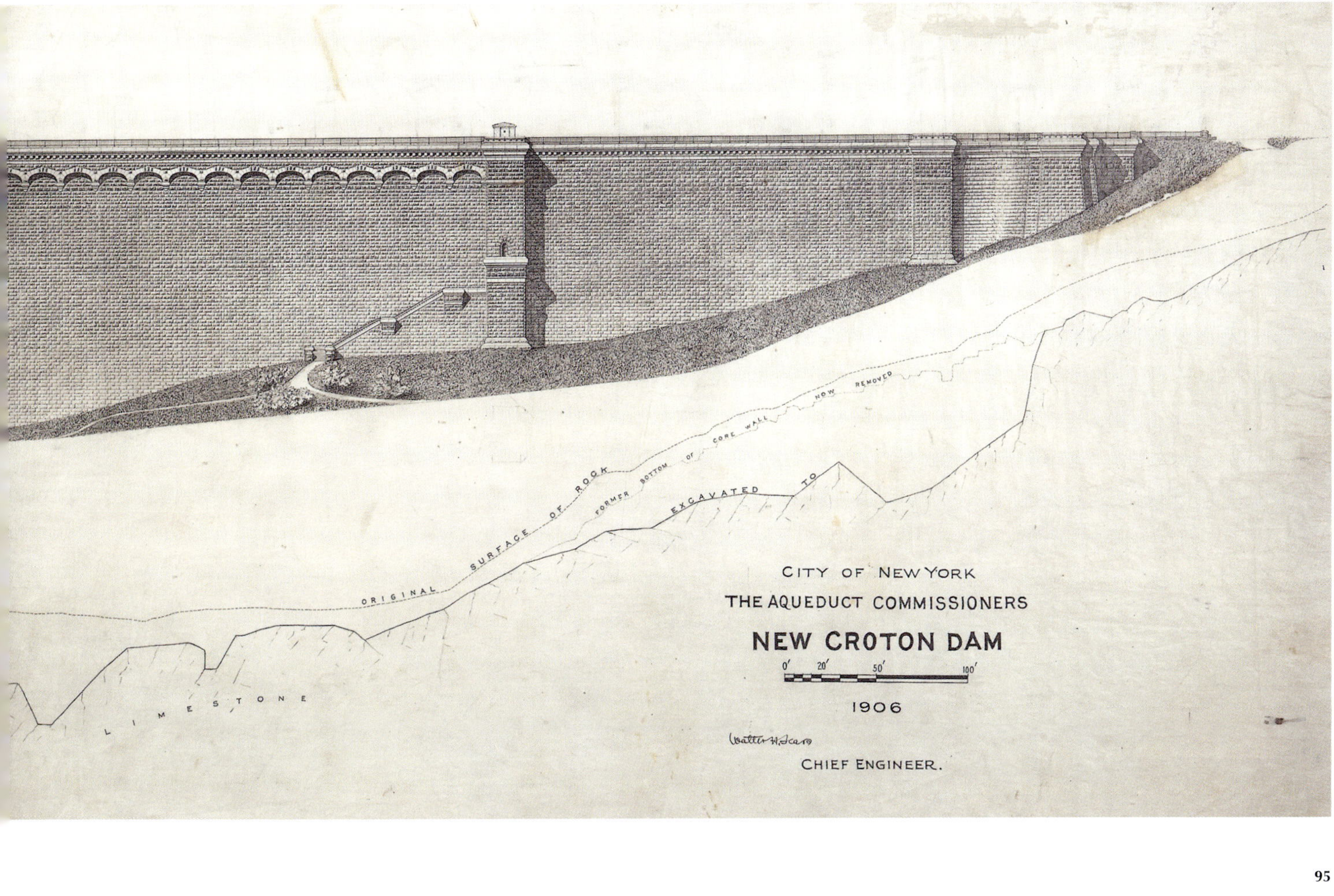

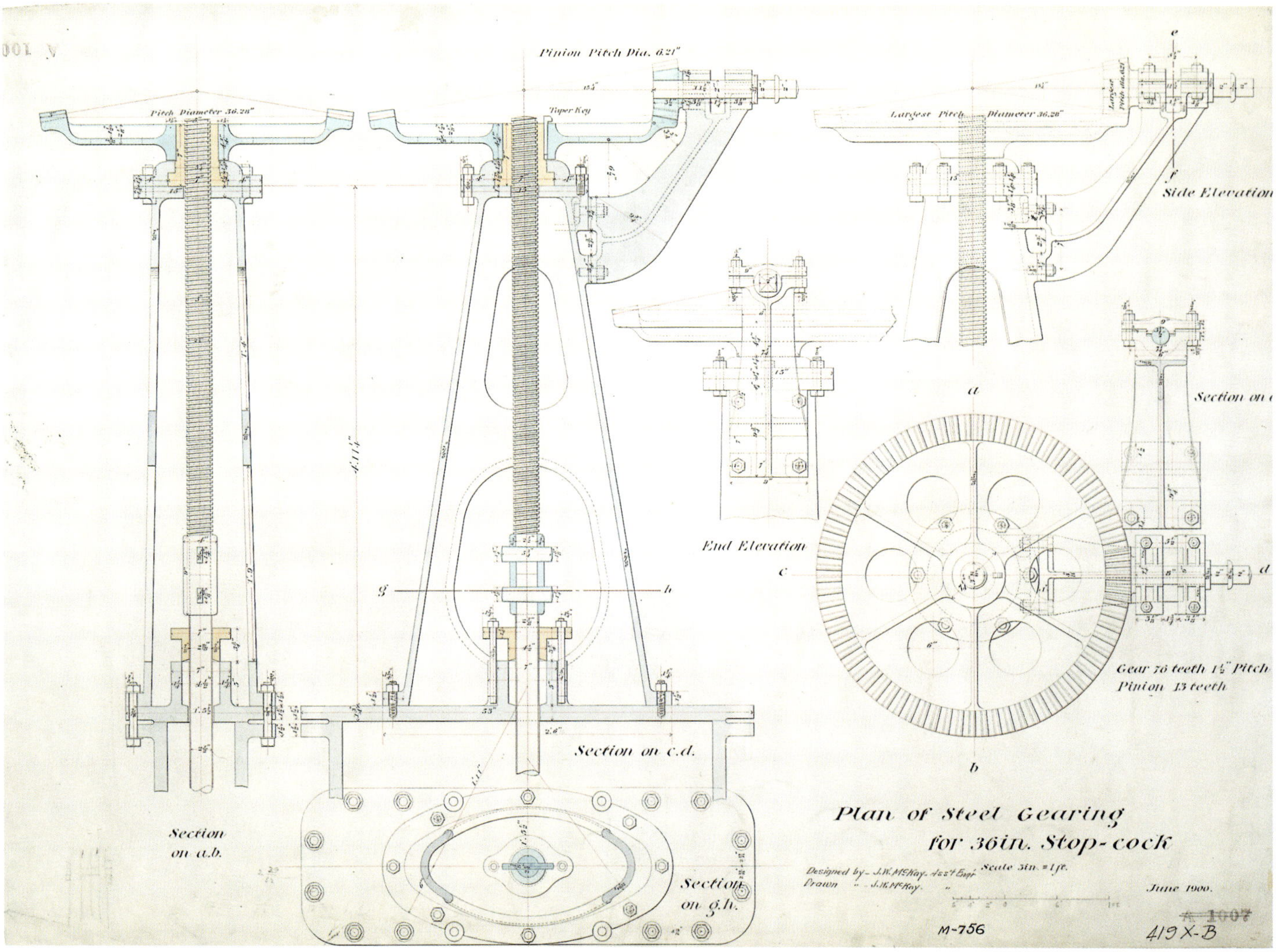

FIG. 63. *Gearing for 36-inch stopcock. Plan, section, and details. June 1900.*

waterproofing of a dam. If the seam failed due to the deep pressure at the base of the reservoir, the dam could be undermined. The surface had to be carefully prepared to receive the masonry of the structure:

> After solid rock had been reached and all the loose and shaky pieces had been removed, the bottom was thoroughly washed by streams of water under heavy pressure and cleaned with brooms. The rock was then painted with a grout of neat cement, applied with brushes, which filled up small cracks and seams. All erosions and open seams were filled with grout . . . large seams were filled by placing small stones in the grout. Springs that were encountered in the foundation were either drained off to a sump hole, or were confined in vertical pipes which were finally filled with grout. [23]

The laying of the masonry began on May 26, 1896. Most of the dam's length was constructed of filled rubble masonry. Where it was exposed, above the back-filled surface level, the dam was faced with cut stone ashlars. The facing stone has a minimum thickness of 28 inches, and the typical stretchers are 3 to 7 feet long. In every course of facing, every third stone is a header, turned back 48 inches into the dam construction to help bond the facing layer to the rubble masonry. Joints at the surface could not exceed one

half inch in thickness. Most of the stone used on the dam is hard, tough dark granite known as "garbo." The stone cutting alone was a huge job, and the quarry employed 851 men, almost as many workers as the dam site itself.

As originally planned, the dam included a section of solid masonry and a section of earth embankment [Figs. 59, 60]. Inspections during construction indicated possible faults in the masonry core of the earthen dam. There was also a general mistrust of high earthen dams, and a Reviewing Committee of Engineers supported these concerns. Work on the earth embankment was suspended, and all work on this portion was ripped out. The revised plans called for a solid masonry structure throughout [Fig. 61].

J. Waldo Smith was named chief engineer of the Aqueduct Commission in 1903 and oversaw the completion of the project. Smith was the primary engineer throughout the development of the Catskill System and one of the key figures in the building of the modern water supply. Smith made his presence felt on the project by revising the profile of the dam and calling for the thickening of the upper 50 feet of the cross section. By 1905, construction on the masonry dam was completed to the point where the gates could be closed [Fig. 63]. The waters rose to

FIG. 64. *Old Croton Dam (at left) exposed due to low water in the New Croton Reservoir. The flowline is indicated on the bridge piers. February 14, 1912.*

FIG. 65. *New Croton Dam. Spillway looking downstream. c. 1905.*

FACING PAGE
FIG. 66. *New Croton Dam. Spillway during overflow. c. 1907.*

FIG. 67. *New Croton Dam. Downstream face. November 8, 1907.*

flood over the Old Croton Dam, gatehouse, and bridge [Fig. 64]. The New Croton Dam, with its superb stonework, elegantly curved and stepped spillway, and arched bridge was then, and remains today, one of the most spectacular works of engineering and architecture in the system, and certainly one of the world's greatest masonry structures [Figs. 65, 66, 67].

In the last years of the Croton era, while the monumental New Croton Dam was lumbering towards completion, and just before the city was to shift away from the Croton watershed and divert all its attention to a completely new system of rivers, two more storage dams were built in the upper reaches of the Croton drainage. Despite the new aqueduct and the massive new dam, some Croton water was still being lost. In January and February 1907, 28 billion gallons poured over the dam to the Hudson and out to sea. New York City wanted every drop of the Croton and devised the last two constructions in that watershed—the Croton Falls Dam and the Cross River Dam—to accomplish it [Fig. 71].

By late June 1905, the contracts for the Cross River Dam were let and work began. Construction of the Cross River Reservoir, the tenth in the Croton System, was completed in May of 1908. The dam was faced with precast concrete blocks and filled solid with cyclopean masonry, a technique where gigantic stones are embedded in concrete [Fig. 72]. When needed, the water from the reservoir is released down the natural channel of the river into the New Croton Reservoir.

Construction of the final reservoir in the Croton System was started in 1906 [Fig. 73]. The Croton Falls Reservoir is actually fed by three sources. It is located near the confluence of the West, Middle, and East Branches of the Croton River, just north of the town of

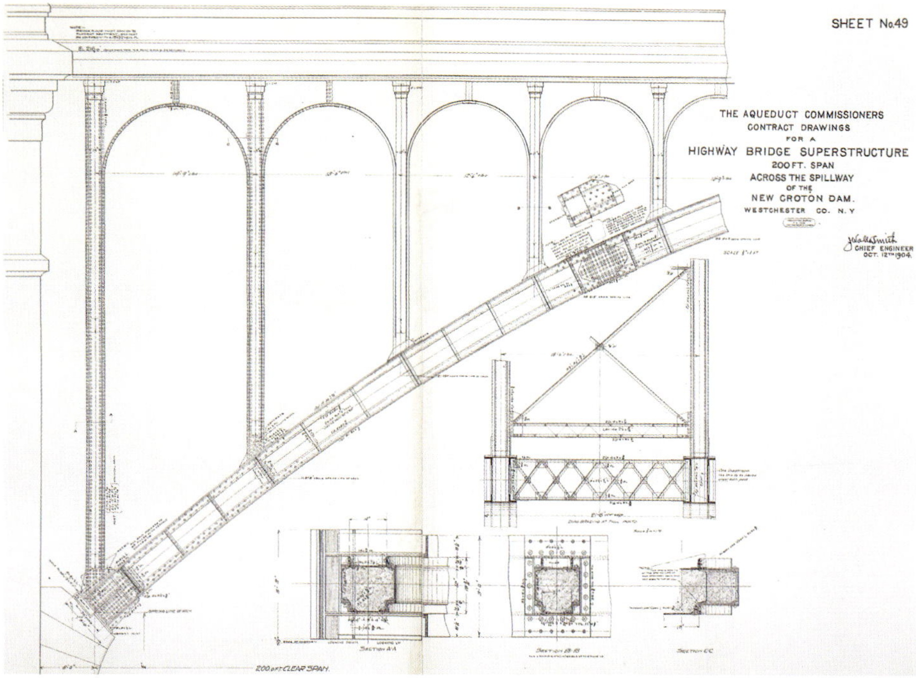

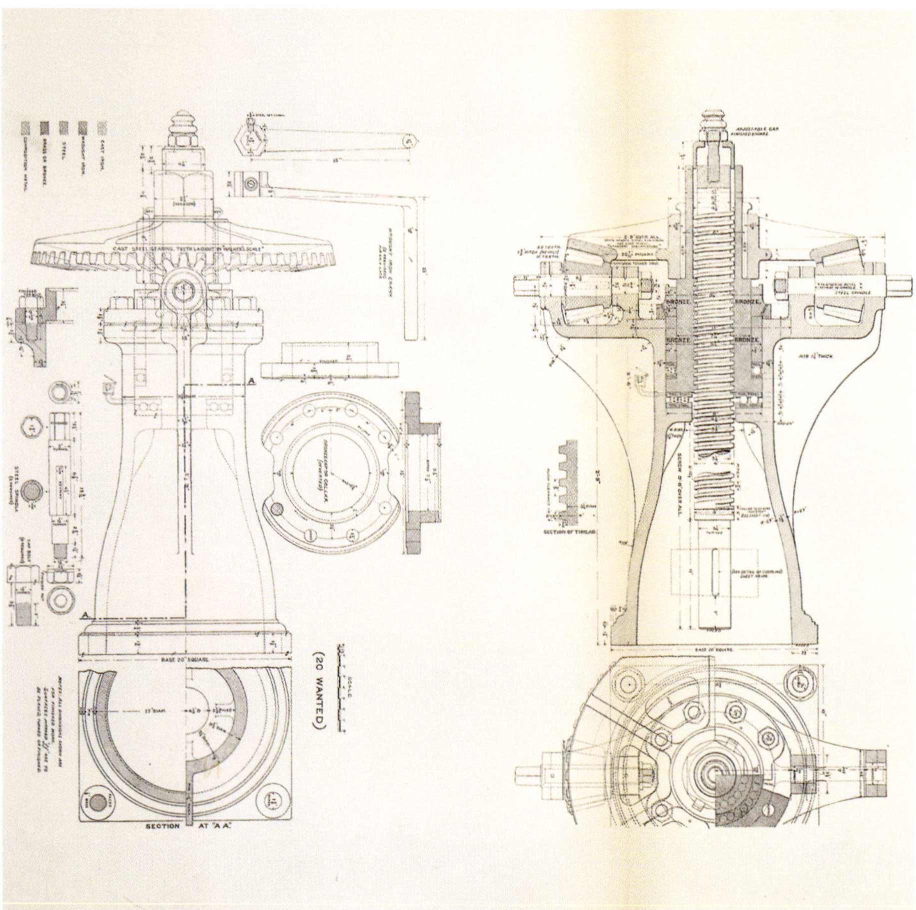

Croton Falls. The largest part of the reservoir system, formed by the Hemlock Dam, is located on the West Branch and the Middle Branch. The water on the Middle Branch backs all the way to the toe of the Middle Branch Dam, a structure farther upstream that had been completed only eighteen years earlier. On the eastern side of the Croton Falls Reservoir is a small body of water whose dam was built primarily for the purposes of diverting the water of the East Branch River to the main Croton Falls Dam. The eastern basin is connected through the intervening hill by a 3,500-foot-long paved open-air channel. The main dam is 1,110 feet long with a maximum height of 173 feet [Fig. 74]. The structure is cyclopean masonry, and was the only dam in the Croton group to use steel-reinforced concrete in the construction. The diverting dam on the East Branch is a low earthen embankment 1,185 feet long, and the spillway is a beautiful stepped concrete channel 1,100 feet long. The New Croton Dam, New Croton Aqueduct, and the other storage reservoirs of the Croton watershed remain in service today and supply about 10 percent of New York City's water [Fig. 75].

A map of Westchester County shows a patchwork of reservoirs that fill almost every mile of the main drainages of the Croton River watershed [Fig. 76]. There are only a few miles where the streams are not stilled by one reservoir or another. The extent to which New York City relied upon this watershed, and the critical role it played in the development of the city cannot be overstated: It was Croton water that made New York City possible. It was this water that took the city through its explosive years of population and industrial growth in the late-nineteenth century, the period in which it became America's premier shipping, manufacturing, and cultural capital.

FIG. 68. *New Croton Dam. Bridge superstructure over spillway. October 12, 1904.*

FIG. 69. *New Croton Dam. Sluice gate lifting machinery. July 13, 1904.*

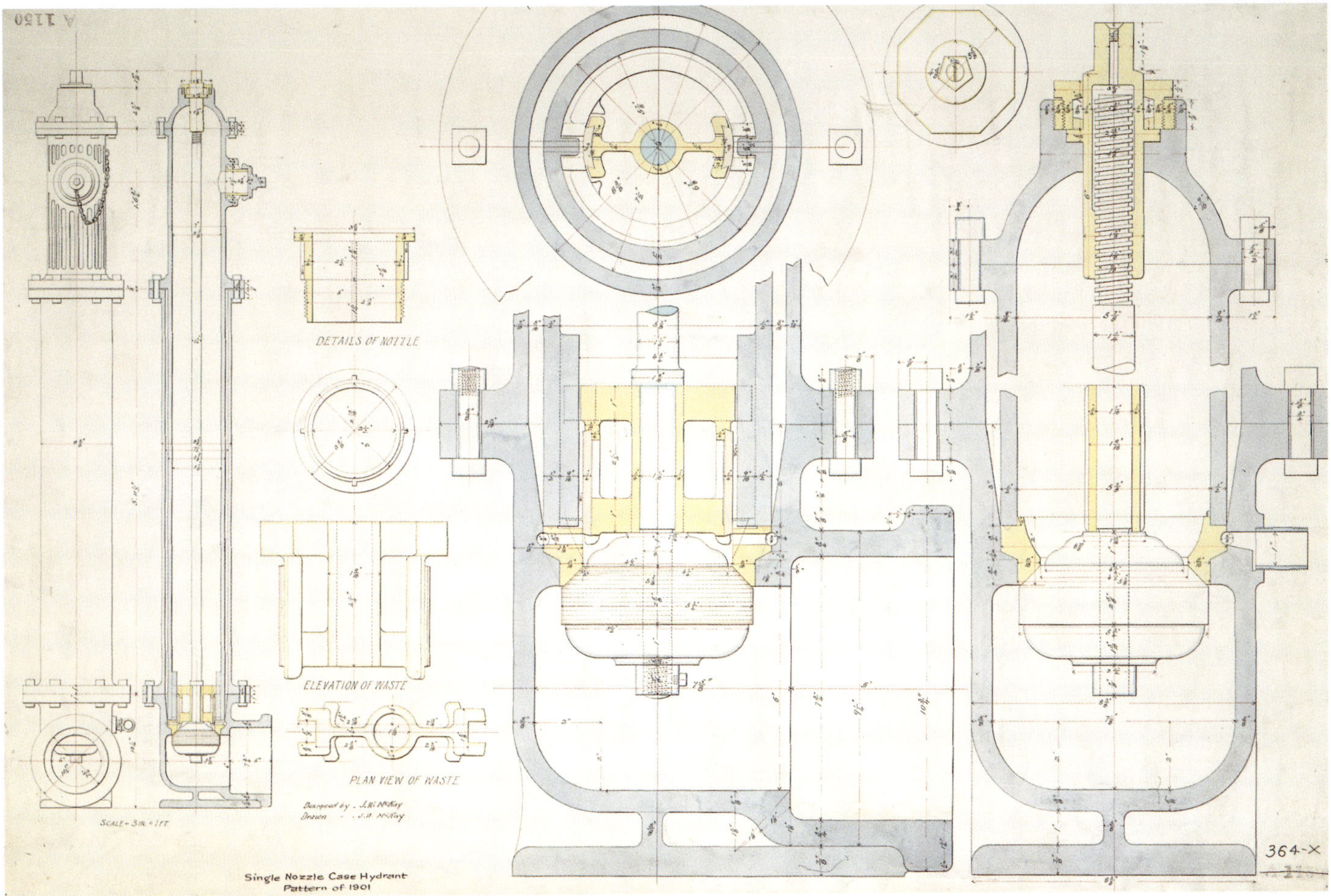

FIG. 70. *Single nozzle case, fire hydrant pattern. 1901.*

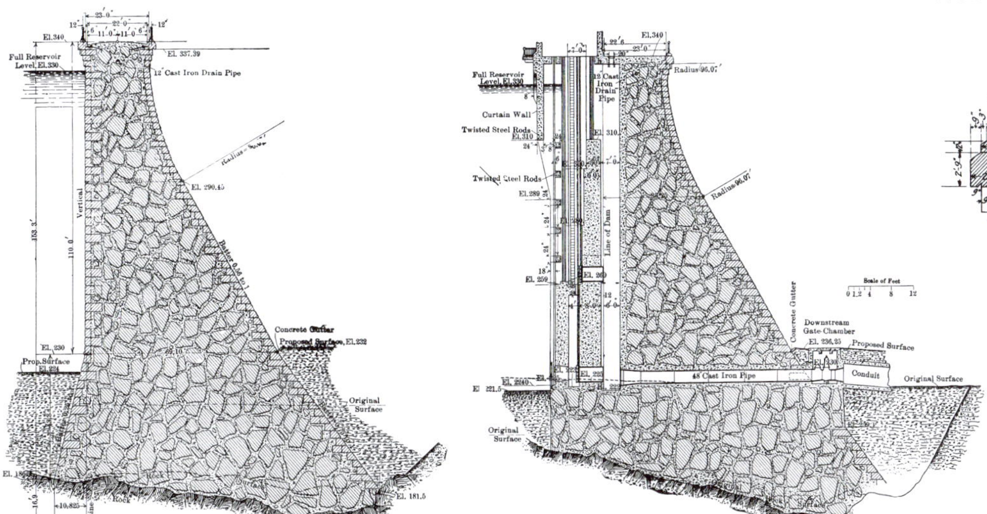

FIG. 71. *Cross River Dam. Maximum cross section through dam and cross section through gate chambers. c. 1905.*

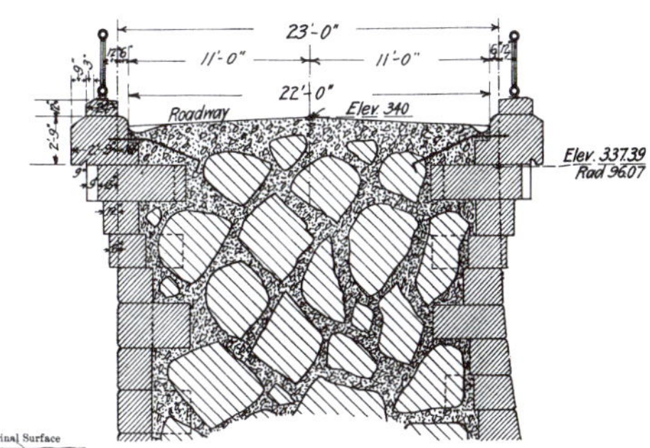

FIG. 72. *Cross River Dam. Detail at top of dam showing cyclopean masonry construction. 1905–8.*

FIG. 73. *Croton Falls Dam under construction, seen from the hill above the spillway. Mark indicates future level of the reservoir when full. December 2, 1909.*

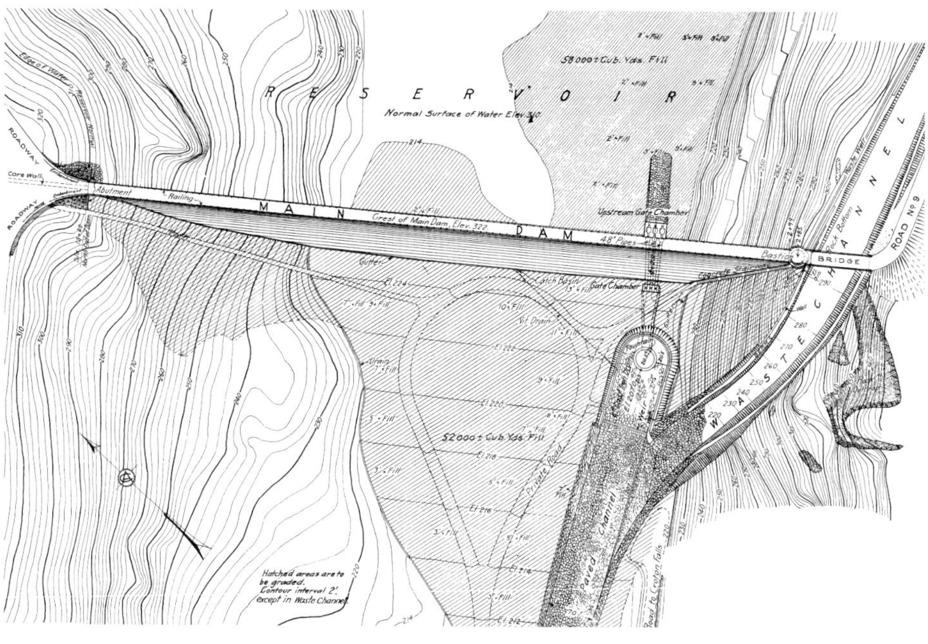

FIG. 74. *Croton Falls Dam. Contour plan of the main dam. June 1907.*

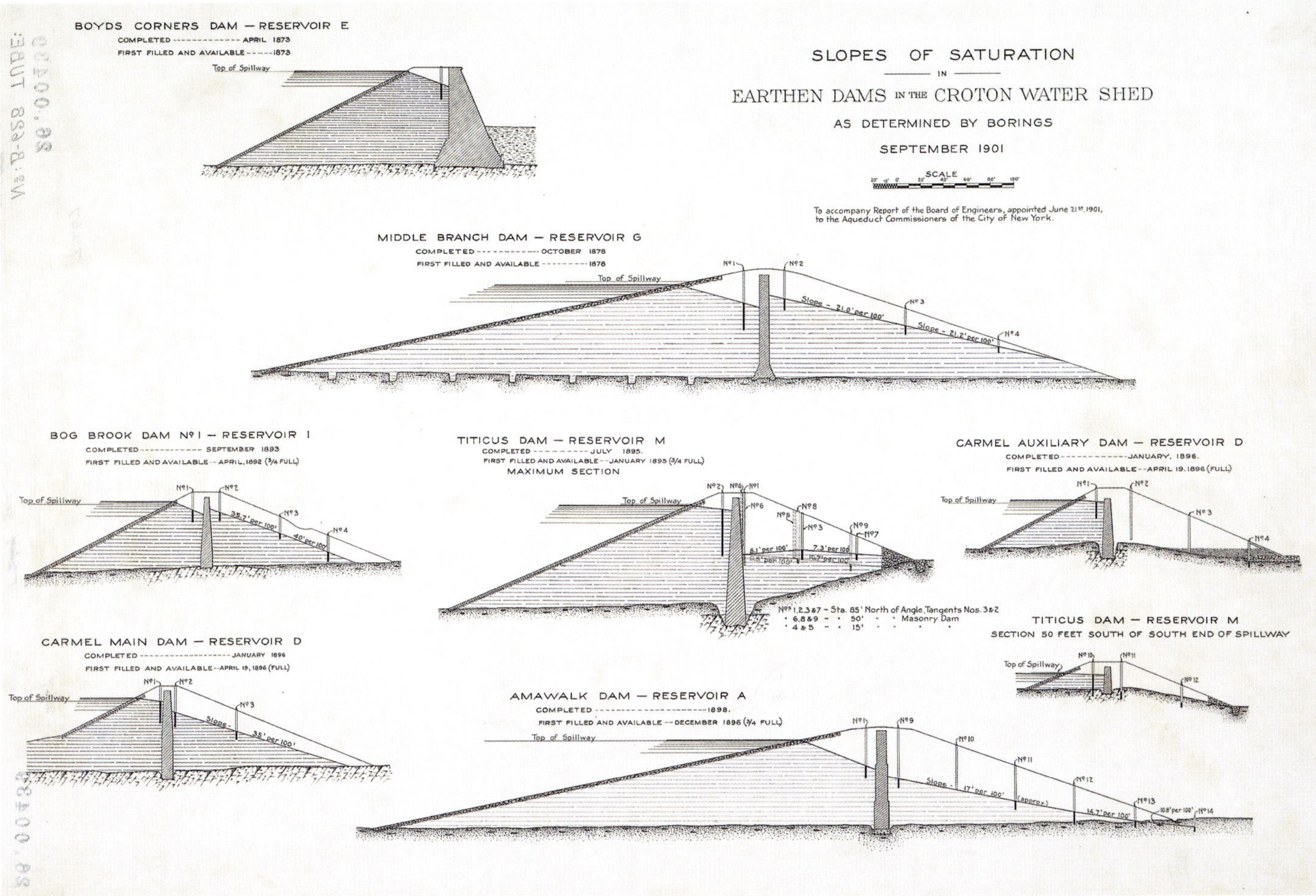

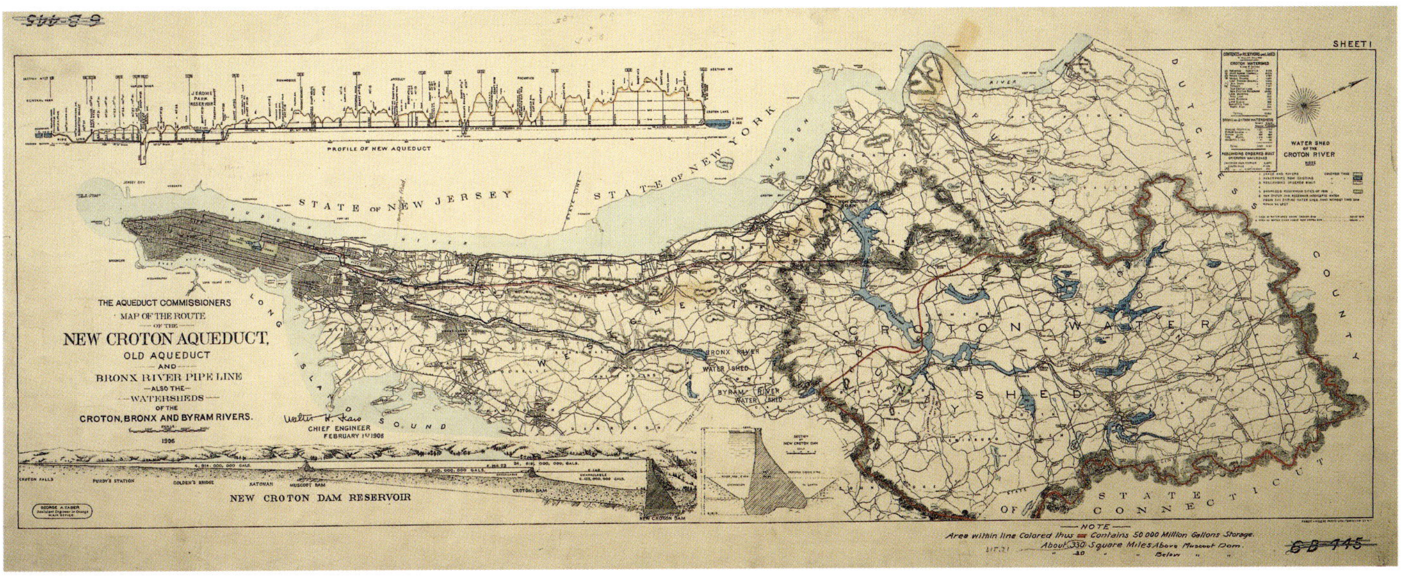

FIG. 75. *Comparative cross sections of earthen dams in the Croton watershed showing slopes of saturation. September 1901.*

FIG. 76. *Map of the Croton River Watershed and route of the New Croton Aqueduct.*

THE HOGHOUSE

Local 147 has been around for over 100 years. From the beginning, most of the men in the union were from either Caribbean or Irish heritage. Everybody else in the world has come through as well, but that's still the background of most of the men today. And they're all men. When I would go to a construction site, one of the first places I'd head was the hoghouse. That's the aboveground trailer where the sandhogs hang out. The jobsites have all these different trailers—the DEP, the General Contractor's, Operating Engineers Local 14/15, the Laborer's, Teamsters, First Aid. The hoghouse—which is kept in order by the hoghouse man—has lockers, showers, toilets, washer & dryer, picnic table with an oilcloth with Santa Clauses on it, coffeemaker and a little black and white TV that's always on. The hoghouse man is a miner who is close to retirement. His job is to make coffee, keep the waterboy's thermos filled, clean up, and be the aboveground contact for the belowground men.

Excerpt from *City Water Tunnel #3*,
written and performed by Marty Pottenger

THE CATSKILL SYSTEM

KEVIN BONE

The year 1881 was very dry. Evaporation from the three upstate Croton System reservoirs, combined with normal reservoir losses due to ground leakage, exceeded the natural flow of the Croton River by 10 million gallons per day (mgd), and 300 mgd of stored water had to be drawn down and diverted into the aqueducts for use in the city. No water flowed over the Old Croton Dam for fifty-two days straight, and the water level in the reservoirs dropped daily. The margin between the rate of consumption and the safe yield had closed. Ten years later, during the drought of 1891, the entire flow of the Croton River had to be diverted into the two aqueducts, and a severe late-fall dry spell left the Croton Reservoirs totally depleted. Almost no water was held in storage for the city. Heavy winter rains did fill the reservoirs, but the message was clear. By the closing years of the nineteenth century, it was evident the additional water made available by the Croton expansions was not enough to supply the city; as New York labored to complete the ambitious final phase of the Croton development, apprehension regarding the adequacy of the water supply system returned once again.

At the same time, the population of Brooklyn was expanding. Because the community drew from the sands and aquifers of Long Island rather than from the Croton system, water was always in short supply. Years before the consolidation of Brooklyn into New York City in 1898, the wells of western Long Island had been drawn down to levels so low they required extensive pumping, a process that lowered water tables and led to well contamination from encroaching sea water. The Brooklyn waterfront and the areas along the Newtown Creek had become important centers for manufacturing, and the diminished water supplies were affecting those industries' ability to compete with the rival factory towns of Philadelphia and London.

In response, the Manufacturers' Association of Brooklyn, under the leadership of Charles Chadwick, appointed a committee to study the problem. Chadwick's group studied two hundred different municipal water supplies, both in the United States and in Europe, and detailed the condition of the New York City system. The report Chadwick presented in March 1897 was significant because it was the first to take several positions that were critical to planning for more water:

> That all plans should contemplate a greater New York; that the State Constitution be amended to separate water debt from the constitutional debt limit; that the project be administered by a commission that would oversee the work from start to finish [thereby freeing long term projects from political manipulation and the changing attitudes of successive administrations]; and that the City's needs for the next fifty years should be considered.

These four recommendations became the cardinal objectives of later plans for enlarging New York City's water supply system.[1]

On January 1, 1898, the city, then consisting of Manhattan and a portion of the Bronx, merged with Brooklyn, Queens, Richmond (Staten Island), and the remaining lands of the Bronx to form its present boundaries. The population immediately increased from 2 million residents to nearly 3.4 million, many of whom supported consolidation because of their desire to share in the water and utilities that New York City was developing. The city was also in the midst of a mass immigration, gaining an additional 135,000 people every year. By 1900 the water consumption rate of the combined boroughs was equal to the safe yield of all the sources then being used: the combined capacity of the two Croton Aqueducts, the 48-inch line from Kensico Lake, and the wells and pumps of Brooklyn. The stage was set for another heroic struggle in the pursuit of water for the metropolis.

FIG. 1. *Catskill Aqueduct. Washington Square Siphon. Siphon house is visible on crest of hill. May 4, 1922.*

West of the Hudson, about one hundred miles upriver from the Port of New York, rises the remnant of a 400-million-year-old plateau that was once sea bottom. This plateau had risen throughout geologic time until its impounded ocean waters could no longer be contained. The escaping water cut great rifts through the plateau, leaving gashes in the land like the Kaaterskill Clove in the Hudson Highlands and the Delaware Water Gap along the southern edge of the basin. At its maximum height the dissected plateau was 5,000 to 6,000 feet above sea level. Over the millennia it eroded down, and the green shale and blue sandstone of the ancient seabed were exposed in what are now the Catskill Mountains. New York City went to the rivers and streams of these mountains for additional water.

Spring snowmelts combined with precipitation of more than fifty inches per year produce runoffs that total hundreds of millions of gallons per day. Water rushes toward the Hudson River from the flanks of features like High Point, Monbucuss Mountain, and Balsam Cap, through Watson Hollow, Stony Clove, and Malyby Hollow. Accumulating in the Esopus, Rondout, and Schoharie Creeks, this water became the heart of New York City's water supply from 1917, when the first Catskill water reached the city, to 1943, when the water of the Delaware River system was added.

The paths to acquiring Catskill water were difficult. Because of the immediacy of the water problems and the potential for opportunity, the situation attracted the attention of a group of speculators called the Ramapo Water Company. Incorporated in 1897 and using what must have been well-established and powerful influence, Ramapo forced through legislation that gave the company an unusual level of authority to acquire lands and water rights. Armed with this mandate, the group developed the first concrete plans for using the waters of the Catskill Mountains for New York City. Much as the railroad companies had been granted rights of condemnation, the Ramapo Company was granted unusually broad powers, allowing it to pursue the privatization of the future water supplies of the city. Shortly after the law passed, the engineers for the company, as required by law, started filing maps that described the company's intended areas of operation. Eight maps, filed in the Ulster County offices in Kingston, covered several hundred square miles of the eastern and southern Catskill Mountains. Although the plans developed by the Ramapo group were the first specific proposals for tapping the Catskill Mountain sources, it was not the first time the idea had been considered.

The Ramapo plan closely mirrored key features of a Catskill water development scheme that was presented in a September 1883 *Scientific American* article. That plan, like the Ramapo plan, proposed seven small storage reservoirs on the Esopus and its tributaries, and an iron pipeline to convey the water to the city.[2] The speculators of the Ramapo Company argued that the city could not afford to build its own Catskill infrastructure, and so private fundraising was needed. In return for raising the capital and funding the project, the company sought a commitment from the city to buy 200 mgd for a period of forty years at a set price of $70 per million gallons of water (a deal that would have ended up costing New York City over $200 million for the life of the agreement).

Tragedy increased the appeal of the Ramapo plan. In March 1899, the Windsor Hotel, a full-block building at Fifth Avenue and 46th Street, was destroyed by fire, taking with it forty-five lives. Firefighting efforts were seriously hindered by water pressure deficiencies. In the aftermath of the Windsor disaster, a group of city officials, hotel proprietors, and merchants met with Mayor Charles Henry Van Wyck to press the case for an increased water supply. The fire chief, the fire commissioner, insurance underwriters, and powerful men like Waldorf-Astoria owner George C. Boldt advocated for the Ramapo plan.[3] By August, the council president and the water commissioner had publicly supported the plan and asked the Board of Public Improvement to accept the Ramapo contract. But from elsewhere within city government there were those who doubted the wisdom of privatizing a major public resource.

Comptroller Bird S. Coler presented a financial report indicating that the city did in fact have the financial resources to build additional infrastructure and that private ownership of the water system would prove more of a financial drain over the long run than any short-term gains could justify. Coler, along with Bridge Commissioner Patrick Shea, mounted a campaign to delay approval of the plan and conduct appropriate diligence. The newspapers began to editorialize their opposition, and public opinion swayed quickly. Businessmen who had originally endorsed the privatizing scheme began to withdraw their support. Armed with the backing of such powerful citizens as William Randolph Hearst, Coler and his delegation convinced the attorney general's office to start an action against the Ramapo Company. As the action moved forward to stop the privatization plan, two important independent studies on additional water for the city were completed.[4]

Coler had hired the engineer John R. Freeman to investigate water supply conditions and to potentially bolster his position against the private enterprise. Coler could not have found a more qualified individual:

> John Ripley Freeman was one of the foremost water supply engineers of his time . . . At the age of seventeen he entered Massachusetts Institute of Technology receiving his degree in 1876. At various times in later life Freeman was consultant engineer to the Chinese government on the improvement of the Grand Canal . . . and an [advisor to the U.S. president] on the Panama Canal. He served on the water supply boards of Mexico City, Los Angeles, San Francisco, Baltimore and numerous other cities . . . He was offered and declined the presidency of MIT.[5]

The Freeman report, submitted to Coler on March 23, 1900, became the guiding document upon which much of the development plans for Catskill water were based. Freeman presented several critical conclusions and recommendations. He produced the specific data needed to support what many intuitively believed: that city consumption was dangerously close to the safe limits of the existing system. Even with strict management, the available sources could not supply the city for more than five years—a very short time in which to bring waterworks infrastructure on line. The situation, as Freeman described it, was dire. Freeman also articulated what previous engineers had suggested: that water waste was huge. He concluded that up to 50 percent of the water that entered the two Croton Aqueducts was not reaching users in New York City. Freeman's report advocated limiting waste by requiring universal metering, by pressuring private owners to reduce loss, and by establishing a program to repair leaks in all city-owned facilities, such as schools and public buildings. Freeman, who had been president of a group of fire insurance companies, added that any future system should be designed to provide enough pressure and volume for effective firefighting. He projected that in order to fulfill current and future water consumption needs, new waterworks would have to supply 150 mgd as soon as possible, and ultimately have the capacity to yield 500 to 750 mgd. This was the first time that supply figures of this magnitude had ever been proposed.

FIG. 2. *Triangulation Station, Plainfield Tower. Plainfield, New York (now Floral Park, Queens). August 6, 1907.*

Several sources for additional water were mentioned in Freeman's report. The Ten Mile River in Dutchess County and the Housatonic River east of the Croton (primarily in Connecticut) were seen as the cheapest and most readily available, and the Esopus Creek water, which the Ramapo Company had targeted, was also noted as a viable option. The Adirondack Mountains, while having abundant supplies, were considered impractical because of the costs and engineering involved in bringing the water such a great distance. Freeman believed the waterworks could be built for $20 per million gallons as compared to the $70 that the Ramapo Company sought. The report concluded that due to the great distance of all these sources, a reservoir should be built near Yonkers as insurance in the event that an upstate aqueduct failed.

While Freeman was at work preparing his report, the Merchants' Association, a private trade group, took up its own study of the Ramapo Company proposal and incidentally addressed the question of an additional water supply for the city. The group formed a committee on water supply, which made a comprehensive analysis of water system finances going back to 1832. The group reached the same conclusion as Coler: private water would be far more costly than water supplied by a public system. The committee made the following recommendations:

1. No contract should be entered into with the Ramapo Water Company. They anticipated increased costs associated with the private water supply company; concluded that the Ramapo model was an ambiguous proposal with regards to water pressure and water quality; noted that the physical plant would be retained by the company at the end of the contract, effectively forcing the city to purchase water from the Ramapo Company in perpetuity; and observed that the city would be deprived of water revenue as a long-term source of income. The bleak history of private ownership of water in New York City was reviewed and the trend toward public utilities in American and foreign cities was cited as a preferable model.

2. All potential private contracts or private suppliers (not just the Ramapo Company) should be opposed.

3. The legislature should grant New York City the power to condemn any water rights needed, except those held by other municipalities.

4. Water-supply bonds should be exempted from constitutional debt limit by legislative action.

5. The city comptroller should present an annual report on water-related finances.

6. Records and reports of the Department of Water Supply should be in a form easily understood by the public.

7. Surplus revenue should be used to pay down bond debt.

8. Immediate steps should be taken toward the development of 250 mgd from Adirondack sources.

9. The city should take action to reduce waste.[6]

These two studies—the Freeman report prepared for the city comptroller and the Merchants' Association report—outlined what would become the doctrine of the expansion of the city water supply. With the help of the comptroller and Governor Theodore Roosevelt, the Merchants' Association introduced legislation intended to prevent the reoccurrence of another private water initiative. The bill limited the power of the water commissioner and the Department of Public Improvement to enter into private water contracts. It also restored powers of condemnation against private companies, allowing the city to confiscate any water to which the Ramapo Company might lay claim. A separate bill was passed that repealed the Ramapo Act of 1895 altogether.

The legislation, and the direction provided by the recommendations in reports, prompted Mayor Seth Low's administration to instigate a detailed and comprehensive investigation of the water supply. The mayor appointed an independent municipal body known as the Commission of Additional Water Supply in December 1902. The commission was charged with making a specific investigation of the current system: to determine how to best protect water quality; to increase its yield and to reduce waste; to make projections of future populations and consumption rates; and to determine the most available future supply, including the preparation of general plans and specifications for structures, and estimated costs and time frames for completion of the system. Finally, the commission was to implement an emergency plan to deliver supplementary water while the primary system was being built.

This commission, consisting of Professor William H. Burr, Rudolf Hering (who helped produce the Merchants' Association report), and John R. Freeman, conducted a comprehensive and authoritative investigation. The Commission of Additional Water Supply was headed by the finest professional engineers of the day, and was supported by more than one hundred assistant engineers working for the Department of Water Supply, Gas and Electricity. In a significant departure from past practice, the professionals gained a new level of authority in the water decision-making process. The recommendations of politicians would no longer carry weight unless they were supported by the city's engineers.

The completed document, the Report of the Commission on Additional Water Supply for The City of New York, consisted of 980 pages that included maps, data, plans, and cost estimates for the construction of dams and aqueducts. The report made specific recommendations supported by hard engineering. Directed by the city's attorneys to avoid interstate water, the engineers recommended developing the Fishkill first, a robust stream system on the east side of Hudson River north of the Croton watershed, and the Esopus Creek in the heart of the eastern Catskills as the second source. The two watersheds, the Fishkill and the Esopus, were thought capable of supplying 300 mgd and would be connected to the city by an aqueduct capable of carrying 500 mgd (in anticipation of additional sources). Dutchess County manufacturers, who relied on these swift streams to power mills, were well represented in state government and moved quickly upon hearing of New York City's ambitions on the Fishkill waters. They lobbied for and successfully passed the Smith-Dutchess County Bill, legislation that prohibited any municipality from taking water that would interfere with the operations of the mills along the three prominent Dutchess County waterways. The bill essentially shut down the possibility of acquiring east-of-Hudson water sources and left the Catskill Mountains as the city's only viable source.[7]

Civic interest remained focused on the water problem. Early in 1904 a group of twelve prominent civic bodies (including the Manufacturers' Association of Brooklyn, the Merchants' Association, and the City Club of New York) were invited by Charles Chadwick to meet and review the prospects for additional water. This group, called the Conference of Commercial Associations, agreed on the need to formulate specific plans and was ready to begin the practical work of bringing upstate water down to the city. The status of this group, representing much of the commercial power of the city, was able to move Mayor George B. McClellan to act. The mayor promptly called for a plan from the Department of Water Supply, Gas and Electricity. The department recommended obtaining water from the Esopus, Schoharie, and Catskill Creeks of the eastern Catskill Mountains—the first time a governmental agency officially supported the development of the Catskill Mountain watersheds.

In a report dated January 16, 1905, the commissioner of Water Supply, Gas and Electricity drew up specific recommendations for a large storage reservoir on the Esopus Creek and for headworks of the planned aqueduct near West Hurley. The commissioner supported the use of a cut-and-cover design for the aqueduct, versus deep tunnels, and defined what was believed to be the most economical line for the aqueduct, with a target capacity of 400 mgd.

New York City's water supply responsibilities were spread over a diverse array of governmental bodies, and this lack of bureaucratic centrality was seen as one of the reasons why the city could not get ahead of its water problem. The Manufacturers' Association of Brooklyn had tried unsuccessfully to introduce legislation calling for the creation of a central water authority in 1902, 1903, and 1904. Early in 1905, with the backing of Mayor McClellan, the association tried again. While past attempts had failed because the governor and the legislature had favored the development of state resources by state agencies, this time a strong delegation that included Mayor McClellan, Charles Chadwick, former Mayor Low, and the committee from the

FIG. 3. *Triangulation station. Constructed by engineers with found material. Yaphank, New York. April 16, 1907.*

powerful Conference of Commercial Associations was able to persuade the governor to support the city's bid for self-determination. McClellan helped craft a fair-minded set of laws including attendant legislation that protected various interest groups.[8] But most important to the city was the McClellan Act, which established the Board of Water Supply (BWS). The preamble to the law reads:

> An act to provide for an additional supply of pure and wholesome water for the city of New York and for the acquisition of lands and interests therein, and for the construction of the necessary reservoirs, dams, aqueducts, filters and other appurtenances for that purpose: and for the appointment of a commission with the powers and duties necessary and proper to obtain the objects . . . Duties of the Board as stated by the law were to proceed immediately and with all reasonable speed, to ascertain what sources exist and are the most available, desirable and the best, for an additional water supply of pure and wholesome water . . . and to prepare maps, plans, estimates and contracts: to acquire real estate and other rights and to construct [these waterworks] upon approval of the Board of Estimate and Appeal and the State Water Commission.[9]

To secure support in the state legislature, two other laws had to be added to the package. Chapter 738 of the Laws of 1905 prohibited taking water from either Westchester or Putnam Counties (although the law did have a provision permitting the city to build distribution works in these counties, thereby allowing for the future Catskill Aqueduct to pass through), and Chapter 726, which gave all communities affected by New York City water development the right to purchase Catskill Mountain water at no greater cost than that charged to city users.

Every expert report on additional water had recommended financing mechanisms to insure the fiscal health of the tremendous Catskill undertaking. In the fall elections of 1905 the public voted to amend the state constitution to "exempt debts incurred by the City of New York after January 1, 1904 . . . for the supply of water" from the constitutional debt limit. This law gave the Board of Water Supply the legal foundation and the economic support to build the new system. (The existing Aqueduct Commission retained authority over the construction of the New Croton Dam and the plans for the final Croton extension—the Cross River Dam and Croton Falls Reservoir—until those projects were completed.) The Department of Water Supply, Gas and Electricity, with the consolidation of the five boroughs into Greater New York, remained the permanent authority for maintaining and operating the entire water supply once the structures were in place.[10]

The new laws granted the Board of Water Supply wide-ranging powers that would combine the functions of condemnation, engineering, design, and construction in a single entity and avoid the bureaucratic burdens of multiple agencies

FIG. 4. *Brooklyn and Long Island Sources. Blythebourne Pumping Station, Brooklyn, New York. c. 1907*

FIG. 5. *Brooklyn and Long Island Sources. Windmill "Belmont" used for triangulation. Babylon, New York. December 18, 1907.*

that had slowed other waterworks projects. The first board was appointed on June 9, 1905 and included Charles Chadwick, who had led the Brooklyn Manufacturers' Association study. The board moved quickly to find an engineer to direct the plans. In August they appointed J. Waldo Smith chief engineer.

J. Waldo Smith (1861–1933) was chief engineer with the New York City Aqueduct Commission from 1890 to 1903 and participated in the construction of New Jersey water supply systems in Newark, Passaic, Montclair, Little Falls, and Jersey City. He served as the chief engineer of the Board of Water Supply for seventeen years and as a consultant to the board until his death. More than any other, he was the individual whose name would be associated with the design and realization of the massive Catskill Aqueduct and its associated waterworks.

The newly formed Board of Water Supply reviewed all the plans that had been presented by various organizations over the past decade. The Ten Mile and Housatonic Rivers, as recommended by Freeman, were prohibited because they were interstate drainages; the Fishkill and Wappinger Creeks were prohibited by the Smith-Dutchess County Bill; and the Adirondacks and Great Lakes were seen as too ambitious and costly. The Hudson River was a possible source, but only if the water was filtered and drawn from above Poughkeepsie where the quality was better. The Suffolk County sources of Long Island, although technically off limits because of protective legislation, were investigated because of their proximity and ease of access, and a Long Island department of the board surveyed and mapped much of the south shore [Figs. 2, 3, 4, 5, 6, 7].

The board focused its energy on the Catskill Mountain sources. The resources—the Esopus, Schoharie, Rondout, and Catskill Creek watersheds—cover over 780 square miles on the eastern and southern flank of the Catskill range and are about one hundred miles north of New York City. The area is heavily forested, gen-

FIG. 6. *Brooklyn and Long Island Sources. West Brooklyn pumping station. c. 1907*

FIG. 7. *Brooklyn and Long Island Sources. Pumping station of the Great South Bay Water Company. Bayshore, New York. April 22, 1908.*

erally free of swampland, and geologically sound for water because of the predominance of bluestone and shale with almost no limestone (which leaches minerals into water). The watersheds are steep, rocky, and for the most part not suitable for agricultural uses—another indicator of good water quality.

The board documented the potential water yields of the drainages and found there was abundant water for New York City. The Esopus Creek, located on the southeastern edge of the Catskill Mountains, drains east and north, joining the Hudson River at Saugerties. The creek is productive, often running at a rate of over 3,000 cubic feet of water per second, a figure more comparable to a river than to a mountain stream. Its watershed (above the anticipated dam site) covers 255 square miles and was first projected to have a safe yield of 250 mgd.

The Schoharie Creek watershed, which is adjacent to and north of the Esopus, drains north toward the Mohawk River and therefore flows away from the city. However, much of its elevation is significantly higher than the Esopus, and a diversion of water through the mountain ridges back into the Esopus was possible. The watershed (again, as measured above the anticipated dam site) is about 218 square miles and was first calculated to have a safe yield of 170 mgd.

The Rondout Creek lies on the south side of the Catskills. Its headwaters are adjacent to the Esopus watershed, and the creek drains primarily east, meeting the Hudson River at Kingston. The surrounding land is forested and the water quality is good, with 196 square miles of watershed. The safe yield was projected to be 125 mgd.

The Catskill Creek is north of the Esopus and east of the Schoharie. It empties into the Hudson River at the town of Catskill. Measured just north of the small community of Durham, this watershed covers 163 square miles and was projected to have a safe yield of 100 mgd. The water, because much of it originated in agricultural land

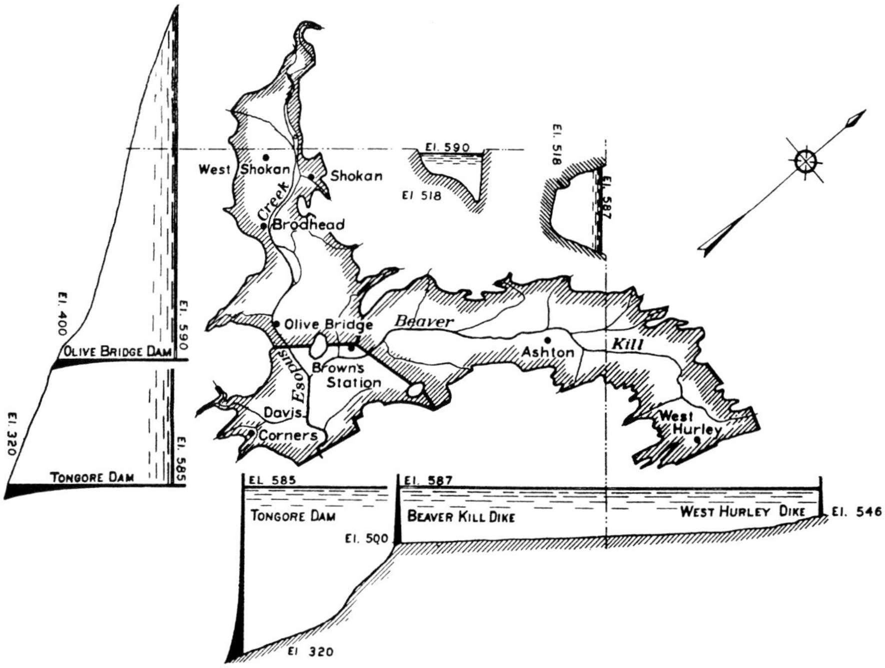

FIG. 8. *Olive Bridge Dam, Ashokan Reservoir. Diagram comparing alternate dam locations. c. 1906.*

and because limestone is present in the local surface rock, was considered the least desirable of the four Catskill watersheds surveyed.

These combined Catskill Mountain sources offered a potential supply in excess of 600 mgd. This was an incredible quantity of water. By comparison, the combined capacity of the great aqueducts of Imperial Rome probably delivered to the capital city no more than 26 mgd. Catskill water was generally clean and low in minerals, and the elevational relationship to New York City allowed the water to be delivered exclusively by gravity to most points within the metropolis. The region was clearly well suited to become the future water farm for America's largest city.

Just four months after its formation, in June 1905, the Board of Water Supply submitted plans to the Board of Estimate and Apportionment outlining its water development strategies for the eastern and southern Catskills. One week later the mayor applied to the State Water Commission for approval. The mayor's package of materials included John Freeman's original report prepared for Coler in 1900, the Burr-Hering-Freeman reports of 1903 and 1905, and the newly generated plans of the Board of Water Supply. After eight months of contentious hearings, where locals fought for the right to stay on their land and businesses attempted to retain their positions in Catskill communities, the inevitable occurred. Driven by huge public need, New York State authorized the city's plans. The board recommended proceeding with the development of the Esopus, Rondout, Schoharie, and Catskill Creeks (in that order). It was decided that the Esopus water would be tapped first because it was the closest of the watersheds and could most quickly provide New York with a much-needed infusion of water. Once Esopus water had checked New York City's looming shortage, building could move forward on the second stage of Catskill development, the Schoharie Creek expansion.

Based upon a more detailed gauging of streams and an updated study of meteorological records, the board revised its earlier estimates of stream flow and concluded that the Esopus and Schoharie could yield, on average, an incredible 875,000 gallons per square mile of watershed per day. Collecting water from the 561 square miles of the two combined watersheds would safely yield an average of 500 mgd (80 mgd higher than the first estimates for these two watersheds).

The Catskill Mountain plan called for a gigantic reservoir to store the waters of the Esopus Creek. From there, an aqueduct would carry water to a new storage reservoir near the White Plains site of the first Kensico Dam and continue down from the Kensico site to an equalizing reservoir on the northern edge of the Bronx at a site called Hill View. From the Hill View Reservoir, the water was to travel through a network of urban aqueducts and deep submetropolis tunnels over thirty-five miles long, ending at a terminal reservoir on the high ground of Staten Island.

The Esopus passes out of the eastern Catskills in a deep valley cut between the Slide Mountain Massif and Ticetonyk Mountain Group. Once out of the mountains the creek merges with the Beaver Kill (a minor tributary north of the main stream), and flows through a gentle rolling landscape called Ashokan, a native term meaning "the place of many fishes." The Ashokan landscape, a remnant of the ancient inland sea, is a natural basin defined by mountains to the south, west, and north, and by a series of low glacial ridges on the southeastern, or downstream, side of the basin. Damming the Esopus effectively recreated an ancient lake that had occupied the site millions of years ago.

The Board of Water Supply engineers debated where best to locate the dam.

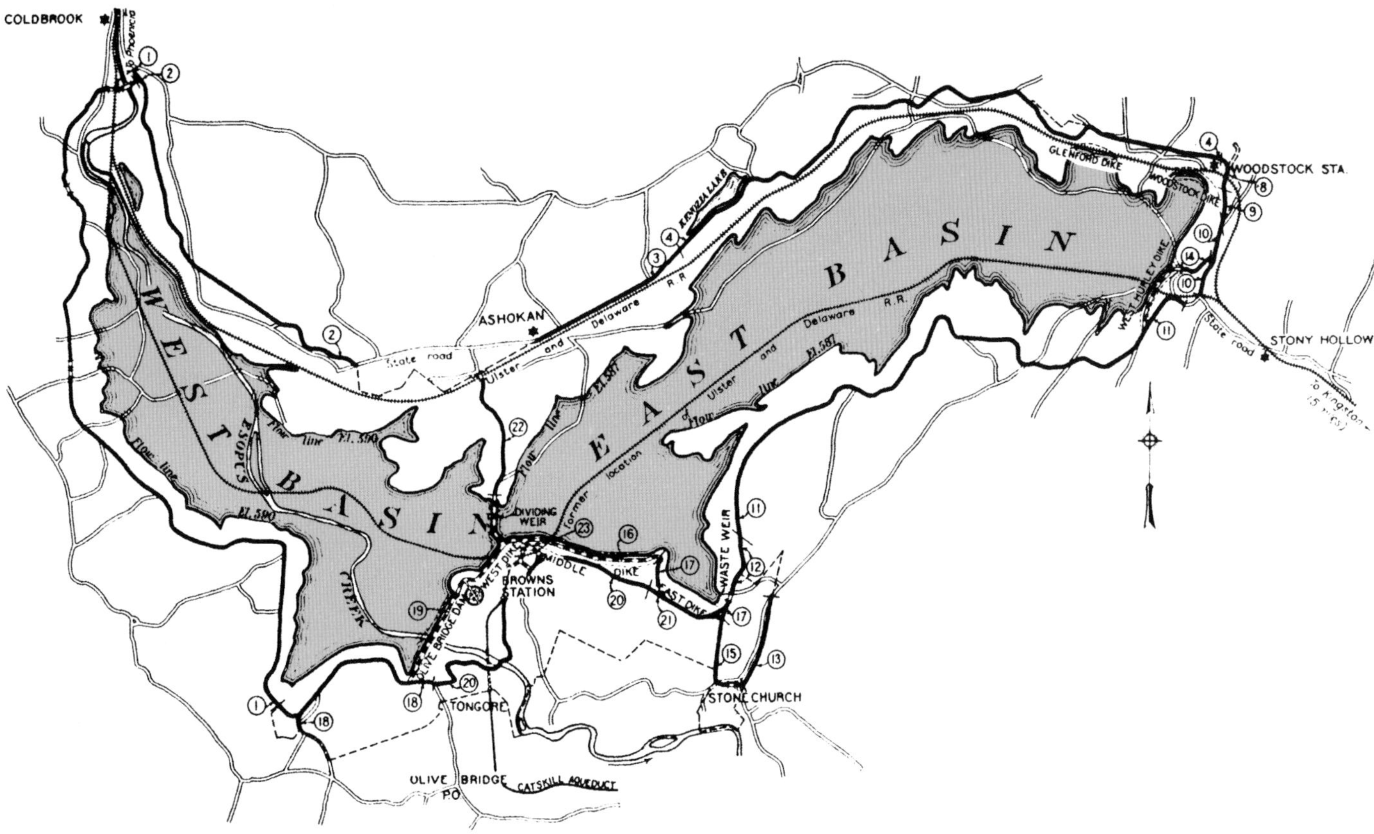

FIG. 9. *Ashokan Reservoir. West and east basins. 1906.*

Detailed surveys of the area done by the board indicated two possible locations: a site near the Olive Bridge just downstream from the confluence of the Esopus Creek and the Beaver Kill (originally identified in the Burr-Hering-Freeman report) and another site two miles farther downstream at a constricted point in the valley, known as the Tongore site [Fig. 8]. This site was attractive because it tapped into a larger portion of the watershed (bringing into the reservoir an extra 5 mgd) and the reservoir itself would have an additional 2½ square miles of surface area, increasing the storage capacity by 10 percent. The Tongore location was also two miles closer to New York City, which would reduce construction time and cost.

Both sites had similar local surface conditions: The Esopus flowed deep in a rock canyon at the bottom of a ravine, with a solid rocky slope to the south and the accumulations of glacial drift to the north. Both sites were cored to determine the subsurface geologic conditions, and engineers concluded both were geologically suitable for a large dam. Initially, it appeared that the Tongore site, with its larger watershed and greater storage capacity, was best. But two factors led the engineers to settle on the upstream site. The Tongore site, being farther downstream and at a lower elevation, required a dam of 270 feet above the streambed to reach the desired elevation for the top crest of the water. This far exceeded the highest masonry dam ever built; the city's own recently completed New Croton Dam was 174 feet high. The engineers were confident they could build such a structure, but the time and cost of building the higher dam, as well as the psychological impact of the enormously high structure on the residents of the towns downstream, were perceived as negative factors.[11]

The Ashokan basin has two shallow valleys within it, one formed by the Esopus and one formed by the Beaver Kill. The two valleys are at an angle to each other and join at one end, creating a V-shaped low area within the larger basin. The Olive Bridge site is near the confluence of these two valleys, and engineers saw that a dam built at this location could create a body of water that was partially divided by the higher land between the two valleys. With a modestly sized division dike built upon the rise of land between the two valleys, the reservoir could be split in half and designed to function as two independent holding areas [Fig. 9]. This two-part configuration would benefit water management and the maintenance of the structures. And because the site was upstream, the dam would not have to be as high. In October 1906, all research and design efforts were directed to the Olive Bridge location and further study on the Tongore site was abandoned.

Planning and preparing for the dam and reservoir included the removal and relocation of towns and farms. The story of Catskill mountain water from the point of view of the people whose homes and communities were lost has been told in two books, Charles H. Weidner's *Water for a City: A History of New York City's Problem from the Beginning to the Delaware River System, 1897–1966* (1974) and Diane Galusha's *Liquid Assets: A History of New York City's Water System* (1999). While the social upheaval that went along with New York City's search for water is outside the scope of this book, it is an important aspect of this story.

Work on the designs for the dam at the Olive Bridge site proceeded with alacrity. In its first year of operation (1906), the Board of Water Supply made 21,400 linear feet of rock borings to determine specific geologic conditions. The board surveyed eighteen thousand acres of land, took one thousand soil samples, and made additional hydraulic measurements of the Esopus Creek and its tributaries.[12] As initially planned, the reservoir would flood 5,978 acres, store about 100 billion gallons of water, and rise to an elevation of +560 feet above mean tide at Ossining. However, the board's hydrological studies concluded that more water was seeping from the watershed than previous measurements had determined. The board reevaluated the plans and revised the design to raise the maximum flow line to an elevation of +590 feet. The revised height increased the surface area of the full reservoir to 8,192 acres and its storage capacity to 132 billion gallons.

The selection of any reservoir site is affected by the subsurface geologic conditions. The pressure of a full reservoir behind a dam can lead to excessive water loss through small fissures and voids in the bedrock, and escaping water can erode rock and undermine the structure. A complete profile of the hidden conditions, based upon thousands of feet of core samples, revealed subtle but important differences in the underlying rock structure at the Olive Bridge site. Subsurface studies showed the main dam site was situated on beds of bluestone, black slate, and green shale, and because the layers of these rocks had remained more or less horizontal throughout geologic time, the rock bedding was impervious to the water above it.

The diversity of the landscape to be dammed (and the subsurface rock profiles) called for different engineering solutions at different locations across the basin. Gaps were closed along the basin's edge east of the Esopus Creek, between Winchell Hill and the rise of land above the village of Stone Church, with a crooked 3.5 miles of moderately high embankments (the West, Middle, Upper, and East Dikes), together called the Beaver Kill Dikes. At the far eastern end of the reservoir, about five miles from the Esopus ravine near the hamlet of West Hurley, three additional embankments totaling 1.6 miles were required (the Glenwood, Woodstock, and West Hurley Dikes). The dikes were earthen mounds built with solid masonry core walls. They were covered with a layer of stone on the sides facing into the water and seeded with grasses on the downstream slope. Most had a brick roadway along the top. The highest of the dikes is 115 feet above the original surface of the land and 185 feet above the bedrock upon which the foundation of the core wall was set.

The dikes along the edges of the basin were relatively routine in structure, similar to the earthen mounds with stone core walls that had been used in some of the Croton System dams and had proved effective for modest-sized embankments. The larger engineering problem was how to close the ravine. Engineers initially examined three structural types for what would be the main portion of the dam: a solid masonry dam (similar to the New Croton Dam), an all-earthen dam, and an earth dam with a masonry core wall.[13] Each offered advantages, but none alone was perfect for the application.

The bedrock at the ravine was covered by deep layers of materials deposited and compacted by the weight of retreating glaciers. This mixture of sand, boulders, rock flour, and clay, called "hardpan," was very dense, relatively stable, and generally impervious. Although this glacial till was dense and hard, it was not solid enough to support the foundation of a masonry dam. Any irregular settlement in the foundation could result in cracks in the rigid masonry structure, and those cracks could lead to leakage, and ultimately failure of the dam. The extreme weight of a masonry dam needs to be situated on solid bedrock, but exposing all the bedrock needed to receive a masonry dam across the entire ravine would have involved the excavation of unrealistically large quantities of earth. An alternate approach was to build directly on the hardpan using a more flexible earthen structure. The use of an all-earthen structure for such a large-scale work was unprecedented and the engineers considered it risky. However, an earth structure in combination with a masonry core wall (similar to the dikes) offered advantages of both the earthen dam and the solid masonry dam. A solid masonry core wall set deep into the bedrock could effectively waterproof an earthen structure, provide the advantage of reducing the huge excavation required for a wide masonry foundation, and still allow for some portion of the dam to be built directly on the ground.

The earthen embankment and core wall combination was suited for some parts of the ravine, but not all. To be stable, earthen embankments, even those with solid masonry core walls, must be significantly wider at the base than they are high, and the slope of the embankment must be gentle enough to resist erosion. The center of the ravine at the Olive Bridge site was deep and the crest of the dam had to rise almost 200 feet. The base of the structure would have been at least 1,000 feet wide and required vast amounts of earth to build. Considering these conflicting criteria, the Board of Water Supply engineers concluded that the Olive Bridge Dam was best designed

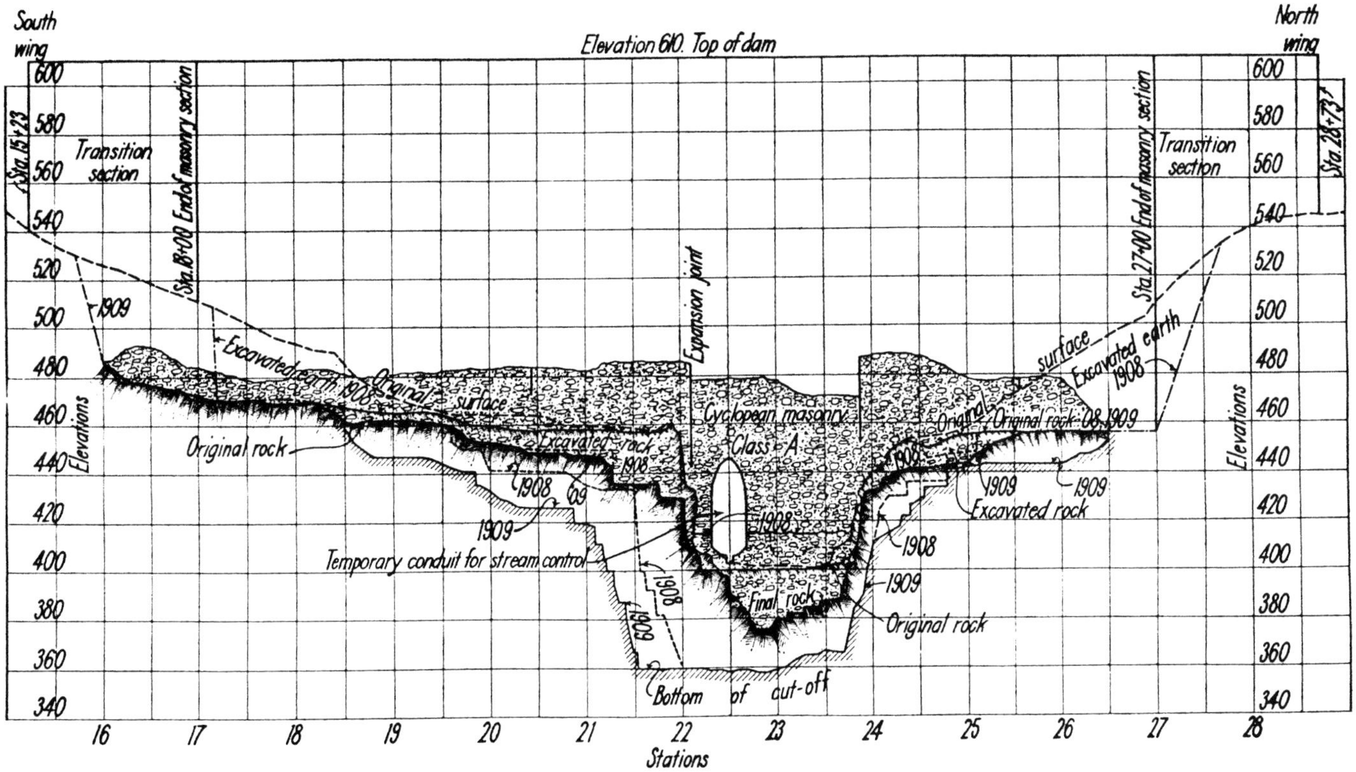

FIG. 10. *Olive Bridge Dam, Ashokan Reservoir. Diagram showing progress of foundation excavation and masonry construction. c. 1909.*

as a composite of different structure types.

In the final design, the deepest part of the ravine was closed with a solid 1,000-foot-long masonry structure. The solid masonry section of the structure stepped up the walls of the gorge at the sides of the ravine, gradually giving way to embankments with core walls.

To build the solid masonry section of the dam, a strip of land was cleared of the glacial overburden to expose the bedrock of the gorge. After removing the cover material, a leveled area 1,000 feet across, 200 feet wide, and averaging about 30 feet deep was blasted and chiseled into the bedrock. Within the excavation area, core holes were drilled another 100 feet down and samples were inspected and tested to locate seams in the stone. This data was used to determine the final depth of the structure. Within the main foundation area, a cutoff trench, designed to stop water from passing through the deepest levels of the structure, was excavated until solid, unfissured rock was reached. The trench was located on the upstream side of the foundation, positioned below the heel of the dam where the water pressures are greatest and where it could best protect the foundation. As a further precaution against water movement deep below the dam, a series of three-inch-diameter holes were drilled into the bottom of the cutoff trench. These holes, averaging 100 feet in depth, were pumped full of grout that forced its way into undiscovered seams and voids, consolidating the layers of bedrock into a solid mass.[14]

The excavation was done with machines called channelers that cut grooves into the rock. Light powder charges were used to break off slabs of stone from the main mass of the rock. As the trench took its final form, excavation was done by hand to avoid any shattering of the bedrock. Leveled benches were established in the bottom of the trench, and it was upon these horizontal surfaces that the foundation of the dam was set. The wall of the dam rises 270 feet from the lowest of the benches. The 30-foot-deep cutoff trench—when filled with solid concrete and cyclopean masonry—keyed the dam's base into the earth. It helped resist the lateral forces of 132 billion gallons of water straining to push the structure down the valley, and prevented water migration underneath the wall of the structure [Fig. 10].

Engineers faced substantial problems of stream water control during the excavation of the dam. The Esopus is capable of flooding up to 38,000 cubic feet per second, and the stream was expected to flood at least once a year at rates of 15,000 cubic feet per second. Flood water had to be diverted from work areas to ensure

FIG. 11. *Olive Bridge Dam under construction with temporary diversion tunnel through dam.*

continuous construction and to limit damage to work in progress. In 1907 temporary stream control structures were set up during the dry season. Two rock-filled timber-framed cofferdams, one above the upstream face of the future dam and one at the downstream face, were built across the ravine, damming the creek. The foundation trench was cut between these protective cofferdams. During excavation operations of the foundation, the water of Esopus was diverted into two 8-foot-diameter steel pipes that channeled the creek past the work site. As the masonry of the dam filled the trench and rose above grade, stream control was shifted through the bypass pipes into an arched passage built into the center of the growing structure. When completed, this conduit was 35 feet wide, 40 feet high, and spanned by an arch 35 feet across [Fig. 11]. The two 8-foot-diameter pipes that had handled the waters of the stream during foundation construction were removed, and the Esopus was directed through the heart of the dam.

This arched passage was the stream-control feature that protected the work site for the next eight years. It was built with a saw-toothed side profile so that when the dam was finally closed, the stepped surfaces locked a massive concrete plug in place. The closing operations began once the dam was about 80 percent complete. While the low waters of late summer ran through a sluice in the bottom of the conduit, concrete was placed into the opening above. The opening was filled to within four feet of the crown of the arch. The remaining work of closing the passage had to be done manually. Workmen crawled into the void to pack it with stones, and grout was pumped in to consolidate the mass. Preparations to close the conduit began in August 1913, and on September 9 the lower sluice was closed and the waters began to back up against the huge masonry wall[15] [Figs. 12, 13].

The Olive Bridge Dam was built using cyclopean masonry, first employed on a large scale at the New Croton Dam. The dam was divided into twelve independent structural zones, each defined by a perimeter wall built of stacked concrete blocks. The precast concrete blocks (called risers) gave dimension and form to the individual sections of the dam. Made at a local plant and allowed to cure for three months, the blocks, varying from 15 to 65 cubic feet each, were stacked like bricks, several courses at a time. Wet concrete and gigantic pieces of rough stone were placed into these boxlike enclosures. These two materials make up the solid hearting of the dam. The concrete riser blocks, because of their size and weight, were able to hold unset concrete and stone, thereby eliminating the difficult and costly step of building formwork for the

FIG. 12. *Olive Bridge Dam. Stream control portal not yet closed. April 3, 1912.*

FIG. 13. *Olive Bridge Dam. Downstream face from Esopus Gorge. September 19, 1913.*

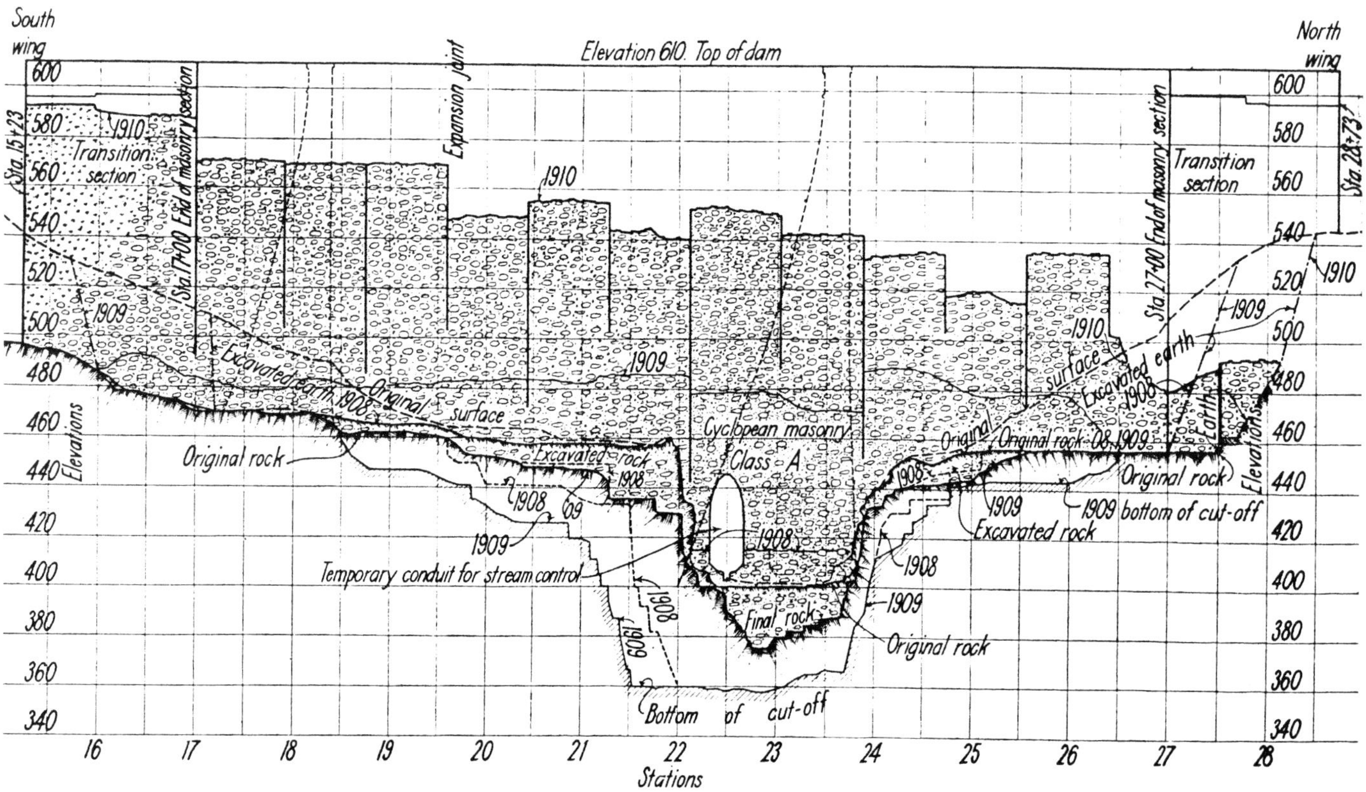

FIG. 14. *Olive Bridge Dam, Ashokan Reservoir. Diagram showing extent of masonry construction in 1910.*

concrete construction. These high-strength concrete blocks served not only as formwork but ultimately comprised the finished material face of the dam.[16]

The cyclopean building method was critically important to the construction of dams of this era and solved several difficult problems associated with the use of large masses of concrete. In concrete construction, the interface that exists between one hardened placement of concrete and the next placement of unset concrete is called a cold joint: it is a seam of weakness in the structure. In a dam it is paramount to resist the lateral force of the impounded water, and these horizontal planes of weakness along the lateral lines of force present a serious problem. Cyclopean construction provided a solution.

The huge stones used in cyclopean construction bridged the layers of concrete and keyed the separate pours together. The cyclopean stones came from the excavation of the foundation trench or from local quarries. Each piece of stone was inspected, brushed with a heavy wire broom, cleaned with pressurized water, cleared of loose spalls, and covered with a coating of grout before being set. The meticulous surface preparation of the stones and concrete insured the strongest possible bond between the component materials. Several cubic yards of concrete were dumped onto the surface of the work area and then the large pieces of stone were placed into it. Each stone was joggled with bars to release any air pockets and to assure a solid mass. Additional loads of concrete were placed around the stones as needed to raise the level of concrete. The large stones were set two or three feet apart and left sticking up at least one foot above the concrete pour to key the layers and eliminate the horizontal planes of weakness.

Prior to the construction of the Olive Bridge Dam, engineers had noted that in temperate climates solid masonry dams were prone to cracking due to the natural expansion and contraction of materials under thermal change. To control this problem and limit water damage within the masonry, the solid portion of the Olive Bridge Dam was constructed with watertight expansion joints. This was the first time that engineered expansion control was used in the construction of a major masonry dam. Above grade, the structure was divided into twelve equal sections, each discreet section ranging from 84 feet to 90 feet wide—a dimension well within the range that expansion cracks on other dams had been observed [Fig. 14]. Rows of the precast concrete building blocks were placed perpendicular to the dam to create transverse joints through the structure. At each of these joints the blocks were set in a staggered tongue-and-groove pattern that locked the independent sections of masonry together across a movable joint. The faces of the blocks along the expansion joint were coated with a

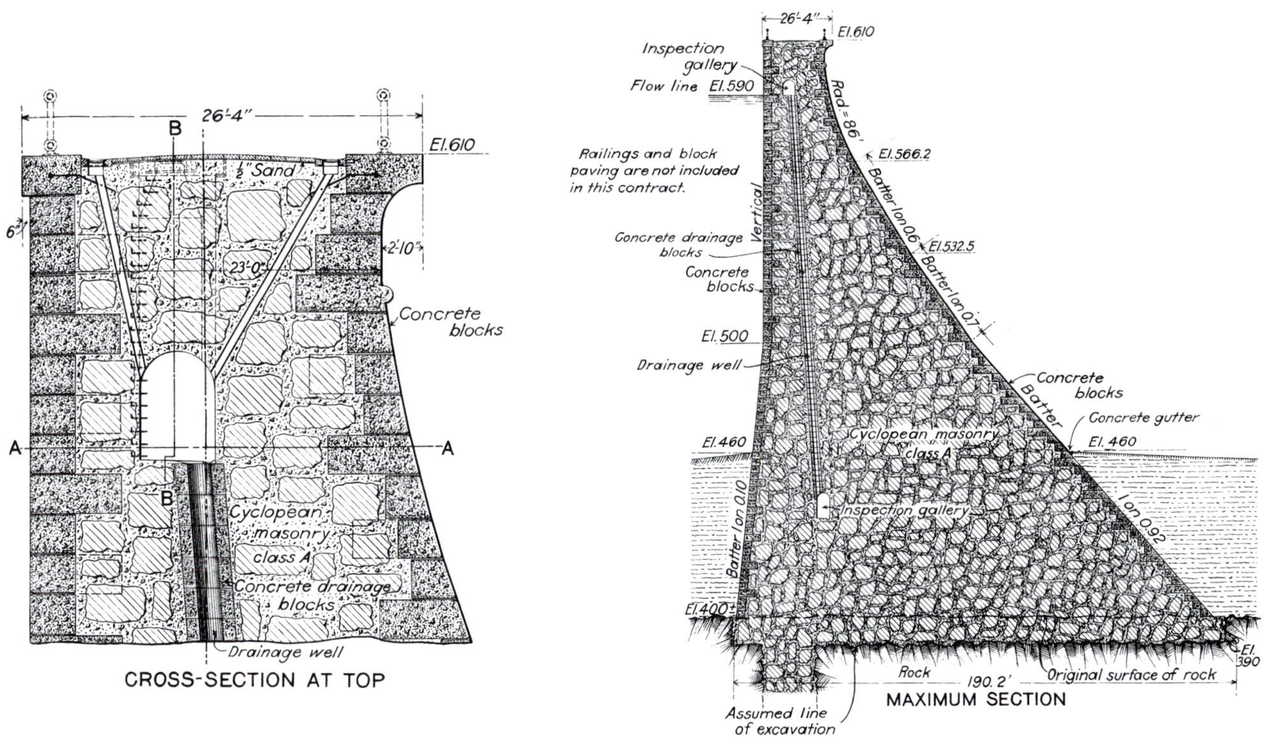

FIGS. 15 AND 16. *Olive Bridge Dam. Cross sections of dam showing concrete block and cyclopean masonry construction, inspection galleries and drainage wells. June 20, 1907.*

one-inch-thick layer of a petroleum-based lubricant that separated the sections of masonry and allowed for movement. A flexible heavy-gauge copper sheet was built into the blockwork across the joint as a water stop. When the dam contracted, the joints opened but remained relatively watertight due to the interlocking bond pattern of the blocks and the copper water stops. Inspection wells were built along the vertical height of the structure at each expansion joint. Engineers could enter the body of the dam by climbing a ladder within these wells to inspect the joints [Fig. 15]. Despite the use of water stops, tremendous pressure still forced some water through the joints. At the time of completion, the dam was leaking a total of about 300 gallons per minute through the joints, a minor amount considering the size of the structure. Water entering through the expansion joints moves through the dam via a system of drainage chambers that discharges the collected

FIG. 17. *Comparative dam profiles showing Wegmann's and Rankine's profiles for the New Croton Dam as well as the Board of Water Supply profile for the Olive Bridge Dam. c. 1906.*

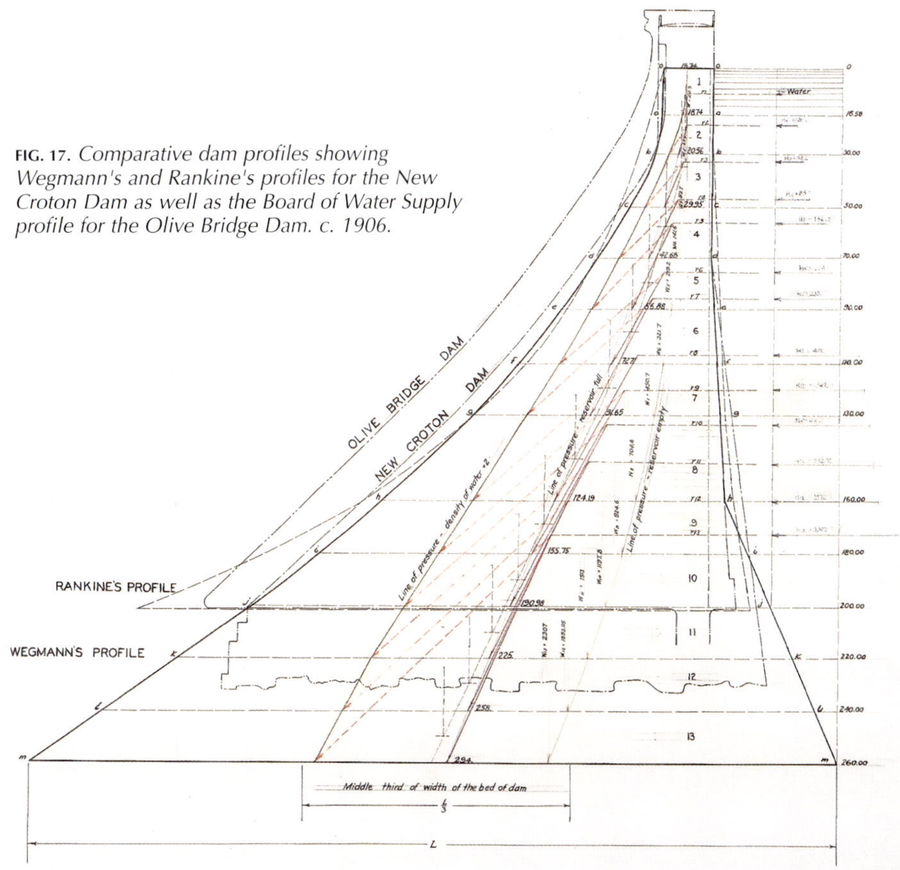

FIG. 18. *Olive Bridge Dam. Downstream face. Transition from solid masonry dam to terraced earthen embankments on sides of ravine. September 25, 1915.*

water in the creek bed on the downstream side of the structure.[17]

In cross section, the finished dam has a near vertical face to the reservoir side. The top surface is 26.3 feet wide and the downstream face curves outward toward a base with a maximum width of 190.2 feet [Figs. 16, 17]. The vertical pressure on the heel of the dam is 28,000 pounds per square foot; at the toe this force diminishes to about 17,000 pounds per square foot. The top of the dam was designed to resist a lateral ice thrust of 47,000 pounds per linear foot.

Two wings, earthen structures with concrete core walls, extend from either side of the solid central portion of the dam [Fig. 18]. The core walls of the embankments extend downward through the glacial overburden to trenches cut in the stone and step up the sides of the ravine following the profile of the rock. At the deepest points the trenches for these core walls extend 50 feet below grade [Fig. 19]. The core walls terminate at an elevation of +596, six feet above high water in the reservoir and fourteen feet below the top

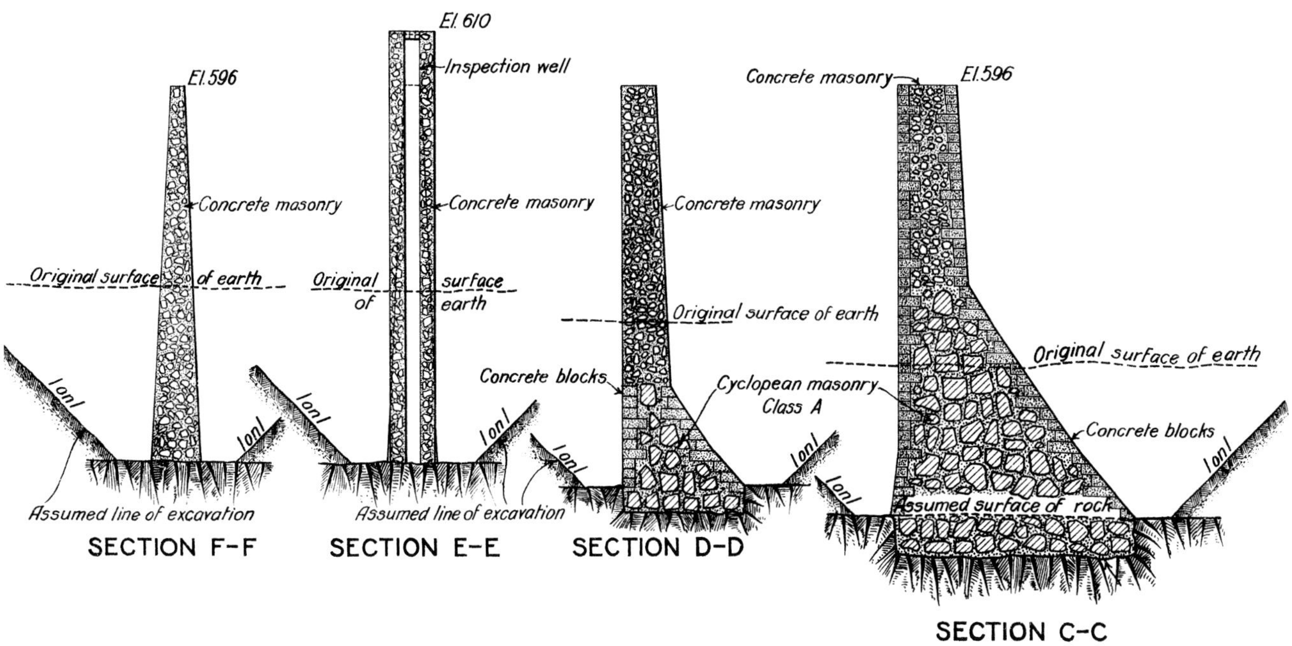

FIG. 19. *Olive Bridge Dam. Different configurations of masonry core walls. June 20, 1907.*

of the finished structure. The embankments are built up of cleaned and graded earthen materials spread on either side of the core wall and compacted with ten-ton rollers. On the upstream side, where the demands on the structure are greater, the earthen materials are laid down in four-inch layers and clad in stone to resist erosion from the wave action of the reservoir. On the downstream side, the materials are laid up in six-inch layers and seeded with grasses to stabilize the sloped surface.

The Board of Water Supply engineers selected the Olive Bridge site in part because the setting allowed the reservoir to be configured as two basins. To create the separate reservoirs, a 2,200-foot-long dividing dike was built along the natural rise of high ground that passes through the reservoir site [Fig. 20]. The dike was built to include a connecting weir, allowing water to spill from the higher west basin (where most of the water enters the reservoir) into the east basin [Fig. 21]. The structure incorporates sluice gates (in the dividing-weir gate chamber) that permit water to be transferred from the west basin back to the east basin and vice versa [Figs. 22, 23]. By dividing the reservoir water, quality could be more precisely controlled. The Esopus is known to be a "flashy" drainage—meaning that large amounts of precipitation can fall in a very short time, causing creek levels to rise and wash down sand, stones, dirt, and organic matter. The floodwater from these "flashes" can compromise water quality. The west basin takes in the water from a very large watershed and is much more prone to the detrimental effects of dirtied floodwaters. In times of high turbidity in the west basin, the dividing weir can be closed and runoff water retained. Water can then be drawn off the east basin only, ensuring the purity of the water that enters the aqueduct. The east basin faces its own water quality problems, as it is built on relatively shallow and flat swampy lands. If the water temperature is high and the reservoir level low, the flooded peat bogs of the submerged wetlands can host algae blooms and high levels of microorganisms. If the

FIG. 20. *Ashokan Reservoir. East basin side of dividing weir with abutments for dividing weir bridge. August 5, 1914.*

FIG. 21. *Ashokan Reservoir. Ashokan Bridge over dividing weir. April 7, 1916.*

FIG. 22. *Ashokan Reservoir. Upper gate chamber on west basin side. October 3, 1923.*

east basin water is compromised, the west basin can be used.

To accommodate floods when the reservoir is at capacity, a waste weir was constructed east of the Beaver Kill Dikes. Waste weirs are often built as part of the superstructure of the dam, allowing water to spill over the top and back into the stream channel. But the design and location of this waste weir had to be approached differently. Spilling water can be a destructive force to a dam and, in times of flooding, can undermine the rock structure of the waterway below. The narrowness of the Esopus ravine and the relative softness of the bluestone and shale creek bed required the spillway be built on a hard rock shelf two thousand feet east of the main dam at the eastern end of the Beaverkill Dikes. The rock ledge drops down into a side ravine that is sound, large enough to accommodate the flow of flood waters, and completely bypasses the stream bed at the base of the dam. By situating the waste weir and spillway at a remote location, the risk to the foundations was eliminated. The waste weir is an elegant curvilinear masonry embankment 990 feet long, with steps designed to dissipate the energy of falling water and limit the scouring effect on the discharge channel[18] [Figs. 24, 25, 26].

The land inundated by the reservoir water required preparation. Trees and brush were cut close to the ground, and all surface vegetation was burned just before flooding. Along the shoreline tree stumps were grubbed out, roots and other organic matter dug up, and topsoil scraped off. These steps ensured that the shores would be presentable if they were exposed when the water level dropped.

Grades were adjusted to eliminate enclosed shallow areas where micro-organisms might flourish. Underlying soft material and pits of organic debris were excavated and backfilled. Perishable materials, buildings, barns, outhouses and other wooden structures, fences, logs, stumps, grass, weeds, and leaves were all gathered and cleared, and the stones of the stone walls scattered.

Sixty-two miles of roads had been eliminated by the time the more than eight thousand acres of land in the Ashokan valley were flooded. To compensate, forty miles of new highways were built, as well as ten new reinforced concrete bridges to transverse the streams, reservoir channels, and rail crossings. The Ashokan Bridge, extending one thousand feet across the reservoir, is the most recognized, and arguably the most beautiful, of the Catskill era bridges. Fifteen spans, each 67½ feet across, are set atop buttresses integrated directly with the dividing weir. Columns bearing on the concrete arches produce open-sided spandrels that create a light and graceful structure supporting the roadway deck. Also notable among the Catskill bridges is the Esopus Bridge, spanning the creek at the northern end of the west basin. The structure consists of five equal arches, also 67½ feet across. Each span was made with two separate arched concrete ribs with vertical concrete cross walls placed over the ribs, again to maintain the lightness of the structure and the overall openness of the profile. The highway surface is a 10-inch-thick reinforced concrete slab supported by cross walls. Concrete balustrades (in contrast to solid parapets) were used to add to the effect of lightness. The bridges of this era were built with expansion joints throughout. Other bridges include the two Bushkill bridges, each with three 25-foot spans bridged with T-shaped concrete girders, and the Traver Hollow Bridge, 493 feet across with a main span of 200 feet, built with three hinged, reinforced concrete ribs [Fig. 27]. The Spillway Bridge and Stone Church Bridge are solid soffit bridges each having a span of 175 feet and a rise of 25 feet. Other minor bridges include the Beaver Kill Bridge and the Boiceville

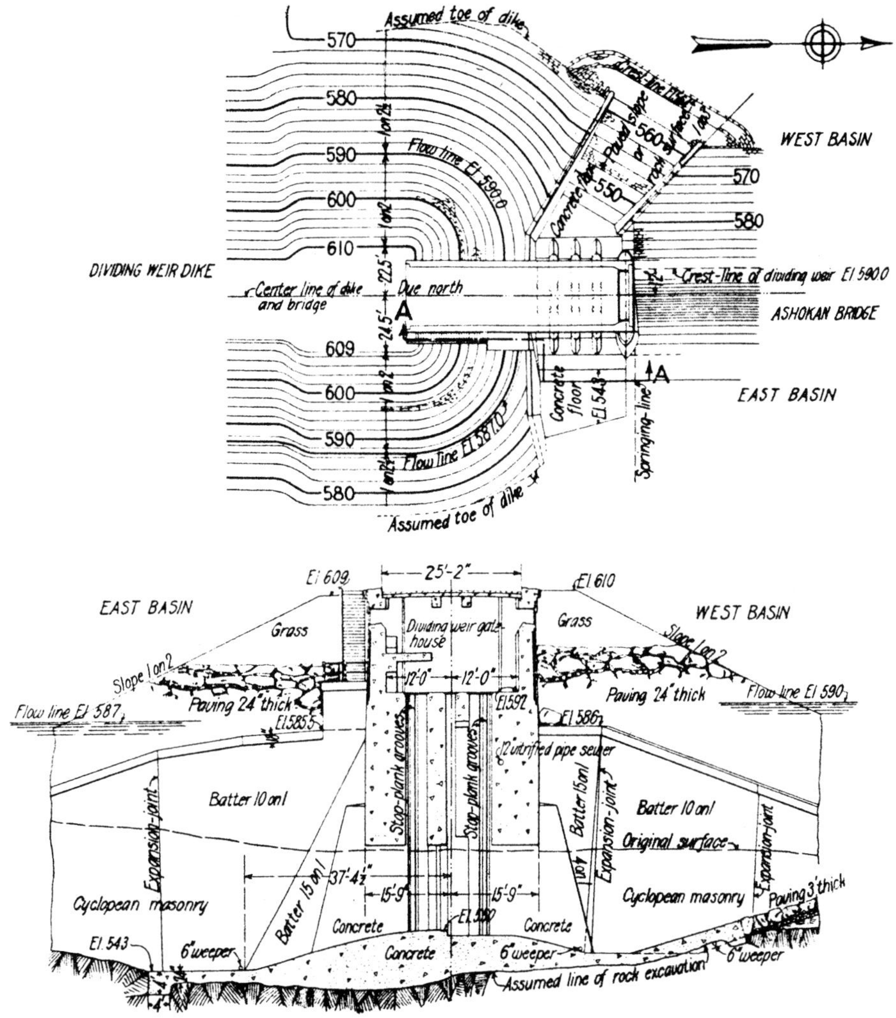

FIG. 23. *Ashokan Reservoir. Plan and section of main gatehouse on dividing weir. c. 1907.*

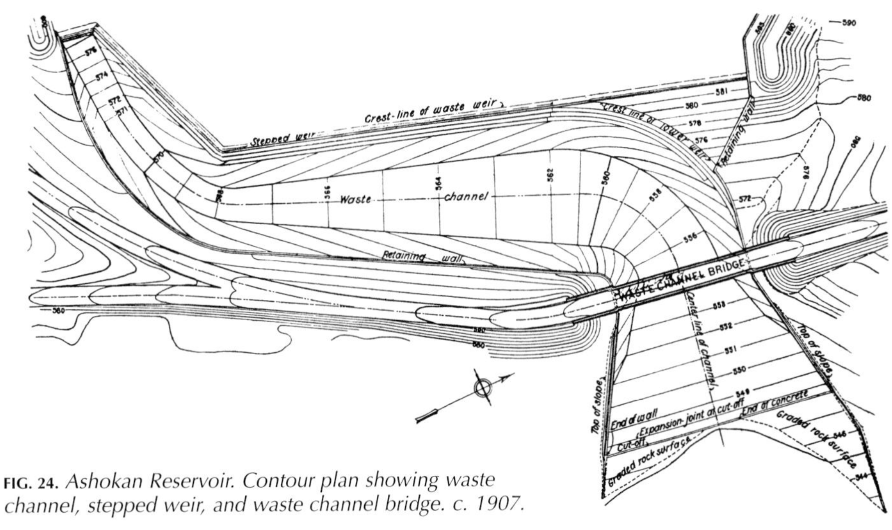

FIG. 24. *Ashokan Reservoir. Contour plan showing waste channel, stepped weir, and waste channel bridge. c. 1907.*

FIG. 25. *Ashokan Reservoir. Spillway, stepped weir, and spillway bridge. August 5, 1914.*

FIG. 26. *Ashokan Reservoir. Full reservoir spilling into channel. Elevation of cresting water at +588.46. Rate of flow is 365 mgd. April 25, 1928.*

FIG. 27. *Ashokan Reservoir. Traver Hollow Bridge on the west basin. June 24, 1916.*

Bridge. The bridges built by the Board of Water Supply are elegant, and functional modern designs that are a visible part of the engineering legacy of the waterworks. The works were typical of the best of the experiments being done in reinforced concrete bridge design at the time, both in the United States and Europe. "Special effort was made in the design of these concrete bridges to express the properties of the materials used and to eliminate inappropriate characteristics of a stone structure."[19]

Sixteen years after the west-of-the Hudson water sources were first seriously considered as a source of additional supply for the thirsty city, the turbulent and rapidly flowing waters of the Esopus became still as they backed up against the Olive Bridge Dam to form the Ashokan Reservoir. The reservoir was filled to capacity for the first time on December 12, 1916.

To release the water out of the reservoir and into the aqueduct required a labyrinth of subsurface tunnels and chambers equipped with the valves, sluice gates, and operating equipment capable of routing hundreds of millions of gallons of water per day [Figs. 28, 29]. Consolidated effluent works in the upper gate chamber are located on the dividing weir and situated to access the waters of either basin. Two pressure tunnels built deep into the rock below the dividing weir connect the gate chamber to the headworks of the aqueduct [Fig. 30]. Since water quality in a reservoir can be poor at one level and fine at another, the upper gate chamber has inlets on both sides of the dividing weir, with vertical control shafts called "wet wells," composed of shutters that allow water to be drawn from any level within the reservoir. The upper gate chamber is used not to regulate the amount of water drawn but only to determine which basin and what elevation the water is

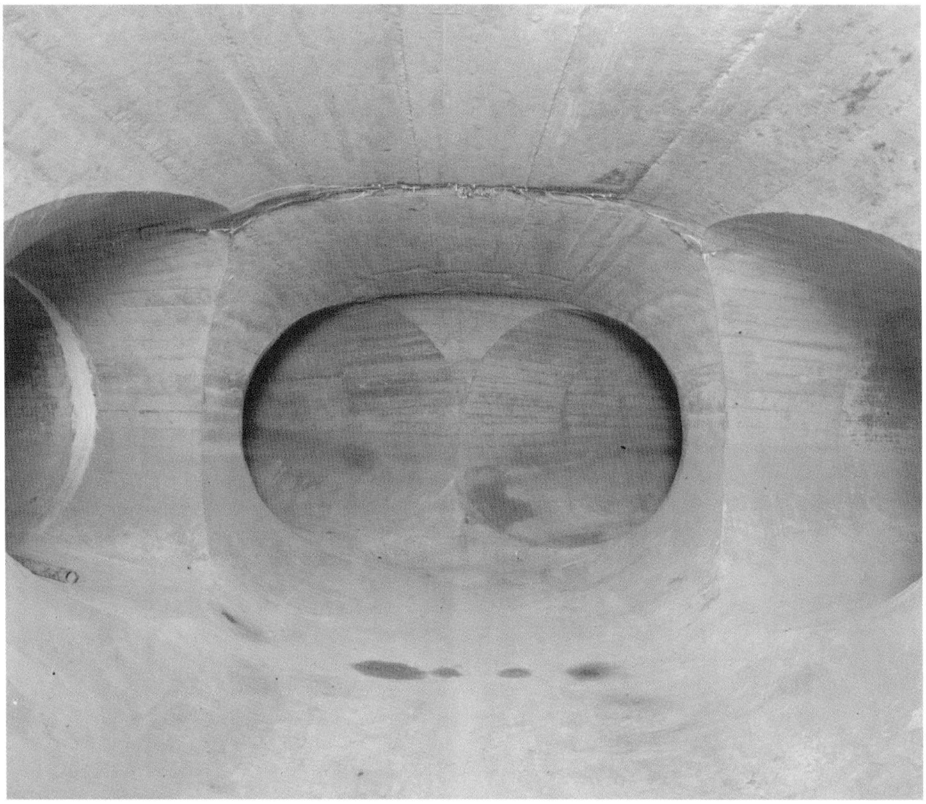

FIG. 28. *Ashokan Reservoir. Upper gate chamber at junction with Catskill Aqueduct. September 19, 1913.*

FIG. 29. *Ashokan Reservoir. Headworks of Catskill Aqueduct at lower gate chamber. September 19, 1913.*

taken from. The chamber inlets are fitted with bronze bars that filter large pieces of debris before the water enters the headworks of the aqueduct. Channels leading to the inlets (5,000 feet long on the west, 3,000 feet long on the east, and both about 100 feet deep) were cut into the rock at the bottom of the reservoir to follow the natural drainage lines of the original streams. These inlet channels insure that water can be taken from the deepest areas of the reservoir.

From the upper gate chamber the water moves through one or both of two pressure tunnels that connect to the lower gate chamber. The pressure tunnels were built one over the other, deep within the bedrock below the center of the dividing dike. The lower gate chamber is part of the headworks on the downstream side of the dam, near the toe of the Beaver Kill Dike [Fig. 31]. It regulates two conduits—the special aqueducts—which take water directly to the screen chamber where fine debris is captured and removed before entering the aqueduct. The lower of the special aqueducts can drain either reservoir basin back into the stream—a protective measure in case water conditions become unacceptable. The water can also be diverted from the lower gate chamber into the aerator basin, where up to 1,020 cubic feet per second of water was sprayed from 1,518 bronze four-inch nozzles. This process, called aeration, kills waterborne organisms by exposing them to oxygen. (Aeration is no longer used, as all New York City water is now treated with chlorine to kill organisms and with alum to remove particulate matter.) The aerator was the central part of a landscape plan below the dam that included the tremendous fountains of the aerator basin, an axial tree-lined approach called "the Mall," and several appurtenant superstructures. The entire headworks assembly is built deep within the bedrock and employs, in many locations, the cyclopean masonry methods used on the dam. These structures were designed to handle twice the capacity of the Catskill Aqueduct, anticipating possible expansion of the system [Fig. 32].

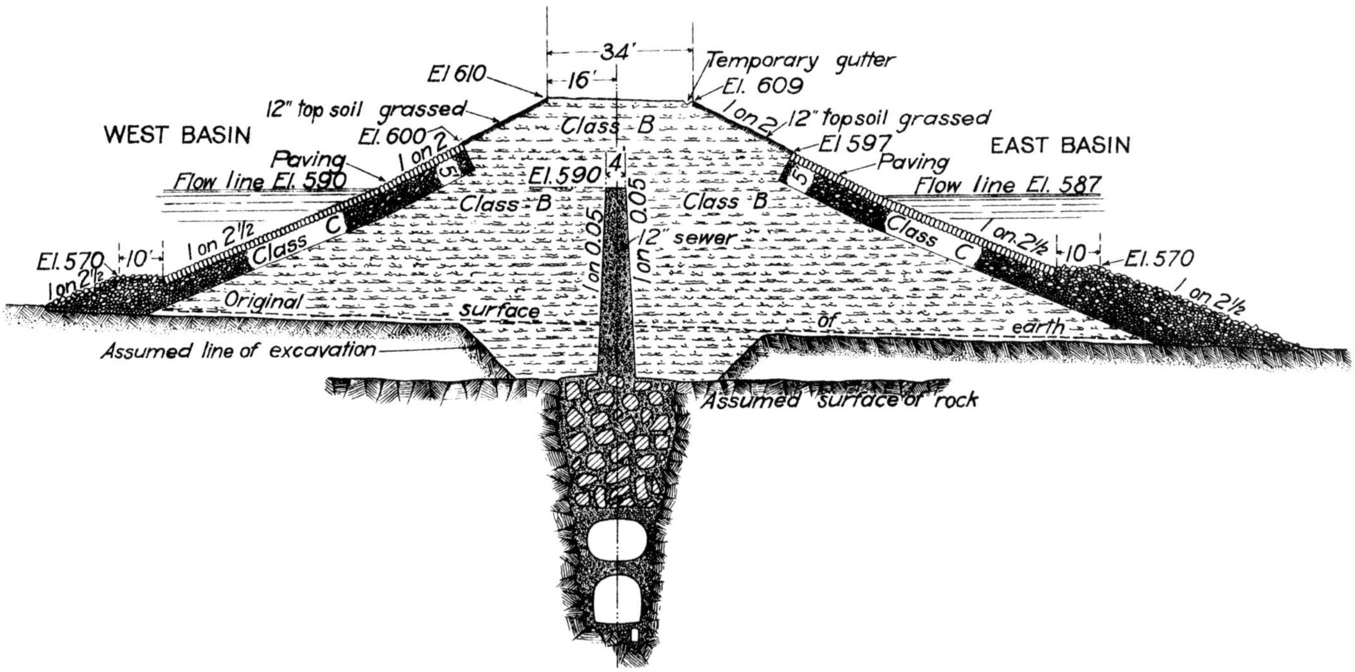

FIG. 30. *Ashokan Reservoir. Maximum cross section of the dividing weir dike showing aqueducts. June 20, 1907.*

During the time when the Olive Bridge Dam and Ashokan Reservoir were under construction, work was also proceeding on the Catskill Aqueduct. Almost ninety-two miles of aqueduct was necessary to negotiate the landscape between the headworks at Ashokan and the receiving reservoir at Hill View. The engineering involved was of unprecedented complexity and the construction was of enormous scale and difficulty. Diverse structural solutions were employed to transverse the hillsides, to pass through the mountains, to cross deep and rugged valleys, and ultimately, in perhaps the greatest engineering achievement of the Catskill Water Supply system, to slip safely under the Hudson River.

Several advances would distinguish the Catskill Aqueduct from its Croton predecessors. The Catskill Aqueduct employs pressure tunnels (siphons) of extraordinary length and dimension to pass under valleys and rivers. The materials used in construction of the system also represented an advance in building technology. Much of

FIG. 31. *Ashokan Reservoir. Headworks of Catskill Aqueduct at lower gate chamber. September 19, 1913.*

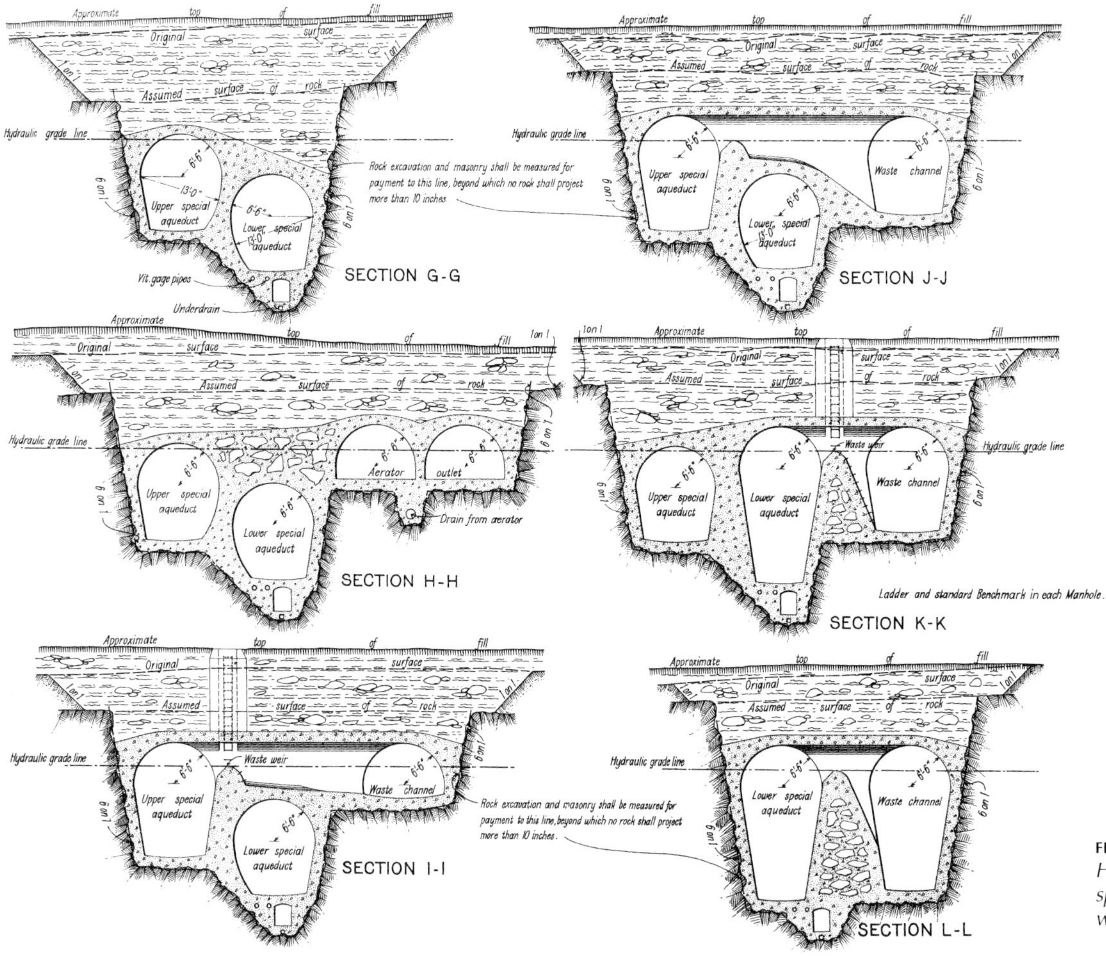

FIG. 32. *Catskill Aqueduct. Headworks. Cross sections of special aqueducts and waste weirs. September 10, 1909.*

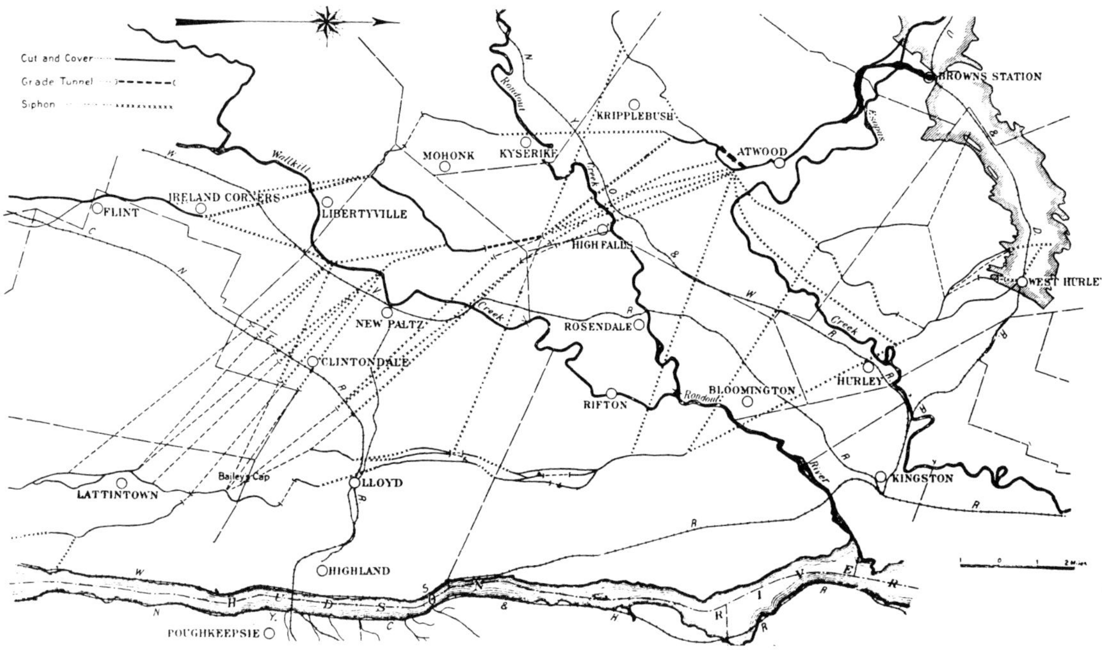

FIG. 33. *Catskill Aqueduct. Map of possible alternate routes for aqueduct. c. 1906.*

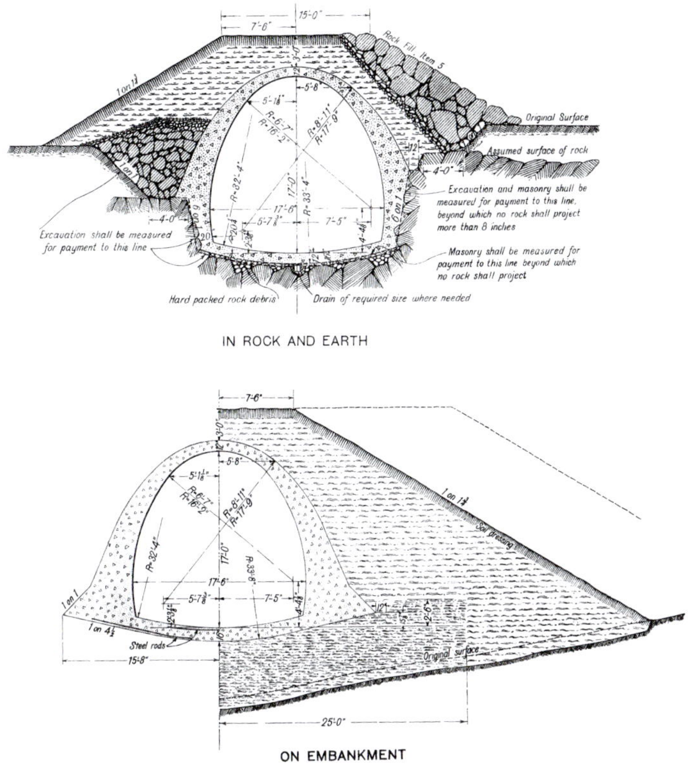

FIG. 34. *Catskill Aqueduct. Alternative types of cut-and-cover construction. September 10, 1909.*

FIG. 35. *Catskill Aqueduct. Concrete cut-and-cover aqueduct before backfilling. Yorktown, New York. August 22, 1911.*

the aqueduct system was built of the material of modern engineering: concrete, not brick and stone masonry, as in the Croton Aqueducts.

J. Waldo Smith and the Board of Water Supply engineers measured, evaluated, and compared alternate routes and structural types [Fig. 33]. In a gravity delivered system, the aqueduct follows the shortest line possible and descends at a uniform rate. Regardless of the route considered, no aqueduct could follow either the shortest line or the perfect slope from Ashokan to the city. To negotiate the diverse path, four principal structural forms for the aqueduct were developed, each suited to different landscape problems.

A cut-and-cover aqueduct is built on the hydraulic gradient descending at a uniform rate to allow the downward flow of the water and is used where the structure can follow the natural contours of the land. The water flows down the slope of the gradient of the structure and is not under pressure. In this configuration, a trench is cut into the earth and the concrete conduit structure is built in the excavated area. The conduit is then completely covered with an earthen embankment. The cut-and-cover conduit is horseshoe-shaped in section. This was the most economical of the basic structure types, and as long as the geography would permit, this was the preferred structure of the aqueduct [Figs. 34, 35].

The grade tunnel aqueduct follows the hydraulic gradient but is built in tunnels cut through the hills and ridges. The water flows down the gradient of the structure and is not under pressure. Grade tunnels also have a horseshoe-shaped cross section, but they are narrower and taller than the cut-and-cover section. In the construction of a grade tunnel, a rough passageway is excavated through earth and rock and then lined with concrete.

The pressure tunnel aqueduct is built below the natural surface of the ground and below the hydraulic gradient. The water is under pressure. These pressure tunnels can be vertical, horizontal, or inclined. Pressure tunnels are typically circular in cross section. A rough tunnel or shaft is excavated through deep rock and lined with concrete. This type of structure, also known as a pressure siphon, was used to pass under deep, broad valleys, rivers, and lakes.

The fourth structural configuration employed on the Catskill line is the steel pipe siphon, which was used when the aqueduct had to pass through shallow valleys. In these siphons the water in the aqueduct

is below the hydraulic gradient and under pressure. The steel pipes are laid in shallow trenches and are inclined to follow the general topography of the valley being crossed. Pipe siphons were most often used in smaller valleys and in locations where the subsurface rock was not sound enough to support tunneling [Fig. 36].

These four aqueduct types were used as required to transverse the diverse geographic features of the line. Of the 92 miles between Ashokan and Hill View, 55 miles of the terrain was suitable for the cut-and-cover aqueduct. The remaining portions of the route required 24 grade tunnels totaling 14 miles in length, seven pressure tunnels totaling 17 miles in length, and 14 steel-pipe siphons totaling 6 miles in length.

The Catskill aqueduct begins at the Ashokan Reservoir headworks. It passes south under the Esopus ravine and continues through the Marbletown Hills and into the Rondout watershed. It dips deep below the Rondout valley, through the Shawangunk mountain range, under the Wallkill River, and through the West Hudson Highlands, where it meets the down-take shaft for the Hudson River Siphon. The aqueducts drops 1,100 feet under the Hudson River, rises up to cross the Croton Highlands, passes under the Croton Reservoir, and continues down to the Kensico Reservoir. The last leg of the aqueduct connects the Kensico Reservoir to the Hill View Receiving Reservoir near the Bronx city line.

The general line of the aqueduct was determined by four key location factors: the takeout at the Ashokan Reservoir, the Hudson River crossing, the intermediate storage reservoir, and finally, the receiving/distributing reservoir in the city. The Burr-Hering-Freeman Commission on Additional Water had recommended that the Ashokan Reservoir takeout be east of the main Ashokan Dam near the town of West Hurley and that the Hudson River crossing be near New Hamburg, about ten miles downriver from Poughkeepsie. The intermediate storage reservoir was planned at the site of the original Kensico Lake near White Plains, and the distribution reservoir was planned for the high ground just north of the Bronx, at Hill View. Early reconnaissance by the Board of Water Supply supported the location of the aqueduct outtake at the eastern end of the reservoir near West Hurley, but as the design work on the Ashokan's two-basin scheme progressed, board engineers relocated the headworks to near the dividing weir, and subsequently the West Hurley designs were abandoned.

With the position of the head of the aqueduct established, the problem of where and how to cross the Hudson River had to be solved. The board considered several structural solutions to get the Catskill water across the Hudson River: a bridge supporting pipes; steel pipes laid in trenches dredged into the river bottom; shield tunnels made of cast-iron plates built into the mud of the river bottom; or a tunnel in deep rock below the river bottom. The tunnel in deep rock promised to be the most durable. It offered the greatest protection to the aqueduct while in operation and was arguably the most economical in the long term. The deep-rock tunnel was also in keeping with the design philosophy that the system should last for centuries. In September 1905, just three months after the appointment of the first Board of Water Supply Commissioners, studies of the geology of the Hudson River began.

Between New Hamburg and West Point, 226 test borings were made totaling 28,000 linear feet. The New Hamburg crossing had been ruled out by 1907, primarily because of unfavorable subsurface geologic conditions, but also because of the relocation of the Ashokan takeout location. Other crossing sites were studied and eliminated. A favorable site near the West Point military grounds was abandoned because work on the West Point reservation would entail an act of Congress and would subsequently slow a project with little room for delay. Eleven

FIG. 37. *Catskill Aqueduct. Hudson River crossing. Drilling operations to determine geologic profile below riverbed. November 26, 1907.*

FACING PAGE

FIG. 36. *Catskill Aqueduct. Hunter's Brook Siphon. Steel pipe laying operations before backfilling. August 25, 1910.*

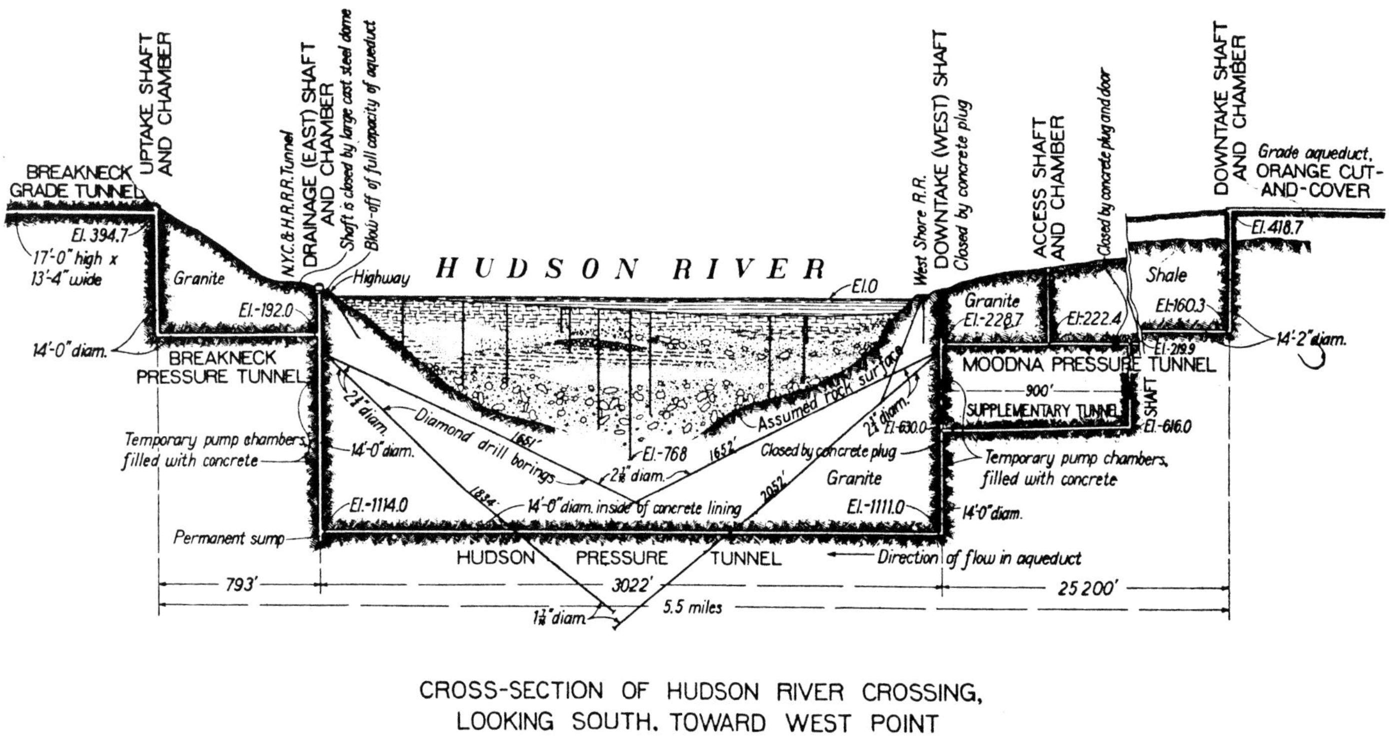

FIG. 38. *Catskill Aqueduct. Profile of aqueduct at Hudson River crossing. c. 1910.*

miles further down river, on the northern side of the Hudson Highlands, two prominent mountains flank the river: Storm King Mountain on the west and Breakneck Mountain on the east. Both mountains are the same hard granitic gneiss, suggesting there might be continuity in the subsurface geology. Based upon the surface observations and limited test borings taken on either bank of the river, the Storm King Mountain site was chosen as the best place to pass the aqueduct under the Hudson.[20]

Somewhere below the bottom of the river was a preglacial gorge of solid rock. To determine the geologic profile of this gorge and its adjacent rock strata was the most complex subsurface exploration ever attempted. Drilling for core samples to determine rock depths below the river were initially done from the water surface on a floating rig [Fig. 37]. It was a very difficult task to drill in a river that had currents flowing in two directions, is an active shipping lane used by large vessels, and required all operations be shut down seasonally because of ice.

It took more than five years to determine the conditions below the river. The first three hundred to four hundred feet of most of the borings went without incident, but numerous problems were encountered as the drilling went deeper into the river gorge. As work progressed, increasingly large boulders were encountered, causing the drill assembly to slip to one side and the bit shaft to bind to the pipe casing walls of the drilling rig. The boulders had to be blasted in order to drill past them. In one incident, the crew unsuccessfully worked for five weeks to free a trapped bit before abandoning a 588-foot hole they had worked on for four months. Such setbacks made progress agonizingly slow. A project engineer called the situation "hopeless." The drilling scow was held in place by six large anchors, but storms, currents, and the wakes of passing ships moved the rig, which put stress on the slender pipes of the drill-shaft assembly, and drilling had to be suspended until water conditions calmed. Floating ice shut down the drill platform for five months a year. One of the holes took six months to put down. Of the fifteen holes drilled from the water surface to explore the bottom of the Hudson River, "two were knocked over by passing vessels, one by its own scow, one fell over from its own weight, two were abandoned because of conditions on the bottom, two were discontinued because [they were] supplanted by other holes, and one, the deepest, was discontinued in December 1910 at a depth of 768 feet as the desired information was obtained by other means."[21]

The difficulties in drilling from the center of the river on a floating platform, and the inability to gather the data they needed in any sort of timely way, prompted the board to take other measures. Two large test shafts were sunk on either bank of the river. Once the shafts reached a sufficient depth in solid rock, chambers were cut from the rock and fixed drilling rigs set at 45-degree angles were built inside the shaft. The test borings were accomplished on these inclined lines. The Board of Water Supply engineers reasoned that if two of the inclined borings could be made to pass each other and set in solid rock the entire way, then the profile of the rock could be reasonably determined.

Although this strategy succeeded, it was also burdened with its own difficulties. In the east shaft, not far below the surface, the first hole drilled took in ninety gallons of water per minute. The pressure of the incoming water undermined the drilling operation, and this hole was lost. The next boring, upon reaching 1,085 in length, produced a 180-gallon per minute flow, and pressurized water sprayed out of the head of drilling machinery. Complications followed, and the drilling equipment seized up, fussing the drill bit into the rock. This hole had to be abandoned as well, and another two months work was lost. But the crews persisted, and finally, almost five years after the first holes were attempted from the floating platform, four inclined borings were successfully drilled, the longest 2,051 feet in length. Based upon the data, the engineers concluded that the solid rock gorge was 900 feet below sea level. The center line of the tunnel was fixed at a depth of 1,100 feet, assuming a minimum cover of 200 feet of solid bedrock[22] [Fig. 38].

The Hudson River Siphon crosses below the river at an elevation 1,114 feet below sea level. The tunnel is a fourteen-foot-diameter concrete-and-steel-lined circular pressure tunnel connecting the vertical shaft at the base of Storm King Mountain with the vertical shaft at the base of Breakneck Mountain, a distance of 3,022 feet [Fig. 39]. The Storm King side of the siphon is 24 feet higher than the Breakneck Mountain side, this being the actual vertical fall of the water after dropping more than 1,000 feet below the river.

Seven years after the Hudson River crossing explorations began, the siphon was "holed through" on January 30, 1912. By the end of the year the concrete lining of the horizontal tunnel was complete and

FIG. 39. *Catskill Aqueduct. Hudson River crossing. Drainage chamber under construction with one shaft cap in place. October 20, 1913.*

the shafts were moving towards completion: "The successful completion of the Hudson River siphon was one of the miracles performed by the Board of Water Supply engineers during the construction of the Catskill system. It was the greatest siphon yet built by man"[23] [Figs. 40, 41].

While the construction of the Hudson River crossing progressed, work was proceeding on other parts of the aqueduct. The Rondout valley floor concealed two deep preglacial gorges that were filled with several hundred feet of glacial drift. Twenty-five thousand linear feet of borings were made in the Rondout valley, indicating twelve different sloping sedimentary rock formations all containing voids that channeled water through the work area. Several of the core-drilling holes produced artesian flows (wells under natural pressure). The northernmost of the ancient gorges was at elevation of +100 feet and buried below 270 of glacial material. The gorge on the southern side of the valley was at +0 (sea level) and covered with 200 feet of drift. The first 200 to 300 feet of rock below the Rondout valley floor is Shawangunk Grit, a hard quartz conglomerate that is very difficult to tunnel and known to contain waterbearing seams.

To avoid this geologic layer and to pass below the bottom of the gorges, the tunnel was set at two different depths. The elevation of the tunnel at the northern shaft was set at –100 feet. Halfway across the valley the tunnel dropped an additional 150 feet, to elevation –250 feet, passing the structure safely under the south gorge. This tunnel, called the Rondout Siphon, was built from nine vertical shafts, one at each end and seven at intermediate locations.

All of the shafts, tunnels, and aqueduct structures were built with concrete. The departure from brick and stone represented, more than any other single innovation, the modernization of construction techniques embodied by the Catskill system [Figs. 42, 43]. The use of concrete as a building material had first been

FIG. 40. *Catskill Aqueduct. Hudson Siphon. Drainage chamber. Crew during operation of stressing shaft cap anchor bolts. February 19, 1914.*

FIG. 41. *Catskill Aqueduct. Shaft cap in railyard. Greenville, New Jersey. September 11, 1913.*

FIG. 42. *Catskill Aqueduct. Steel reinforcement for cantilever floor of a chamber. January 26, 1915.*

FIG. 43. *Catskill Aqueduct. Lower effluent chamber at Kensico Reservoir. January 29, 1914.*

FIG. 44. *Catskill Aqueduct. Formwork for concreting operations in place. May 12, 1910.*

FIG. 45. *Catskill Aqueduct. Cut-and-cover construction showing completed concrete arch before backfilling. September 15, 1908.*

FIG. 46. *Catskill Aqueduct. Cut-and-cover aqueduct. Backfilling operations in progress. November 28, 1911.*

presented as part of the Burr-Hering-Freeman investigation:

> The combination of steel rods and concrete required for this type of construction [the concrete steel aqueduct] has already been the subject of much study and experimental investigation by engineers. There are a number of effective forms of steel rod sections available for this purpose, and experience with works already completed show that aqueducts of the highest excellence can be made with reinforced concrete.[24]

The basic structure of the aqueduct, the cut-and-cover, was a cast-in-place concrete enclosure that allowed the thickness of the walls to be incrementally varied according to the different stresses resulting from the forces of flowing water and retained earth [Figs. 44, 45]. The bottom of the conduit is an inverted arch that supports the downward weight of the sidewalls of the aqueduct and the weight of the flowing water. The parabolic sidewalls bow slightly outward, resisting the inward push of the retained earth and outward force of the water inside the aqueduct. As the depth and weight of the water decreases, the sidewalls diminish in thickness, allowing for the conservation of building material. Across the top is a continuous semicircular vault, completing the enclosure and supporting the earthen cover. The top portion of the conduit, which needs to resist only the weight of the earthen cover, is the thinnest section of the shell [Fig. 46].

Grade tunnels were used to pass the aqueduct through ridges and hills. To decrease cost and construction time, the cross section of the grade tunnel was reduced to two-thirds the width of the cut-and-cover aqueduct and built on a steeper grade, allowing the smaller conduits to maintain the required water-delivery capacity (a steeper gradient created a faster rate of flow, which compensated for the smaller cross sectional area). Additionally, a tunnel with a narrow profile involves less demanding excavation procedures. The tunnels, regardless of whether they passed through earth or rock, were fully lined with concrete. Grade tunnels were not designed to resist pressure from ground water outside the tunnel. Instead, permanent "weep holes" —small pipe-lined openings in the wall of the tunnel—allowed the external ground water to flow into the aqueduct, which relieved pressure on the outside of the tunnel wall lining. The exact profile of the construction varied according to subsurface conditions. In the most stable rock, the rough-cut opening was simply lined with concrete varying from three inches thick on the bottom to eight inches thick at the arch. With less stable rock, timber or steel shoring was built in place and the concrete lining enveloped the frame. In poor rock or loose soils, a concrete lining as thick as twenty inches was used.

The aqueduct traveled through large valleys in tunnels below the hydraulic gradient; in these structures the water is under pressure. In addition to the two structures previously described—the Hudson and Rondout pressure tunnels—there are five other pressure tunnels along the Catskill Aqueduct line—at Wallkill, Moodna, Breakneck, Croton Lake, and Yonkers. The typical pressure tunnel is constructed deep in sound rock connected to the aqueduct grade structures by vertical shafts at either end [Fig. 49]. These deep rock tunnels were the most difficult aspect of Catskill Aqueduct construction. The Harlem River Pressure Tunnel, built to take the New Croton Aqueduct below the Harlem River, was the only tunnel-building precedent that would approach the problems encountered on the Catskill line. The pressures that the Harlem River tunnel had to resist were the equivalent hydrostatic head of 130 vertical feet of water. The Catskill Aqueduct's operating pressures were more than three times greater.

The tunnels were built deep in solid rock to resist the outward force of pressurized water [Fig. 50]. Calculations on projected forces and profiles of the geologic conditions at various sites concluded that the tunnels should be covered with a minimum of 150 feet of sound rock. At the Hudson River crossing, where pressures were the highest in the system, the tunnels were built with a minimum sound rock cover of 200 feet. The dead weight of these rock covers was enough to exceed any upward pressure exerted on the tunnel by the water within. All the pressure tunnels were round in cross section and fully lined with concrete, which varied in thickness according to local rock conditions [Fig. 51].

In addition to the outward forces from the pressurized water within the tunnels, the vertical shafts and horizontal portions of the pressure tunnels had to resist the inward pressure from ground water. When the tunnels are full and pressurized, the internal force is so great that the external forces are offset and have little impact. But when the tunnels are empty—during construction operations and servicing—these external forces can be enormous. In places where the surrounding rock is porous or has water-bearing seams, the external pressure on the concrete tunnel lining was calculated to be as high as 2,000 pounds per square inch. The circular cross section was selected to oppose the compression from these external

FIG. 47. *Catskill Aqueduct. St. Elmo Siphon. May 4, 1911.*

FIG. 48. *Catskill Aqueduct. Completed cut-and-cover aqueduct showing the hole through which a boat would be lowered into the tunnel for inspections. June 19, 1912.*

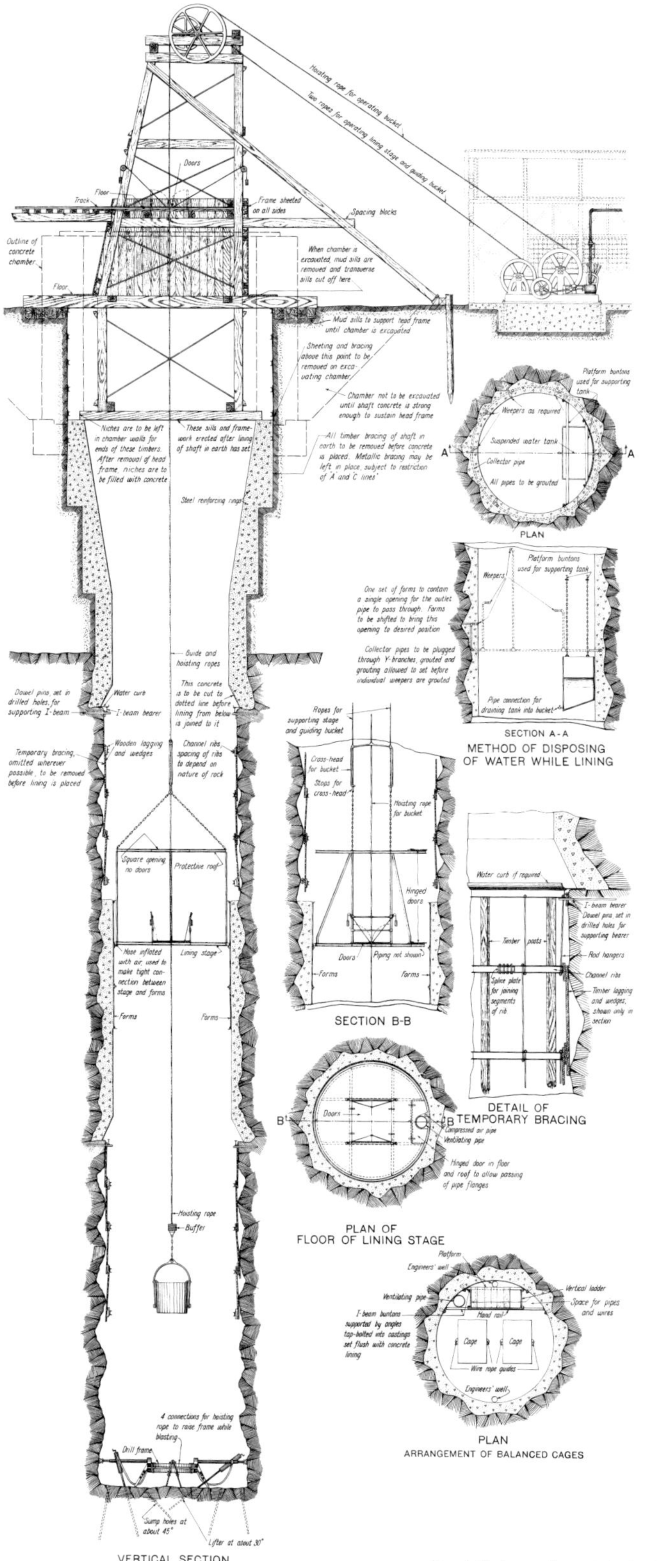

FIG. 49. *Catskill Aqueduct. Methods for sinking and lining permanent shafts. September 1, 1908.*

hydrostatic forces: the round concrete form acts as a continuous compression ring that can resist forces from any direction.

The thickness of the concrete lining on pressurized tunnels varies from a minimum acceptable thickness of 17 inches at the Hudson crossing to a minimum acceptable thickness of 10 inches on the smaller tunnels. The minimum acceptable thickness refers to an assumed circle into which no rock is allowed to project. Because the excavated surface of the rock was irregular, the actual average thickness of concrete was often 50 percent greater than these minimum prescribed values. As with the grade tunnels, the wall assemblies differed according to the demands of the geology and employed various types of reinforcing [Fig. 52]. Steel frames, timber frames, and steel sheet liners, fully imbedded in the final concrete lining, were incorporated in different stretches of the tunnels [Fig. 53]. The inside diameters of the pressure tunnels were typically 14½ feet. The tunnels curved gently from the horizontal sections to vertical shafts, providing a smooth transition for the watercourse, reducing disturbance to the flow and scouring action against the walls of the structure [Figs. 54, 55]. All the pressure tunnels were constructed to permit dewatering. Wherever possible, one of the shafts was placed near an existing streambed so that the tunnel could be pumped out to facilitate inspection and service[25] [Fig. 56].

Steel pipe siphons were used for crossing short valleys, or in locations where the rock was of poor quality and unsuited for tunneling. Each of the fourteen siphons used in the system was made with three pipes laid side by side across the valley. There were a number of advantages to the three-pipe configuraton. With three pipes the system had enough redundancy to allow for one of the lines to be taken out of service for maintenance or emergency. If necessary, two lines, operating under slightly higher pressures than normal, could carry the full capacity of the aqueduct. The shaping, drilling, and machining of the pipe ends required a large specialized factory that was impractical to set up

FIG. 50. *Catskill Aqueduct. Rough excavation for pressure tunnel before concreting. March 9, 1910.*

FIG. 51. *Catskill Aqueduct. Cross section of typical pressure tunnel with concrete lining in progress. c. 1910.*

at each siphon location, but transportation of the pipe sections was problematic since the dimensions of pipe could not exceed the clearance along the railroad tracks. But by employing three pipes in the siphon scheme, the diameter of each pipe could be kept small enough to fit on flatbed train cars. In addition, the mortar lining that was to used to protect the inside of the steel could be more stable and securely bound to the pipe if the overall diameter was a tighter circle.[26]

The pipe was made from riveted sheet steel, $\frac{7}{16}$ inch to $\frac{3}{4}$ inch thick. The lower the pipe was below gradient, the greater the water pressure, and hence the need for a thicker pipe wall. The diameters are either 9 or 11 feet. The larger pipes were used for siphons closer to the city in case they had to carry Croton water as well. To assemble the siphons, the pipe sections were placed on concrete support cradles in open trenches excavated across the land [Fig. 57]. The sections were positioned, cleaned of rust, and riveted and caulked together [Fig. 58]. When the full length of pipe on a given siphon was completed, the pipe was closed at the downstream end and filled under pressure to test the joints for water tightness. The pipe was drained and repairs were made as needed. Once the pipe was refilled and under pressure, it was enveloped in a protective layer of concrete 6 inches thick at the top and 18 inches thick at the sides. The pipe remained filled during the concreting process to prevent the metal from deforming under the weight of the concrete. After the concrete cover had set, the pipe was dewatered and the interior lined with a 2-inch-thick layer of cement mortar [Figs. 58, 59]. Both the exterior layer of concrete and the interior mortar lining retarded corrosion of the metal. Once the concreting and the mortar operation were complete, the whole pipe assembly was covered by an earthen embankment and seeded [Fig. 60]. Concrete siphon chambers at both ends of each siphon form the connection between the grade structure (either a cut-and-cover or grade-tunnel) and the pipes [Figs. 61, 62]. Sluice gates in these chambers control the flow of water into the siphon pipes.[27]

In order to send water to the city as soon as possible, the steel pipe siphons were brought online with only one of the three pipes completed. The remaining two parallel lines were added later. By 1924 the Catskill aqueduct was complete to capacity and delivering 650 million gallons of water per day from the Catskill Mountains to the city.

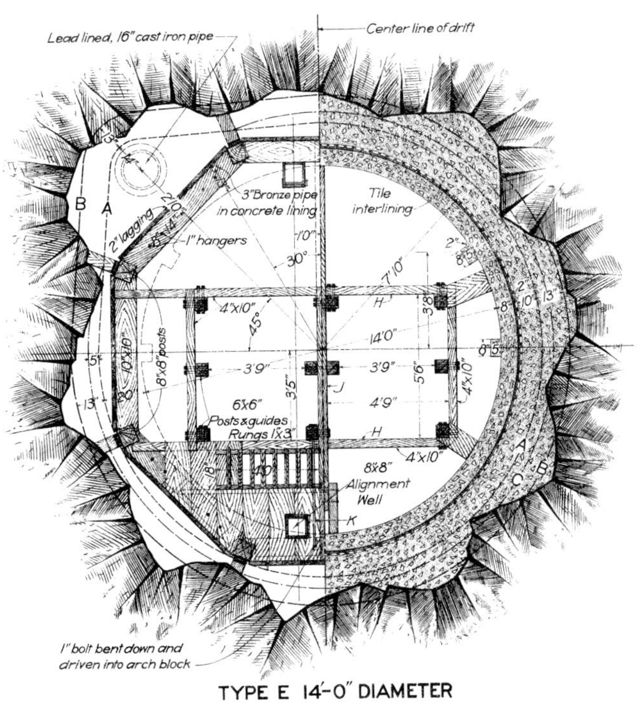

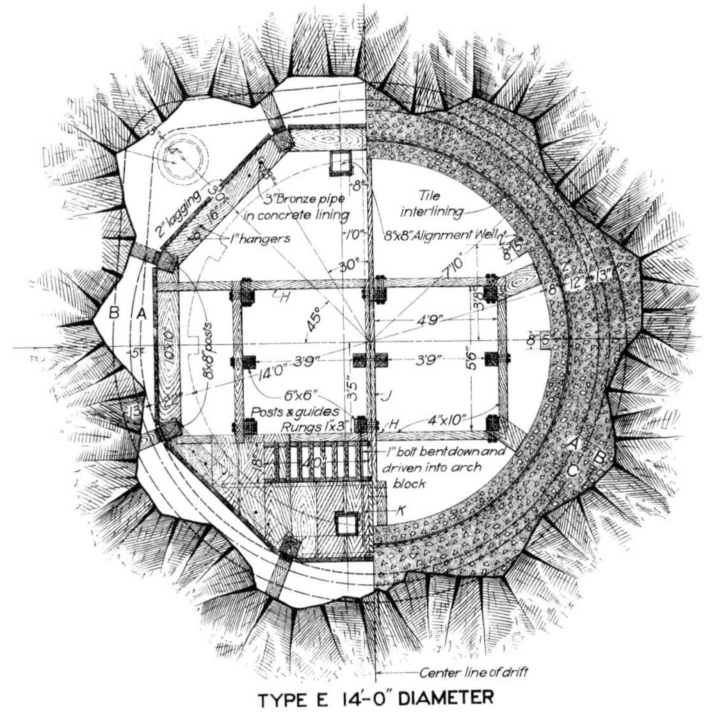

FIG. 52. *Catskill Aqueduct. Standard sections of drainage shafts for pressure tunnels showing methods of construction. February 15, 1908.*

FACING PAGE

FIG. 53. *Catskill Aqueduct. Rondout Siphon. Steel formwork with finished concrete tunnel in foreground. Grouting operations in progress. January 23, 1913.*

FIG. 54. *City Tunnel #1. Framework of concrete form for the transition from a 14-foot diameter shaft to a 12-foot diameter tunnel. June 8, 1914.*

FIG. 55. *City Tunnel #1. Completed concrete formwork, which was subsequently sawed into pieces and lowered into place. June 26, 1914.*

FIG. 56. *Catskill Aqueduct. Rondout Pressure Tunnel. Discharge conduit to Rondout drainage chamber. September 11, 1915.*

The Catskill water supply was at risk of interruption if any part of the 92-mile aqueduct had to be closed because of structural failure, or even for routine repairs. Freeman, in his report to the comptroller in 1900, recognized this weakness and suggested the use of the old Kensico Dam and the Rye Ponds as a possible site for a large storage reservoir to hold emergency water reserves close to the city [Fig. 63]. Similarly, the plans submitted in 1903 by the Commission on Additional Water Supply recommended the development of a large storage reservoir at the old Kensico site. The Board of Water Supply agreed.

The Kensico location is sixteen miles from the Hill View Reservoir, where the Catskill aqueduct terminates and where the city distribution aqueducts originate. The separation of these two bodies of water was not considered problematic, because the line between Hill View and Kensico was relatively short and planners knew that a backup aqueduct could be built at a reasonable cost to connect the two reservoirs. That backup came in the form of the Delaware Aqueduct, completed in 1945.

Surveys for the New Kensico Dam began in 1906, and preparation of the work area started in 1909. The site, in the rolling hill country of southern Westchester, is in a flat glade where the valley sides draw together. It is on the Bronx River at the junction with Davis Brook. A single masonry dam, 1,850 long, 307 feet from the base of the foundation to its crest, and 128 feet above grade, closed the main valley to create the storage reservoir. A modest dike about 1,450 feet long and 25 feet high was required one mile north of the dam to fill a shallow gap on the west bank. Combined, the structures would create a 2,081-acre reservoir that held 30.6 billion gallons of water—enough to meet the city's needs for about two months.

The Old Kensico Dam was situated in the middle of the area most suitable for the new dam. The Board of Water Supply considered locating the new dam just downstream of the old Kensico structure. Using this approach, the old dam could be left intact and the water kept flowing during

FIG. 57. *Catskill Aqueduct. Fort Hill Siphon. Steel pipe siphon on concrete saddles under construction with siphon chamber in background. October 25, 1922.*

FIG. 58. *Catskill Aqueduct. Peekskill Siphon. Manual cleaning operations of steel pipe siphon interior prior to the placement of two-inch mortar lining. May 1, 1923.*

FIG. 59. *Catskill Aqueduct. Bryn Mawr Siphon. Two-inch mortar lining of steel pipe siphon in progress. May 19, 1912.*

FACING PAGE
FIG. 60. *Catskill Aqueduct. Sprout Brook Siphon after pipe has been covered. November 15, 1912.*

construction. (The water that old Kensico provided was a small but critical part of the supply and needed to be maintained.) But the plan was regarded as risky: to build a new dam just downstream from an aging earthen structure seemed unwise. A failure of the old dam could inundate the construction site and severely delay the whole Catskill project. The best location was just above the old dam (in the middle of the old reservoir), where the sides of the valley were steepest and where the new dam could be set squarely—an advantage for both engineering and for architectural effect. To accommodate this plan, the old dam was drained and two temporary timber-reinforced earthen dikes were built in the valley, diverting the water of the Bronx River into wooden sluices that were connected directly to the 48-inch Bronx Conduit. This arrangement kept the Bronx River waters flowing into New York City taps during the seven years it took to build the new dam.[28]

FIG. 61. *Catskill Aqueduct. Yonkers Siphon. Completed east leg of elliptical branch tunnels. November 21, 1913.*

FIG. 62. *Catskill Aqueduct. Yonkers Siphon. Junction of three elliptical branch tunnels with standard circular pressure tunnel. November 21, 1913.*

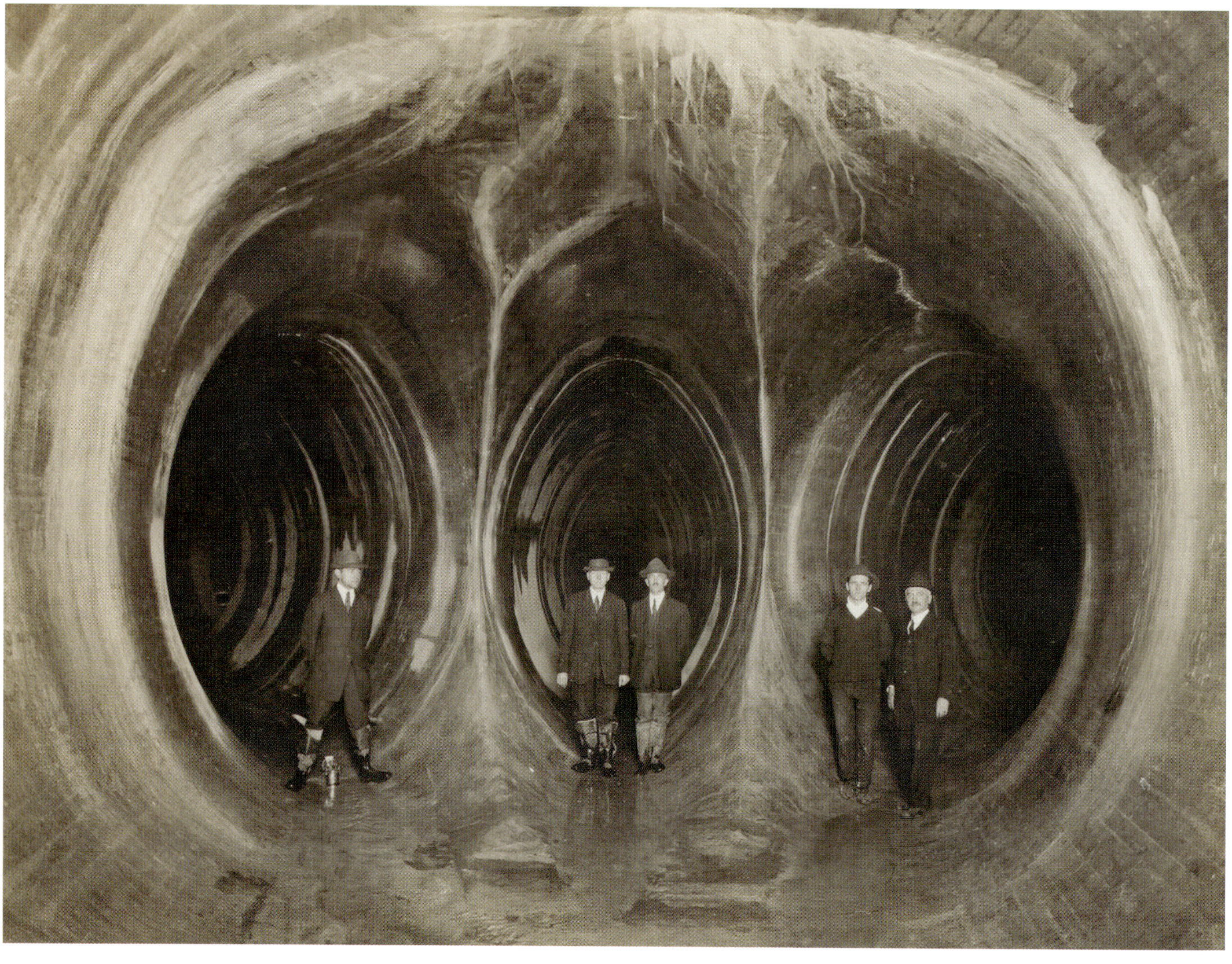

By 1911 work had begun on the foundation. Surveys revealed substrata of Manhattan schist, Fordham gneiss, and Inwood limestone. The upper layers of schist and gneiss had bands of severely decayed rock, but the limestone had almost become marble. Despite the inclination to avoid a water-soluble rock like limestone for waterworks foundations, this limestone was considered suitable. To prevent percolation beneath the dam, an exceptionally deep cutoff trench was dug 130 feet below the base excavation and 179 feet below the valley floor. The limestone was pumped with grout to fill any undisclosed seams or voids, and the sidewalls and bottom of the trench were cleaned and brushed with grout. The trench was then filled solid with cyclopean masonry.

East of the Bronx River Valley were outcroppings of gneiss that served as good building stone. The stone requirements on the project were huge: 635,000 cubic yards of concrete aggregate, 235,000 cubic yards of cyclopean stone, 22,000 cubic yards of dimension stone, and 8,000 cubic yards of roughly squared stone. The contractor purchased fifty acres of land slightly uphill from the dam site (the downhill slope from the quarry to the building site made moving the huge stones easier) and set up a quarry and stone-cutting plant. The quarry was situated so that the scar on the landscape was not visible from either the area highways or from the top of the dam.[29]

Once the cutoff trench and foundation excavations were filled, construction proceeded with precast concrete blocks (as at the Olive Bridge Dam). The blocks were laid up to define the perimeter of each structural bay, and each discreet structural area was filled with the cyclopean masonry. By any measure, the New Kensico Dam was built at an incredible pace: 300,000 cubic yards of masonry were placed in 1913 and 490,000 cubic

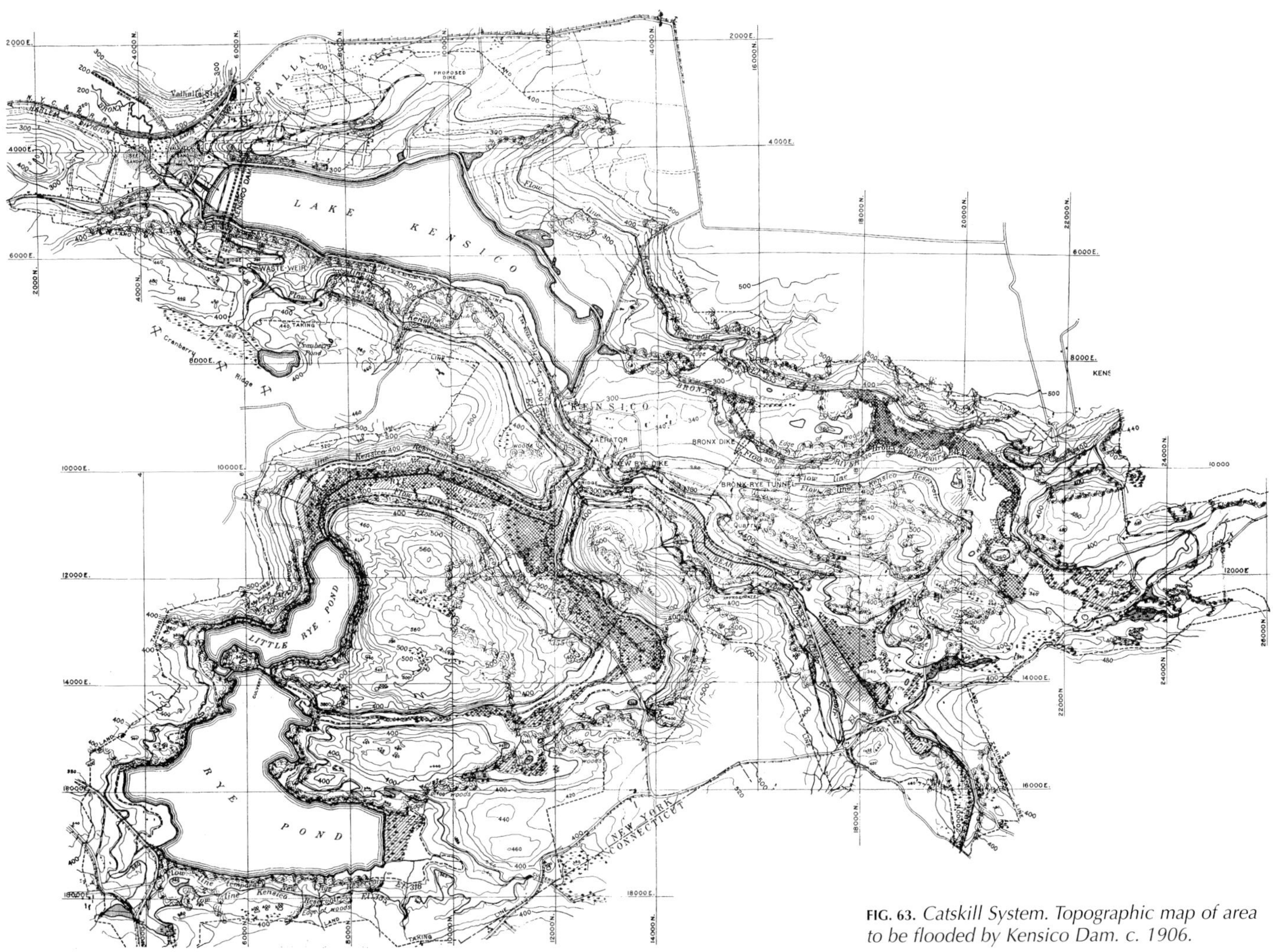

FIG. 63. *Catskill System. Topographic map of area to be flooded by Kensico Dam. c. 1906.*

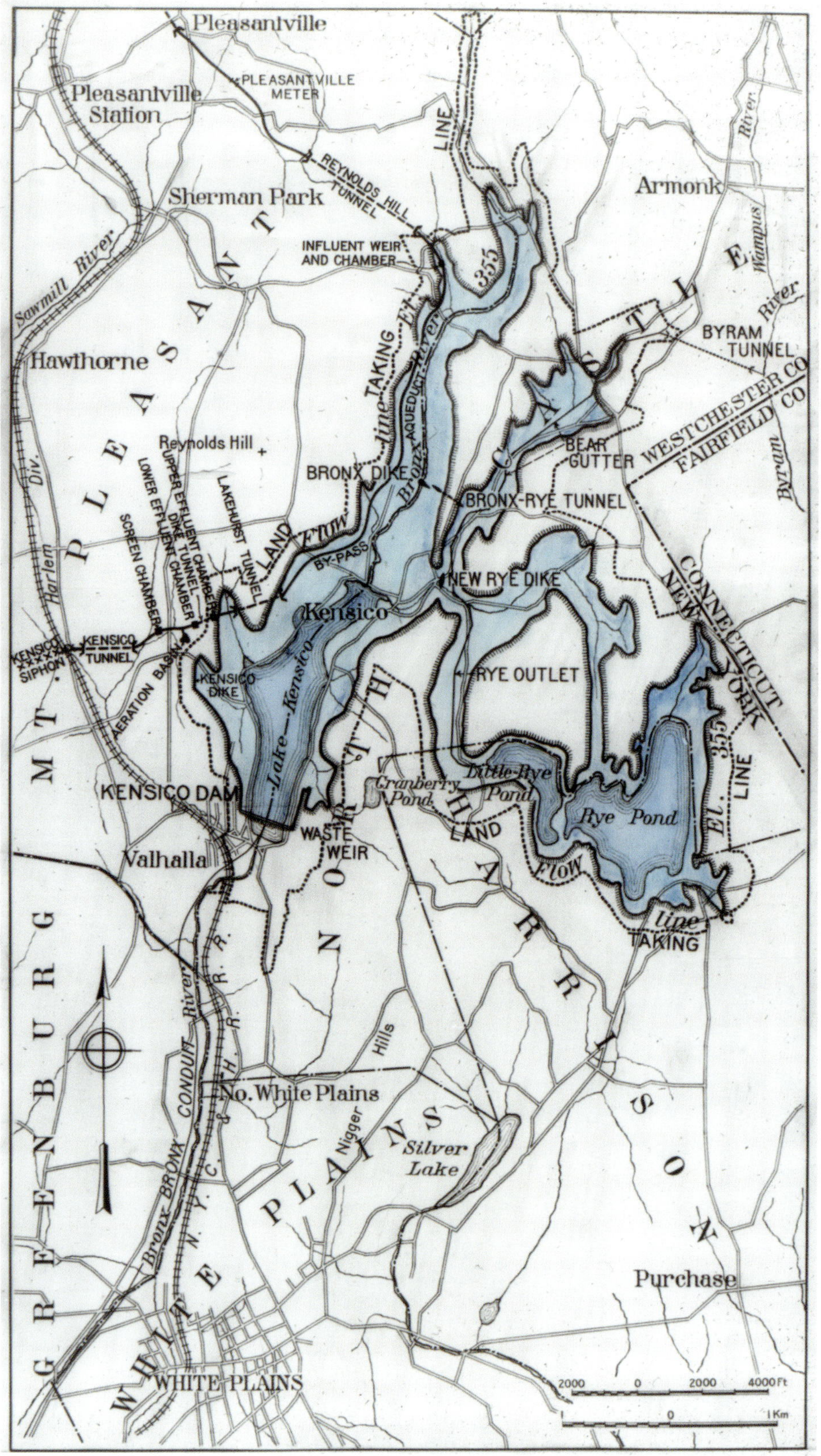

yards placed in the 1914 season—84,450 cubic yards were placed in one month alone. (In contrast, the single best month on the Olive Bridge Dam was 35,300 cubic yards of masonry.) The contract stipulated that by November 14, 1917, the dam be capable of storing (and drawing off) water at a minimum elevation of +315 feet and be completed by February 14, 1920. The dam reached full height in October 1915, more than four years ahead of schedule.[30]

All land within the boundaries of the reservoir above elevation +320 was grubbed of tree stumps, stripped of organic matter, and cleared of peat and sediment. Because it was also an equalizing reservoir, the water level in the reservoir was expected to rise and fall quickly, exposing wide segments of the bottom when levels dropped. Thirteen months were required to grub 502 acres of land. Stumps were blasted using dynamite and loose organic material was scraped off. The gates of the dam were closed and the reservoir filled with Catskill Mountain water in the winter of 1916, completely submerging the Old Kensico Dam [Fig. 64].

The New Kensico Dam, reservoir, bridges, and surrounding landscape plan are among the major architectural and engineering achievements in the Catskill system [Figs. 65, 66, 67]. Indeed, the dam ranks among the great masonry structures of the world. In cross section, the structure was very close to the ideal theoretical profile proposed by Wegmann and employed on the New Croton Dam [Fig. 68]. The upstream face is almost vertical. The downstream face is a true parabolic curve that spreads out from a 28-foot-wide top to a 233-foot-wide base [Figs. 69, 70].

The dam is divided into structural zones separated by expansion control joints [Figs. 71, 73, 74]. Kensico, like the Olive Bridge Dam, incorporates a network of inspection galleries and wells within its interior [Fig. 72]. At the base, the sloped surface of the downstream face meets the

FIG. 64. *Catskill System. Plan showing old Lake Kensico and Rye Ponds subsumed by the new Kensico Reservoir. c. 1916.*

FIG. 65. *Kensico Reservoir. Rye Outlet Bridge. Timber centering and formwork for concrete arches. May 2, 1911.*

FIG. 66. *Kensico Reservoir. Rye Outlet Bridge. Concrete arch and open spandrel construction. December 1, 1911.*

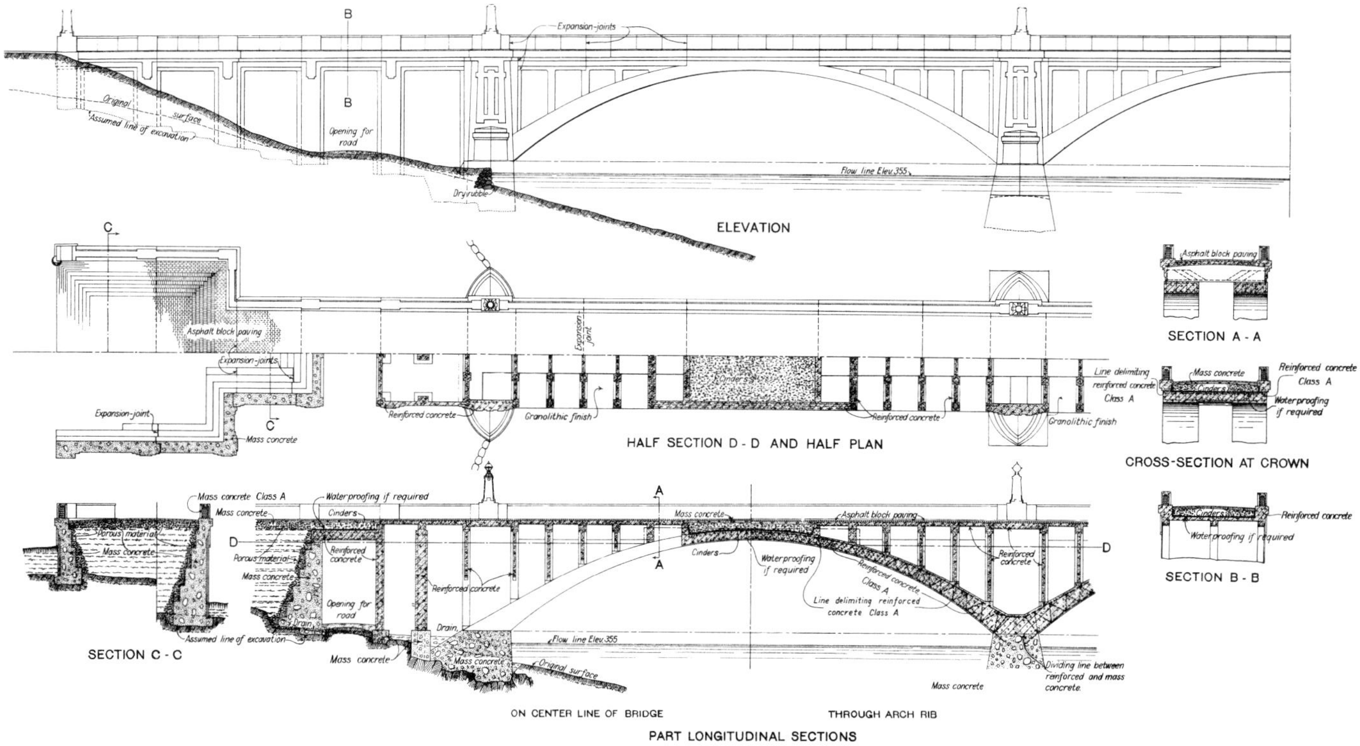

FIG. 67. *Kensico Reservoir. Rye Outlet Bridge. Plan, section, elevation. November 5, 1909.*

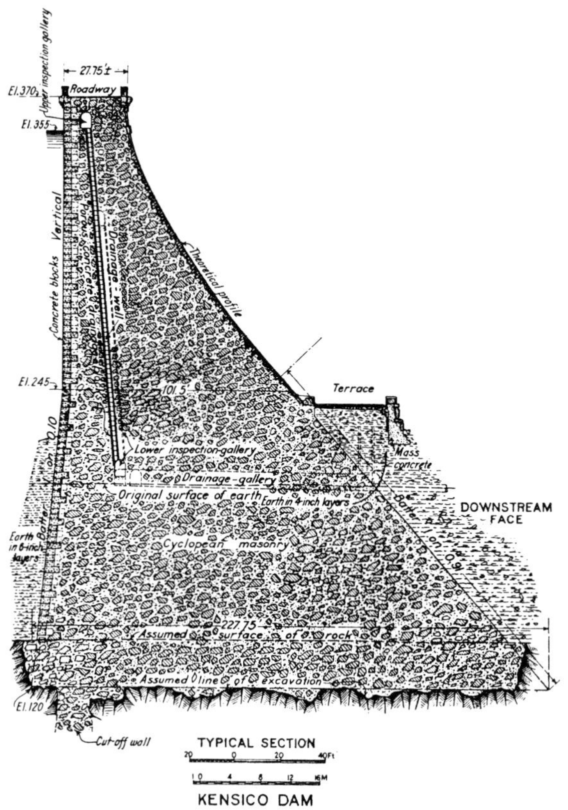

FIG. 68. *Kensico Dam. Cross section. c. 1910.*

ground at a masonry terrace 32 feet across and 10 feet high. This terrace houses the lower gate chambers, blow-off pipes, and connections to the Bronx conduit (the 48-inch line).

The architectural treatment of the downstream face is richly detailed. It is clad in rusticated granite blocks that express the dam's structural order and reveal the location of the expansion joints. The face is divided into panels that correspond to the 21 structural sections of the dam [Fig. 75]. Each panel is bordered with 15-foot-wide raised stone bands along the expansion joints. The entire face is topped by a continuous projecting entablature with a course of rough-hewn medallions, a projecting cornice, a coping assembly, and a simple parapet. Within the face of the panels is a grid of dimensional stone headers, each stone one-foot square and projecting six inches from the surface. The headers create a decorative diamond pattern, but also served a functional need during construction. The projecting stones provided

FIG. 69. *Kensico Dam. Dam construction showing precast concrete block supporting cyclopean masonry infill. July 23, 1914.*

ledges upon which to rest temporary scaffolding required for working on the masonry. The cladding stones for the downstream facade are themselves cyclopean in scale, some weighing as much as sixteen tons [Figs. 76, 77, 78].

At the top of the dam, at either end of the structure, are two rotundas. These terminal pavilions are on terraces located symmetrically about the central axis of the dam, projecting beyond the face to overlook the dam and park below. The pavilions mark the extremities of the dam and establish an architectural connection between the axis of the superstructure and the axis of the landscaped mall.

The spoils from the foundation excavation were used to build up and level the valley floor, upon which a landscaped mall was developed. The mall is organized around a central axis that incorporates an elevated terrace at the base of the dam, four square pavilions that flank an aerator fountain, decorative stone paving and walkways, and extensive formal plantings. The axis and walkways of the mall extend south and meet the traffic circles at the termination of the Bronx River Parkway. The grandeur of the new dam, the parklike

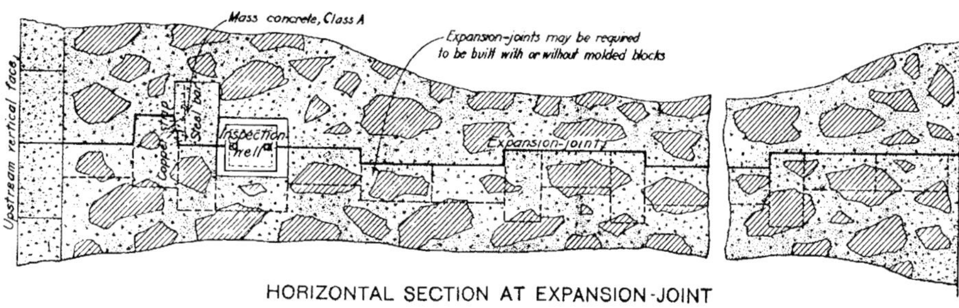

FIG. 70. *Kensico Dam. Downstream face of dam with stepped masonry substrate prior to completion of stone facing. October 29, 1914.*

FIG. 71. *Kensico Dam. Detail of expansion joint and inspection well. November 5, 1909.*

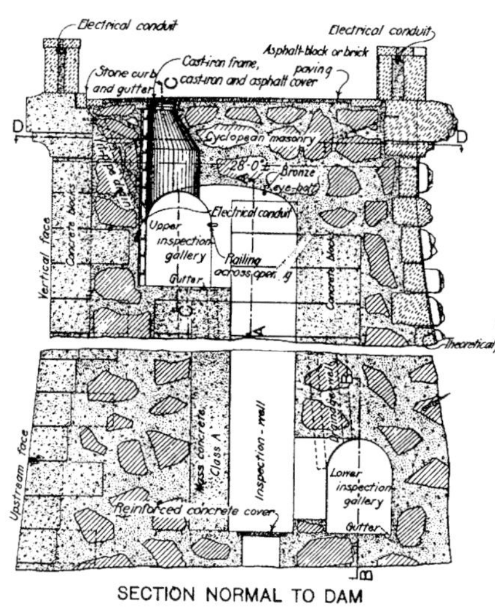

FIG. 72. *Kensico Dam. Detail of access to inspection galleries. November 5, 1909.*

FIG. 73. *Kensico Dam. Cross section of the dam showing vertical upstream face and interlocking concrete block wall construction. December 18, 1914.*

FIG. 74. *Kensico Dam. Precast concrete block walls forming expansion joints. Cyclopean masonry infill. September 9, 1914.*

FIG. 75. *Merritt H. Smith Memorial at Kensico Dam. Smith (1862–1926), as Department Engineer for the Board of Water Supply, directed the construction of the dam. October 18, 1929.*

FIG. 76. *Kensico Dam. Storage yard for dimensioned stone in foreground with block yard and camp in middle distance. September 9, 1914.*

setting of the mall with its magnificent fountain, and the elegant landscaping became a destination for city residents.

The New Kensico Dam impounded the waters of the eastern Catskills in the 30.6-billion-gallon capacity Kensico Reservoir [Fig. 79]. Although the Kensico Reservoir's primary function is as a storage reserve in close proximity to the city, it also figures in the day-to-day managing of the supply. At times of high water use (typically in the mornings and during the day) the reservoir releases as much water as daily use demands, an amount that often exceeds the supply coming in from the Catskill Aqueduct. At night, when usage in the city drops, the amount of water leaving the Kensico Reservoir is cut back, and water from the Catskill Aqueduct is diverted back into the reservoir to replenish the supply for the next day. This process is known as equalizing.

The Catskill Aqueduct arrives at the northwest extremity of the Kensico, where the hydraulic gradient of the aqueduct corresponds to the flow line of the reservoir. An open weir on the side of the aqueduct allows some or all of the water to spill over into the reservoir [Fig. 80]. Mechanical sluice gates within the influent chamber control the amount of incoming water and can, when required, completely direct the flow of the Catskill Aqueduct water around the reservoir, diverting it into an eleven-foot diameter, 2½-mile-long tunnel known as the Kensico Bypass. The bypass tunnel joins up with the waterworks of the lower effluent chamber, where the waters can be directed south to the city through the last segment of the Catskill Aqueduct [Fig. 81].

The Catskill Aqueduct leaves the Kensico outlet chamber and travels south through White Plains and Yonkers in a line that utilizes a combination of all the basic aqueduct structure types. The original Burr-Hering-Freeman plans, as well as the BWS's own plans, called for a filtration facility between Kensico and the city limits. The board acquired 314 acres of land near Mount Pleasant, and plans were developed for a plant of rapid sand filters at a site called Eastview. This facility

FIG. 77. *Kensico Dam. Stone dressing shed. Squaring up and hand dressing surface of stones. January 22, 1915.*

FIG. 78. *Kensico Dam. One of several full-scale models executed to test stonework and masonry technique. April 9, 1914.*

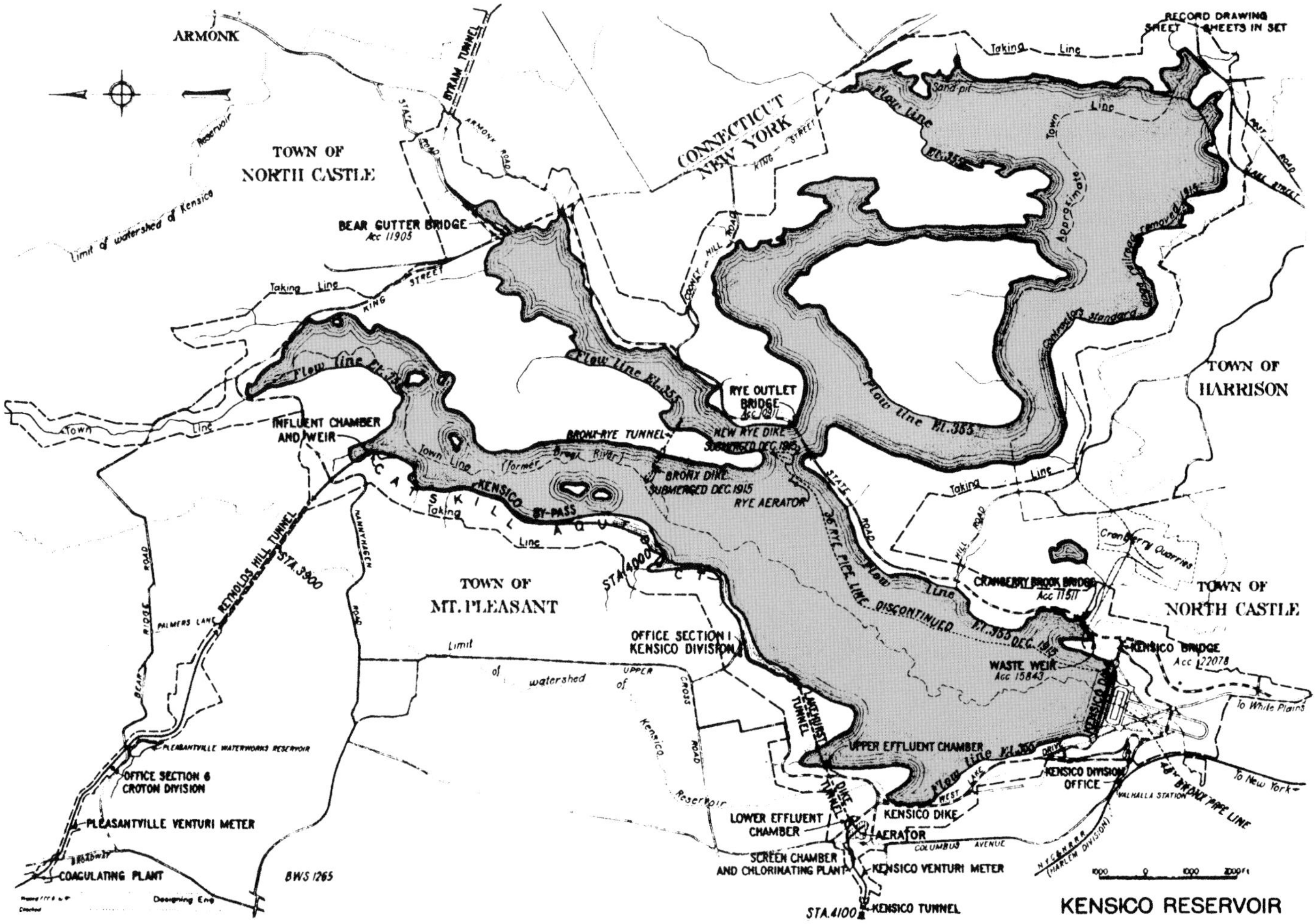

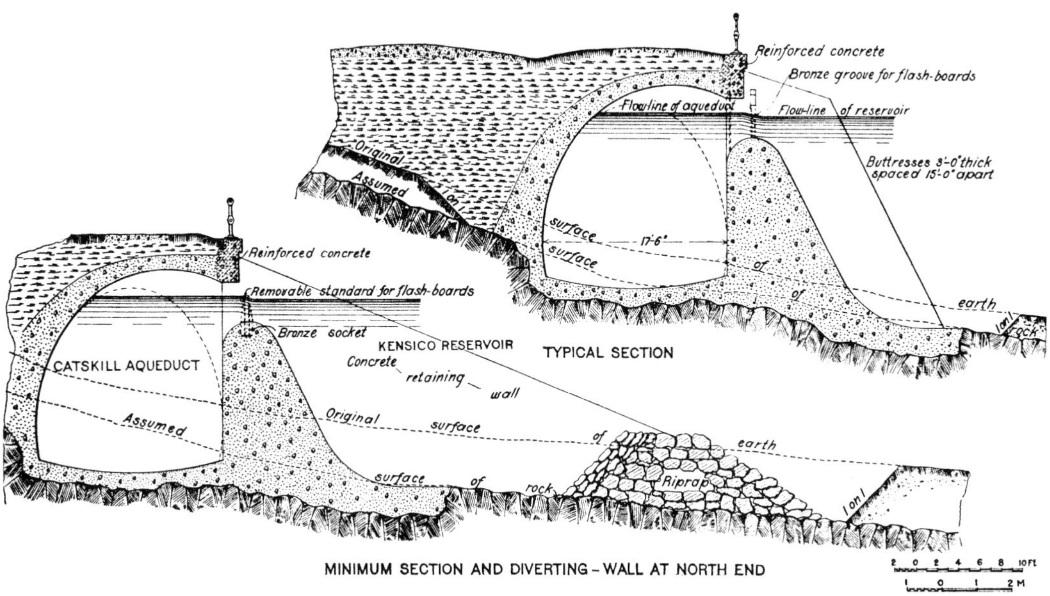

FIG. 79. *Kensico Reservoir. Plan. c. 1916.*

FIG. 80. *Catskill Aqueduct. Cross section of influent weir at Kensico Reservoir. c. 1910.*

would have been the largest water filtration project ever attempted, but building the essential components of the system (the reservoirs and aqueduct) was prohibitively expensive, and the project was abandoned. The water filtration question has been under consideration ever since and remains a core issue in New York City water planning and politics. The Eastview site was sold, and in 1998 the city announced plans to build a facility in the Bronx to filter Croton water. The larger question of filtering the entire system is still being debated.

The Catskill water makes one last pause before entering the tunnels and pipes of New York City. The aqueduct heads from the lower effluent chamber of Kensico south for seventeen miles to the Hill View Reservoir [Fig. 82]. The facility at Hill View is used strictly for equalizing. Hill View stores very little water, only 900 million gallons. Its location, about one-half mile north of the New York City line, is on the highest ground in the area, which maximizes the available pressure and increases the delivery potential in city pipes. (The location was first identified by Burr-Hering-Freeman for an equalizing reservoir.)

The Hill View Reservoir blocks no waterway and has no dam. The ninety-acre basin, roughly rhomboid in shape, was excavated out of the top of a knoll using 70-ton steam shovels, locomotives, and horse-drawn wagons [Figs. 83, 84]. Although the excavation effort was not completed until December 1914, the embankments, concrete sidewalls, shafts, chambers, and dividing dike were all phased to allow the project to move forward while the work of digging the basin continued [Fig. 85]. Construction of these appurtenant structures began in 1910. The basin was fully lined to facilitate cleaning, reduce water loss, and prevent pollutants from leeching in. A 6-inch-thick concrete slab covers the entire bottom, and 8-inch-thick concrete slabs line the embankment below the water level. Stone paving covers the sides and top of the embankment. The reservoir is separated into two basins by a 2,740-foot concrete dividing dike. Either

FIG. 81. *Kensico Reservoir. Upper effluent gate chamber from the inlet channel. October 13, 1913.*

FIG. 82. *Hill View Reservoir. Uptake chamber from Catskill Aqueduct. August 6, 1912.*

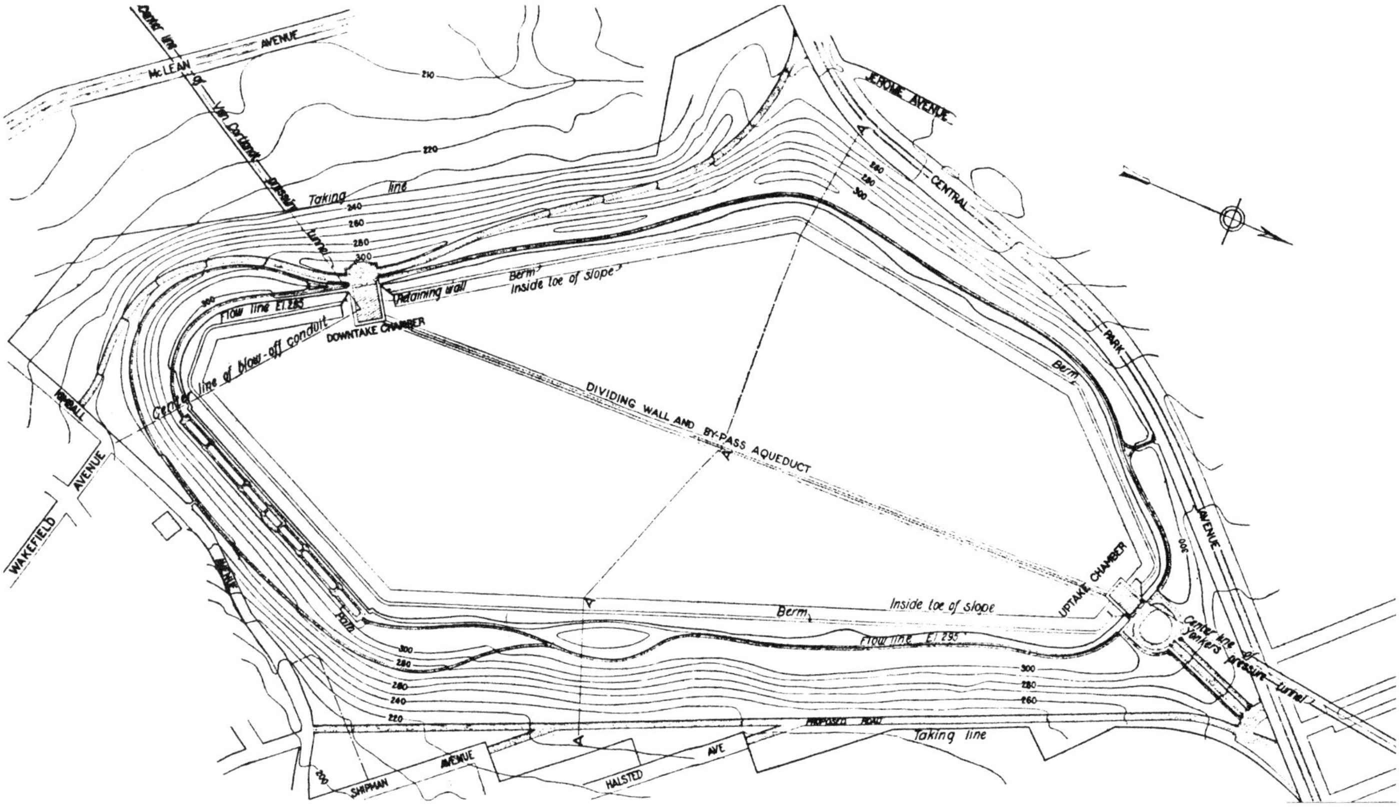

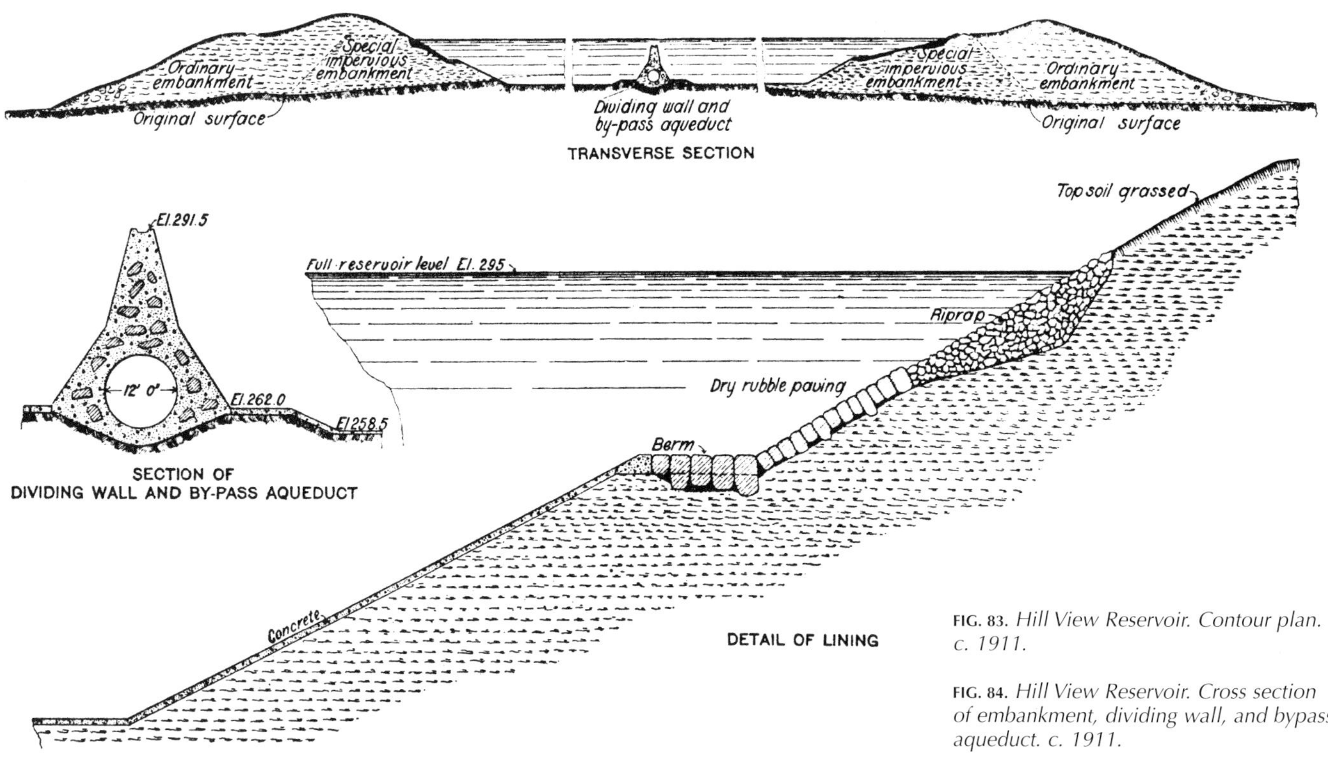

FIG. 83. *Hill View Reservoir. Contour plan. c. 1911.*

FIG. 84. *Hill View Reservoir. Cross section of embankment, dividing wall, and bypass aqueduct. c. 1911.*

basin may be used separately and either may be bypassed completely. The dividing dike contains a 12-foot-diameter bypass tunnel connecting the uptake and downtake chambers where the water passes into City Tunnel #1 [Figs. 86, 87]. The flow line is at elevation +295. Water filled the west basin in December 1915.[31]

The water entering Hill View is tested, chlorinated, and when required, coagulated (a process to remove particulate matter). Engineers designed a roof structure to cover the reservoir to protect the water from pollutants in the air and from sunlight (which promotes algae growth), but by 1909, views on filtration and water protection had changed. Along with the plans for the Eastview filtration plant, the plan to enclose the Hill View Reservoir was put aside.

In the course of building the reservoirs and associated waterworks of the Catskill System, numerous landscape and environmental problems were encountered. Aesthetic requirements at the building sites called for some landscape work, particularly at the malls and along the roadways, but the primary motivation for planting and forestation was the protection of the watersheds. The building process disturbed huge tracts of land, removed vegetation, and altered topography. Excavations for the foundations, sites for the contractor's facilities, spoil banks, borrow pits, temporary roads, and work-staging areas all left behind severely degraded land that threatened the quality of the water the system was built to collect. To address these problems the Board of Water Supply developed an enormous landscaping and forestry program. New trees and plantings were established on all the disturbed land. Existing forests were cleared of deadwood, diseased trees, fallen timber, hedges, scrub brush, and fruit trees (because of fruit spoilage). Thin areas in the forest were underplanted with young pines. Nurseries were established in Ashokan valley and in Westchester at Brown's Point to supply the seedlings for the gigantic reforestation effort. Grids of trees on six-foot squares can still be seen in the watershed around the margins of the reservoir. The benefits of

FIG. 85. *Hill View Reservoir. Basin under construction with shaft headhouse in foreground. May 6, 1915.*

FIG. 86. *Hill View Reservoir. Downtake chamber. The three rectangular openings are waterways from the east basin. The high arched opening allows boat access. August 8, 1913.*

FIG. 87. *Hill View Reservoir. View across forebay of uptake chamber into east basin of reservoir. November 21, 1913.*

FACING PAGE
FIG. 88. *Kensico Reservoir. Screen chamber. View upstream with circular aqueduct beyond. January 29, 1914.*

a stand of trees around a reservoir are many. The roots stabilize the soil and minimize erosion. The organic-debris mat on the forest floor shades the ground and slows evaporation by keeping temperatures low. Evergreen trees were used for much of the planting because they are hardy and indigenous to the region. In addition, pine needles produce a less rotten mulch than deciduous trees. Between 1911 and 1919 the BWS planted more than 1.5 million trees.[32]

The last part of the aqueduct is within the city itself. Eighteen miles long, this section connects the five boroughs. The original Board of Water Supply plans of 1905 proposed pipelines from Hill View Reservoir. Further investigations revealed that a total of 32 48-inch-diameter pipes would be required to deliver all the water in the Catskill Aqueduct. However, laying so many large distribution pipes in the already packed substructure of the city streets would cause enormous disruption to traffic and would likely interfere with other utilities and inhibit future construction of subways, sewers, and the expansion of the water supply itself.

The engineers concluded that a deep-rock pressure tunnel was preferable. This would minimize disruption of existing street and sewer infrastructure, provide the added protection of rock cover to the water conduits, and allow for greater flexibility in route selection. Surprisingly, the tunnel was also found to cost significantly less than the pipelines.

Existing geological records and test borings were reviewed to determine the best route. The line selected followed the West Side to lower Manhattan and crossed the East River at South Street into Brooklyn, where branch conduits connected the water to Queens and across the Narrows to Staten Island. With the line generally selected, 224 borings were put down at the most critical points: the deep gorges of the Harlem River, East River, and Manhattan Valley, and at known planes of weakness within the rock formations that lie below the city. The final route permitted direct delivery of Catskill water (via vertical riser shafts) to the Jerome Park Reservoir and the Central Park Reservoir, and (via lateral conduits) to the High Bridge Reservoir and the 135th Street Gatehouse of the Croton Aqueduct.

The appropriate diameter of the tunnel was carefully considered. The capacity of a tunnel increases exponentially as its size increases, but a larger diameter causes a decrease in pressure if it is not filled to capacity. The board developed a tunnel size that provided some extra capacity so that supplemental emergency water could be added to the tunnel without undue loss of head (static pressure). The tunnel was designed to gradually decrease in size as water is withdrawn along the line, from a

FIG. 89. *City Tunnel #1. Shaft with riser pipe in place. Shaft cap set and riveted to riser. Bronx, New York. November 5, 1914.*

FIG. 90. *City Tunnel #1. Shaft cap and valve assembly at site near High Bridge. May 21, 1914.*

maximum 15-foot diameter upon leaving Hill View Reservoir to an 11-foot-diameter tunnel below the East River. Graduated conduit lines, from 66 to 36 inches in diameter connect the main tunnel's terminal shaft at Fort Greene Park in Brooklyn to south Brooklyn, Queens, and Staten Island.

Like the pressure tunnels of the upstate aqueduct, the tunnels of the city aqueduct were set at depths that provided for a least 150 feet of solid rock cover. The tunnel below the Harlem River is 331 feet below tide; at the East River, the tunnel is 704 feet below tide. Passing below the Manhattan Valley, the tunnel is 352 feet below the surface. The rest of the tunnel in Manhattan is about 200 feet below the surface.

The tunnel, like other deep-rock pressure tunnels along the aqueduct, was built from a series of vertical shafts. There were four basic shaft types used in the city portion of the aqueduct. The first is the construction shaft, which gets workers and equipment down to the tunnel headings. A construction shaft can be converted to other uses upon completion, or plugged and filled. The second shaft type—the city shafts—serve for distribution. They tap the deep-rock tunnel and bring the water up to street level. These shafts are constructed to be permanent. They are lined and equipped with devices that control the flow of water [Fig. 89]. The third shaft type is used for dewatering the tunnel. It is equipped with pumps and located near a natural drainage site where the water can be discharged. The fourth shaft type provides access to the cutoff valves, which are used to shut down and isolate sections of the tunnel. Shaft locations are determined by local geological conditions and surface restrictions [Fig. 90]. Twenty-four shafts were sunk along the length of the City Aqueduct [Fig. 91].

The Brooklyn Conduit, the Queens Conduit, and the Bay Ridge Conduit added another 12½ miles of secondary aqueduct to the Catskill water distribution network. City conduits draw water from the terminus of the deep-rock City Tunnel #1 at Shaft 24 in Fort Greene Park, Brooklyn. They are located under the

FIG. 91. *City Tunnel #1. Delivery of a 23,576 pound, 72-by-48-inch bronze shaft cap. Myrtle Avenue and North Elliot Place. Brooklyn, New York. January 21, 1914.*

streets and are composed of either 66-inch steel pipe or 48-inch diameter cast iron pipe (the exception being the Narrows crossing to Staten Island). City conduits did not exceed the 66-inch diameter because of severely limited space below street level. Cast iron was the most economical pipe material of the day and used wherever possible, but it was considered unsuitable for a pressurized pipe over 48-inches in diameter. For the larger conduits where the internal pressures were higher, 66-inch riveted steel pipe was used.

The 66-inch Brooklyn Conduit extends $3\frac{1}{2}$ miles to 36th Street in Bay Ridge, where the water continues down the 48-inch Bay Ridge Conduit for a distance of just over 3 miles. The Queens Conduit, another lateral arm of the aqueduct, originates at the uptake shaft in Fort Greene Park in Brooklyn and is $6\frac{1}{3}$ miles long. The first $2\frac{1}{3}$ miles are 66-inch steel pipe and the remaining 4 miles are 48-inch cast iron pipe [Figs. 92, 93]. The line divides into several 30-inch branch connections that serve the eastern limits of Queens.

The city conduits were laid in trenches that were manually excavated to avoid damage in the constricted and crowded spaces below the street [Fig. 93]. The sides of the trench were shored with timber sheeting to prevent collapse, and the pipes were lowered from two movable derricks designed for the operation. Each section of pipe was placed in the trench, leaving the leading end blocked up in the air to receive another section of pipe. Each successive section of pipe was slid into the last, until the upper bolt holes could be aligned using temporary drift pins that guided the pipe sections together. This assembly sequence used the weight of the pipes to force the joint closed. The sections were then bolted together, and the supporting blocks removed, allowing the entire pipe to be lowered into the bed of the trench. These conduits included all the saddles, valves, gate connections, and venturi meters (for measuring flow) that were required to divert the water into the secondary mains, and provided provisions for future connections to the system. The fittings were built into masonry chambers to protect the hardware elements.

The Board of Water Supply specified a number of treatments to extend the useful life of the iron and steel conduits. Many of the pipes were dipped in a burlap-reinforced, mineral-fortified rubber coating. Some were painted with a more traditional bitustatic coating (a petroleum- or coal-based product). BWS specifications spelled out handling techniques to minimize damage to the coatings during shipping and pipe placement and required a final inspection and recoating of rivets and other exposed metal components. All pipes were tested under pressure exceeding the anticipated operating pressure before any backfilling was permitted.[33]

The 130-mile journey of the Catskill waters to the communities of the five boroughs encountered one last design and construction challenge: the crossing of the Verrazano Narrows to Staten Island. Although submarine pipe construction was not new, the scale and the unique environmental conditions at the Narrows made this siphon crossing particularly difficult. During the early twentieth century, New York Harbor was the busiest port in the world, and the Narrows, the entry to the harbor, was the primary anchorage for merchant vessels awaiting Coast Guard clearance to dock at the city. It was also an important military holding area.

Shipping regulations required that any obstruction be at least 45 feet below low tide on the side anchorages and 60 feet below low tide in the shipping channel. The board had considered a deep-pressure tunnel like the one that had been used to cross under the Hudson River. However, geologic uncertainty and construction problems combined with the relatively small quantity of water to be carried made it hard to justify the tremendous cost involved in building a deep-rock tunnel.

Methods for securely connecting the pipe pieces together, laying the pipe, and burying the pipe all presented problems that landbased pipe did not. Merchant ships, while holding in the Narrows, dropped huge anchors capable of digging deep into the ocean floor. The sand is soft and an anchor could drag the bottom, catch on the pipe, and rupture the line. The board ran tests with various large anchors and concluded that a minimum cover of 8 feet was required to protect the pipe. With this protective cover and the 60-foot restriction in the center of the channel, the grade for the trench was established. Depending on the depth of water, the trench needed to be between 20 and 30 feet deep in the ocean floor.

The Hudson River discharges into the Narrows, and the currents there are strong. This water typically flows about three miles per hour in both directions, depending on the tide, and at times flows in opposite directions at different depths. The axis of the trench and pipe is perpendicular to this flow. The current could be expected to push the soft debris on the bottom and fill the trench before a pipe was laid. Two months were required to complete a section of trench, including setting and testing the pipe, and the trench had to remain stable throughout that time. Test excavations to determine the trench profile began in April 1914. The first tests, done on a relatively narrow, steep-sided trench, showed the embankments were prone to collapse. The width of the trench bottom, initially set at 10 feet, was increased to 30 feet. The slopes of the side embankments were adjusted from near-vertical to a rather gentle one-on-two slope. This configuration proved stable. As the trench approached either shore, the cut had to be as deep as 60 feet. In these areas the cut was sheeted and braced with timber to prevent collapse.

FIG. 92. *City Tunnel #1. Truck hauling a 30-foot length of riser pipe through city streets. Brooklyn, New York. July 21, 1914.*

FIG. 93. *City Tunnel #1. Williamsburg conduit under construction in Hooper Street. Taken from the elevated train on Broadway. Brooklyn, New York. July 24, 1925.*

The pipe itself presented the final technical challenges. Stresses on the connections brought on by the current could only be countered by a flexible pipe assembly (rather than the more common rigid submarine pipe). Existing flexible pipe joints were essentially rigid pipe with a cast ball-and-socket terminus. After coupling together, the two sections were filled with molten lead and the lead was hammered tight after the metal had set (a process known as caulking). The technique worked for sewers and low-pressure water applications, but tests showed that there was far too much leakage in a high-pressure water delivery application. The lead shrank after cooling, and the joints were deformed and loosened by both the placing of the pipe and the force of the currents against the pipes once they were set. The alternative was to caulk the pipes after they were in position, but the strong tidal currents made the detailed work of underwater lead placement extremely difficult. Divers were limited to doing the work for an hour or two at slack tide. A joint was needed that required no underwater caulking and met the pressure requirements for water delivery.[34]

Machinists were brought in to improve the existing designs for flexible pipe joints. Precision standards were established that allowed a deviation of no more than

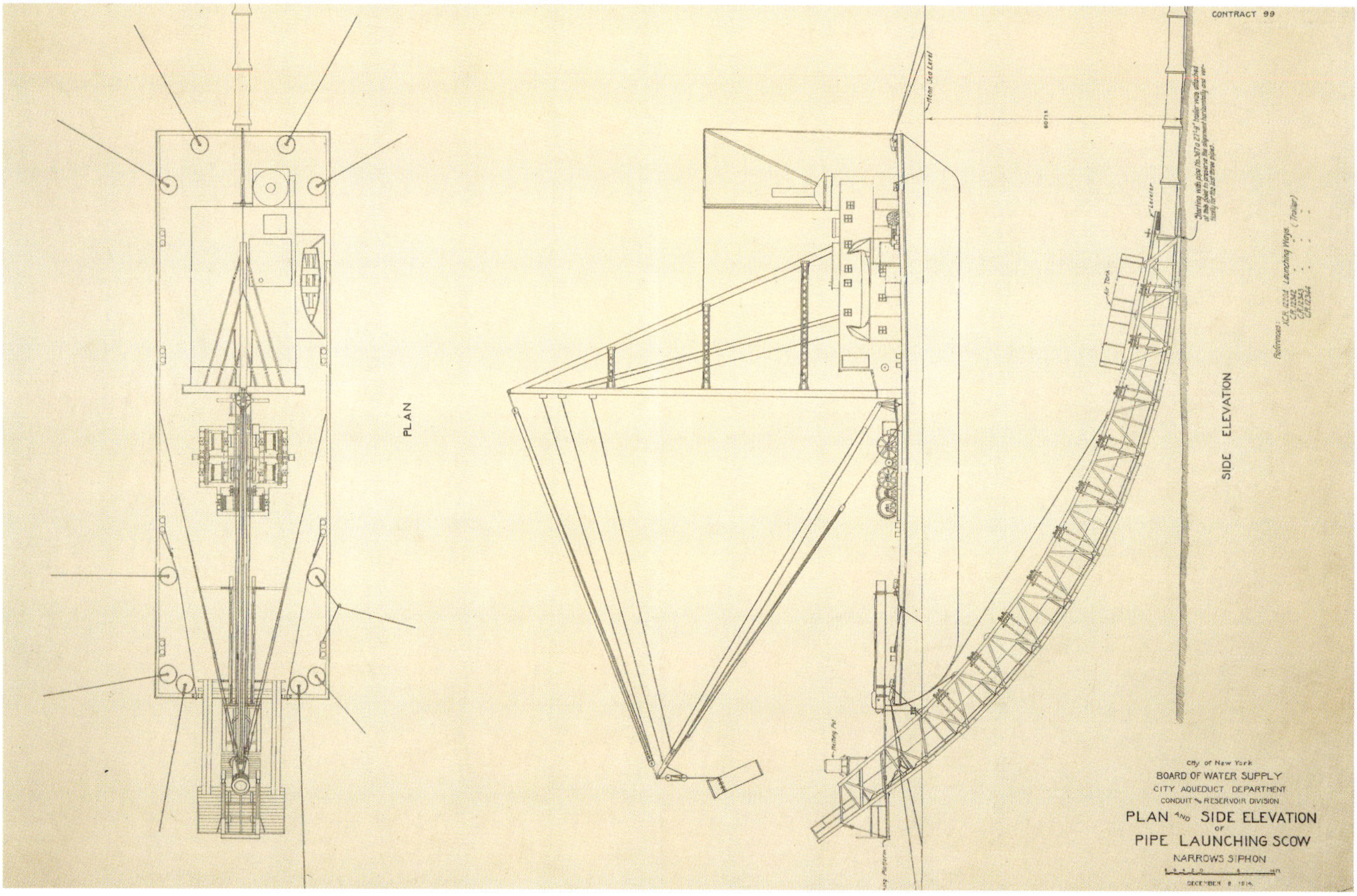

FIG. 94. *Narrows Siphon. Pipe launching scow for laying the siphon across lower New York Bay to Staten Island. Plan and side elevation. December 8, 1914.*

1/100 of an inch from the specified dimensions of the connection: this was an incredible level of precision for such a large manufacturing operation. Alternative lead alloys for caulking were tried and various methods developed to force the lead into the joints under pressure. The pipe was made of 12-foot sections. That length enabled the pipe to curve so that it could be draped into the trench on the sea floor. The assembly procedure allowed all the joints to be caulked above water level. The design was finalized, tested, determined to be generally watertight, and found to be strong enough for the job.

The pipe was laid from floating scows fitted with specially designed derricks [Fig. 94]. A giant structural steel cradle and skidway in the form of a curvilinear truss hung from the scow down to the ocean floor [Fig. 95]. The pipe was positioned and assembled vertically in the cradle, caulked above water level, and lowered down the skidway [Fig. 96]. Each joint used almost three hundred pounds of lead and required about three hours to pour and complete. As each piece was joined, the growing length of pipe was gently lowered along rollers and slid into the water and the trench below. At the beginning of the construction process every joint was tested at pressures of one hundred pounds per square inch before being lowered into the water. After 129 joints tested without failure, spot inspections [Fig. 97] were deemed sufficient.[35]

The bottom of the trench was dragged to establish a smooth pipe bedding before pipe placement to assure there were no surface irregularities. Stone and tunneling debris from the building of the New York City subways was floated out to the site and dumped from barges to cover the pipe and backfill the trench. Despite numerous delays related to floating ice, fog, low temperatures, and rough water, the 10,300-foot-long Narrows Siphon was completed in twenty-six months:

> On July 10, 1916, a hydrostatic test was made on the pipe between the 36-inch gate on the Brooklyn shore and the 36-inch gate on the Staten Island shore, a total length of 10,330 linear feet of 36-inch pipe, of which 9,740 feet were the sub-marine type. At a

> pressure of 130 pounds per square inch the leakage for the entire line averaged 5.6 gallons per minute . . . less than is considered allowable for land lines of similar size.[36]

A fifty-six-acre reservoir was built in the hills of the high ground of Staten Island at the site of Silver Lake. This reservoir provided an equalizing and distribution center on the island and stored a reserve of water for the community in the event that the Narrows Siphon failed or required closure for servicing. The reservoir was located in two shallow basins, one that held the existing Silver Lake and the other a low-lying wetlands area [Fig. 98]. Geologists found that the ground was fairly impervious, and engineers concluded that a fully paved storage reservoir (as had been built at Hill View) was not needed. The reservoir basin was made of compacted earth with riprap slopes on the embankments of the dikes [Fig. 99]. The reservoir, like the Hill View Reservoir, was divided into two separate holding areas to facilitate water quality management and reservoir maintenance, and the dividing dike held the gate chambers and distribution works. Catskill water was first turned into the reservoir in January 1917 and by late February the water had reached the flow line, at elevation +228.[37]

In the first year of operation, it became evident that the rate of water loss in the Silver Lake Reservoir was greater than expected. Investigators concluded that the leakage was general in nature and could not be specifically assigned to any particular structural deficiency. The costs of paving the reservoir to address the problem of widespread seepage seemed unwarranted, and further investigations of the water loss were dropped. The Silver Lake Reservoir marked the completion of phase one of the Catskill Water Supply system and served Staten Island until 1970, when water loss and quality issues prompted the city to build its replacements: the largest enclosed water-storage tanks in the world. Built on higher ground, the reinforced concrete structures hold 100 million gallons. The Silver Lake Reservoir was retired from the water supply system and has reverted to park and recreational use.

FIG. 95. *Narrows Siphon. Steel truss cradle used in laying the siphon. June 7, 1915.*

FIG. 96. *Narrows Siphon. View from the deck of the pipelaying scow* Comptroller *of work crew fastening sections of the pipe together. September 17, 1914.*

The Catskill Aqueduct was built to carry at least 500 mgd, but the Esopus Creek and its tributaries could only be relied upon to supply half that amount. To complete the supply, the original Board of Water Supply plans of 1905 called for using the Rondout Creek in the southwestern Catskills, and the Schoharie Creek in the northern Catskills. These two additional sources could provide more than enough water to fill the aqueduct to capacity.

Surveys had been made of these two drainages. Extensive drilling operations revealed that the bedrock in the Rondout Creek basin was hundreds of feet below loose glacial fill, and design engineers concluded that the geologic conditions were not favorable for the construction of a large dam. By 1915 the hydrological investigations made by the board concluded that the potential yields of water from the Rondout Creek were relatively modest. While these investigations diminished the feasibility of the Rondout Creek, studies showed that the Schoharie basin water yields were significantly larger than early field studies had indicated. The 236 square miles of watershed above the original proposed Prattsville dam site on the Schoharie Creek were estimated to yield up to 200 mgd.

The Schoharie is a large stream located north of the Esopus drainage in the heart of the high northern Catskill region. The upper reaches of the Schoharie drainage are just over the ridge from the northern side of the Esopus watershed, at elevations above two thousand feet near Hunter Mountain, Windam Mountain, and the town of Tannersville. The Schoharie Creek and its tributaries flow north, and the main stream connects to the Mohawk River at

FIG. 97. *Narrows Siphon. Diver employed on the pipelaying scow. November 27, 1914.*

FIG. 98. *Silver Lake Reservoir. Staten Island, New York. c. 1917.*

Fort Hunter near the town of Amsterdam, about seventy miles north of the Ashokan Reservoir.

Gravity takes water in the opposite direction it needs to go to reach New York City. The Board of Water Supply's Schoharie plans addressed that fact by building a dam across the Schoharie Creek, reversing the direction of the flow, and sending the water south via a diversion tunnel built through the intervening Shandaken Mountain Range. The water is released into the natural channel of the Esopus Creek about eleven miles above the Ashokan Reservoir.

The investigations for a dam on the Schoharie Creek began at Prattsville, a location first identified in the early schematic plans of the BWS and approved by the State of New York. But local geologic conditions proved difficult, and subsurface studies moved farther downstream. Five alternate locations were investigated before a suitable building site was identified just below the village of Gilboa [Fig. 100]. The downstream location, besides being geologically desirable, also collected water from a larger watershed, including the waters of Manor Kill, Bear Kill, Johnson Hollow Brook, and Huntersfield Creek. The resulting watershed is 314 square miles. Projections for the dependable yield of the Schoharie sources were upgraded to 250 mgd (and would later prove to be significantly more than that). The amended proposal for a dam location was reviewed and approved by both the New York City Board of Estimate and the State Conservation Commission. The Board of Water Supply completed the plans for the Gilboa Dam and Shandaken Tunnel in November 1917.[38]

The Schoharie Valley near Gilboa is steep and narrow, and it could not accommodate a large storage facility. To deal with the limitations of storage, the eighteen-mile-long Shandaken diversion tunnel was built with a capacity up to 650 mgd—well beyond the projections for the Schoharie stream's average daily flow (and more than the total original design capacity of the Catskill Aqueduct). The water is diverted into the Esopus Creek, where it is captured by the Ashokan Reservoir. At peak capacity, the reservoir created by the Gilboa Dam could hold only 14 percent of the total annual flow of water coming down the Schoharie Creek, but the huge Ashokan Reservoir, which can hold 103 percent of the average annual flow of the Esopus, was more than adequate to handle the extra water.[39]

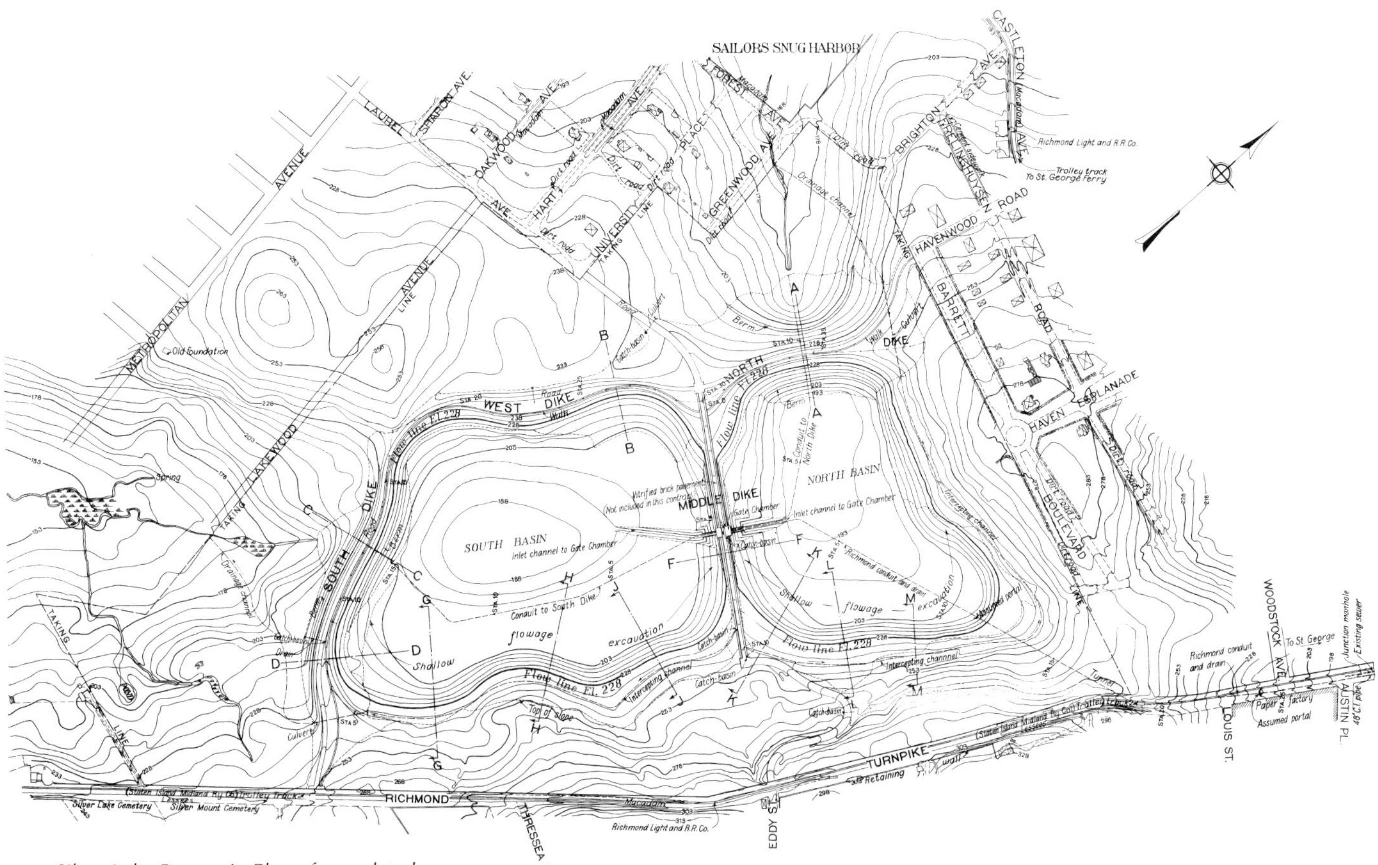

FIG. 99. *Silver Lake Reservoir. Plan of completed reservoir. Staten Island, New York. December 16, 1912.*

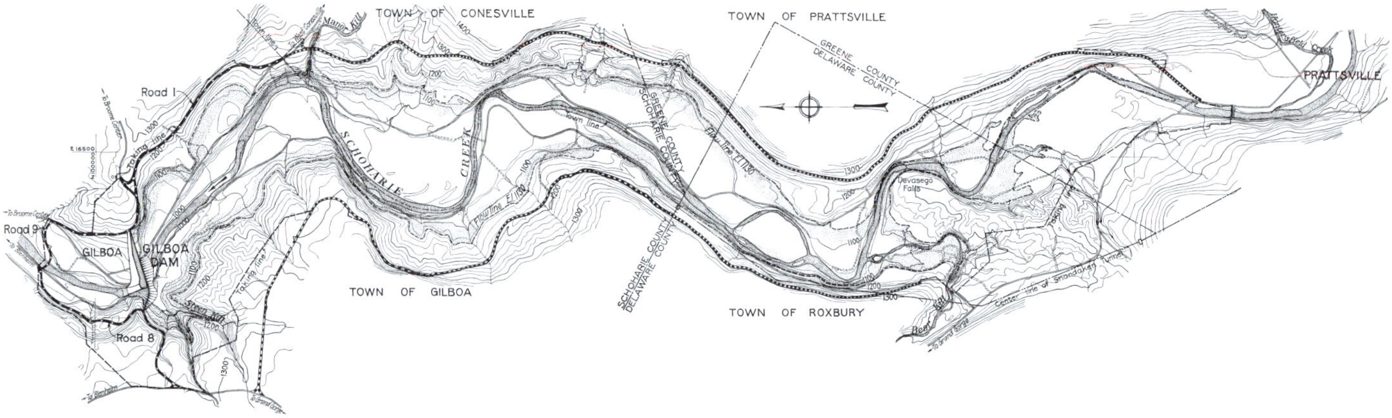

FIG. 100. *Schoharie Reservoir. General plan of reservoir showing Gilboa Dam and path of original Schoharie Creek. October 25, 1917.*

FIG. 101. *Shandanken Tunnel. Intake gatehouse. The tunnel connects the northern Catskill Schoharie watershed to the eastern Catskill Esopus watershed. July 25, 1926.*

The Shandaken Tunnel, from the intake works on the Schoharie Reservoir to the outlet portal on the Esopus Creek, was, at the time of construction, the longest tunnel ever built anywhere in the world [Fig. 101]. It is horseshoe-shaped in cross section, 11 feet 6 inches high, and 10 feet 3 inches wide. Starting at the intake portal elevation of 1,950 feet, the water drops down an 174-foot intake shaft about four miles upstream from the dam, and thereafter the waters flow south through the uniformly sloped tunnel, dropping downward one inch per hundred feet. The water spills from the south portal into a stone-lined channel and connects with the natural streambed of the Esopus Creek at an elevation of 970 feet. It is a natural-gravity tunnel in which the water is not under pressure. The Shandaken Tunnel was built from eight shafts, varying in depth from 320 to 630 feet, and sunk into the steep mountain slopes along the route. The deepest shaft is 629 feet and at it greatest depth the tunnel is 2,215 feet below the top of the range. The 95,180-foot tunnel was blasted out of the bluestone and shale of the mountain range and lined with a minimum of twelve inches of concrete. By February 1924, a temporary diversion dam had been erected across the Schoharie Creek and the tunnel was sufficiently complete to allow the stream to be turned down into the portal and sent south to the waiting city. The creek has, except for brief periods when the tunnel is serviced, followed this route ever since.[40]

The Gilboa Dam presented unique design problems. The large and exceptionally productive Schoharie watershed was prone to severe flooding. Even the diversion tunnel's huge capacity could not handle all the water expected under extreme conditions, and the tightness of the valley left little room for excess storage in the reservoir. A sudden increase in stream flow could overwhelm the tunnel and stress the dam. To remove excess floodwaters quickly, the Gilboa Dam was designed to overflow across the entire top of the structure. Steps in the downstream face of the dam absorb and diffuse the intense energy and destructive power of water falling 180 feet in elevation and control the potential for erosion at the foundation. To understand the hydraulics of water spilling over the stepped dam face, the Board of Water Supply engineers constructed working models in the Schoharie Creek, testing different scales and different step proportions [Fig. 102]. In the final design, the overflow section, constructed of cyclopean masonry faced with 2-foot-thick slabs of local bluestone, was 1,324 feet wide with eight giant steps that averaged 17 feet in height and 13 feet in width. Building started in June 1919 and the dam was first placed in service in 1927 [Figs. 103, 104, 105]. The resulting reservoir is 5.8 miles long, less than three-quarters of a mile wide at its widest point, and floods 1,145 acres.[41]

During the years that the board worked to capture the waters of the Schoharie Creek and build the Shandaken Tunnel and Gilboa Dam, it also worked to expand the capacity of the Catskill Aqueduct to carry the increased flow of water. The Catskill Aqueduct first came into service in 1915. The fourteen major siphons were initially built with only one pipe each. To accommodate the additional water, each siphon had two more 108-inch steel pipes placed alongside the original siphon pipes.

The building of the Catskill System required an enormous workforce—at one point the number reached 17,243 men [Figs. 106, 107]. The Board of Water Supply recognized early on that the welfare of its workers and their families was

FIG. 102. *Hydrologic model of Gilboa Dam to test stepped weir profile. September 24, 1919.*

FIG. 103. *Gilboa Dam. Central stepped masonry portion of dam and core wall for west dike. Town of Gilboa on upstream side of dam. November 19, 1924.*

vital to the progress of the project. Camps with suitable housing, built by contractors under supervision of the board, were established with consideration for the health and welfare of the workers. The larger camps evolved into small towns with water supply and sanitation, electric lights, telephones, banks, stores, bakeries, hospitals, post offices, churches, and police and fire protection. (It should be noted that despite this social progress, the camps remained segregated into sections for "Italians, Negroes, and Whites"). The board understood the need to foster a spirit of cooperation between the agency, the contractors, and the workers and encouraged self-help and personal betterment. Many of the laborers were newly arrived immigrants, and multilingual day and evening schools were established for them and their children. The attention paid to the human side of the construction process produced a positive and productive atmosphere essentially free of labor strikes.

FIG. 104. *Gilboa Dam. Steel arch forms in place for the downstream half of the stream-control conduit through the dam. September 20, 1922.*

FIG. 105. *Gilboa Dam. Stepped waste weir and spillway channel. November 20, 1926.*

FIG. 106. *Ashokan Reservoir. Olive Bridge Dam. Celebrating storage of water in reservoir. Excursionists at luncheon on top of dam. October 11, 1913.*

The building of the Catskill System marked the beginning of a progressive labor movement, particularly in regard to large municipal projects. By contractual obligation, builders had to maintain a decent quality of life for workers and their families. Camp hospitals with staff doctors were required, and a core of sanitary police employed by the Board of Water Supply inspected food and water.

Workers were required to undergo periodic checkups for various communicable diseases. By the standards of the times, the workers were treated well. The Catskill System also marked the beginning of mandatory worker's compensation in New York State.

The Catskill System redefined the meaning of public water in New York City, not only structurally and proportionately but also organizationally. The positive solutions to many former organizational problems created a unique atmosphere during the time of construction. The unity felt among those involved in building the Catskill Water System is perhaps best expressed in the words of Chief Engineer J. Waldo Smith, in the last issue of the *Catskill Water System News*, December 20, 1913:

> The work we set out to accomplish has been very largely completed, and practically every difficulty has been overcome. With relatively little remaining to be done, it is but inevitable that many of us must seek elsewhere for places where our activities may continue and where we can go on doing good and useful work. To none, more than to me, is this parting a sorrow, as I look back over what we have together succeeded in doing and as I recall our many pleasant, useful and happy associations.
>
> To those of us who remain the parting is as hard as for those who go, for, one and all, we realize the extreme improbability, in the few years that are left to each of us, of ever again being connected with an organization of such magnitude, of such cohesiveness and unity of spirit and, withal, of such intense honesty and loyalty of purpose.
>
> As you all seek far and near for fields in which to continue your efforts, it is my earnest wish that the fullest measure of success will be yours and that you will carry abroad and spread wherever you go the same spirit of truth, of loyalty, of common honesty and of regard and respect for the rights of others.
>
> And I trust, too, that you will never forget, and with me ever bear in mind the happenings, events and many pleasant associations which were in reality ours during the days of the Catskill Aqueduct.[42]

FIG. 107. *City Tunnel #2. Completed section of steel interlining ready to be lowered by derrick to bottom of shaft. July 15, 1932.*

UPHEAVAL

In order to build New York City's magnificent system of 16 reservoirs, we condemned, bulldozed, razed and flooded 12 towns out of existence. And like most big cities, we did so primarily through the use of bullying, bribery and general skullduggery. And the people up there haven't forgotten it. Every year the people who come from those towns hold annual gatherings at the reservoirs' edges, where the old roads run right down into the water. They come with their children, their grandchildren and their great grandchildren and stand there, pointing: "See, over there. That's where your grandparents had their store. And I think over there, that's where your great grandparents on your father's side were married. And I think that's where your great-great grandmother was buried . . . wasn't that it, Honey?" If the water is really low, you can sometimes still see the foundation lines . . . I never met or interviewed anyone up there who didn't, at some point in our talking, stop and say to me: "Don't forget. It's our water too."

Excerpt from *City Water Tunnel #3*,
written and performed by Marty Pottenger

ENTERING
CANNONSVILLE
RESERVOIR
THIS VALLEY TO BE FLOODED
BY WATER IMPOUNDED BY THE CANNONSVILLE
DAM NOW UNDER CONSTRUCTION
BOARD OF WATER SUPPLY
CITY OF NEW YORK

THE DELAWARE SYSTEM

KEVIN BONE

As work on the Catskill System advanced toward completion, the Board of Water Supply (BWS) predicted that the combined capacity of the Catskill and Croton supplies would be outpaced by the growing demands of the city's water users. In his annual report to the board for the year 1920, Chief Engineer J. Waldo Smith warned that in spite of the monumental and heroic effort being undertaken, planning for the next stage of expansion to the water supply needed to begin immediately. He cited the annual increase in the rate of consumption of 42 million gallons per day (mgd) since the first part of the Catskill system was introduced in 1917. He calculated that at that rate, the 600 mgd supplied by the Catskill and Croton system would only meet demand until 1932: "Thereafter the water supply situation will, in general, be the same as that which led up to the development of the Catskill project in 1905 . . . there will be a period of only seven years (from the completion date of the entire Catskill project in 1925 to 1932) during which the available supply will meet actual requirements. This situation, while perfectly satisfactory from the view-point of the present, is, nevertheless, one of serious import for the welfare of the City's future."[1]

In response to these dire predictions, the Board of Estimate and Apportionment directed the Board of Water Supply to move quickly in studying the problem and proposing solutions. The laws of 1905 that first created and empowered the BWS, and ultimately led to the construction of the Catskill System, were amended to cover this new mission. Once again, the BWS engineers were charged with the task of seeking additional water for New York City.

Engineer Thaddeus Merriman, who was the senior authority on new sources of supply, stated in his annual report at the end of 1926 that "the most available sources . . . [to] provide sufficient water for the City for the next 25 years were the streams tributary to the Hudson River on the eastern side of the Hudson River north beyond the Croton watershed."[2] Despite the Smith-Dutchess County Act, which specifically prohibited New York City from entering Dutchess County for water, Merriman issued a report, map, and plan for the development of new sources in Columbia, Dutchess, and Rensselaer Counties.[3] The proposed east-of-Hudson scheme also included plans for a new aqueduct. This aqueduct would divert East-of-the-Hudson waters to New York City, as well as squeeze more water from the upper Croton River watershed, including the Muscoot River and the East and West Branches of the Croton. This time, unlike in 1905 when he successfully championed the development of the eastern Catskills, Merriman's recommendations were not accepted.

The commissioners of the Board of Water Supply had other ideas. They submitted plans to the Board of Estimate and Apportionment that proposed the development of the headwaters of the Delaware River in the western Catskill Mountains. They believed that the rivers and streams of this remote, sparsely populated, and heavily wooded area offered the best sources for new water. The long, narrow valleys of the western slope drain the wet, high peaks region of the Catskill Mountains, an area that receives as much as sixty inches of precipitation annually. This region, like the eastern Catskills, was an ideal place to build facilities to gather and store water. The upper Delaware River basin was first recognized as a possible source for New York City water in the land filings made in 1895 by the Ramapo Water Company. But the city would have to wait forty-four years, build the eastern Catskill System, suffer through the Depression, prevail in major legal challenges, and weather most of World War II before any water from the Delaware River system would be tapped to help meet its needs.

FIG. 1. *Neversink Reservoir. Diversion tunnel. Centerline of excavation at intersection with the inclined spillway. August 13, 1942.*

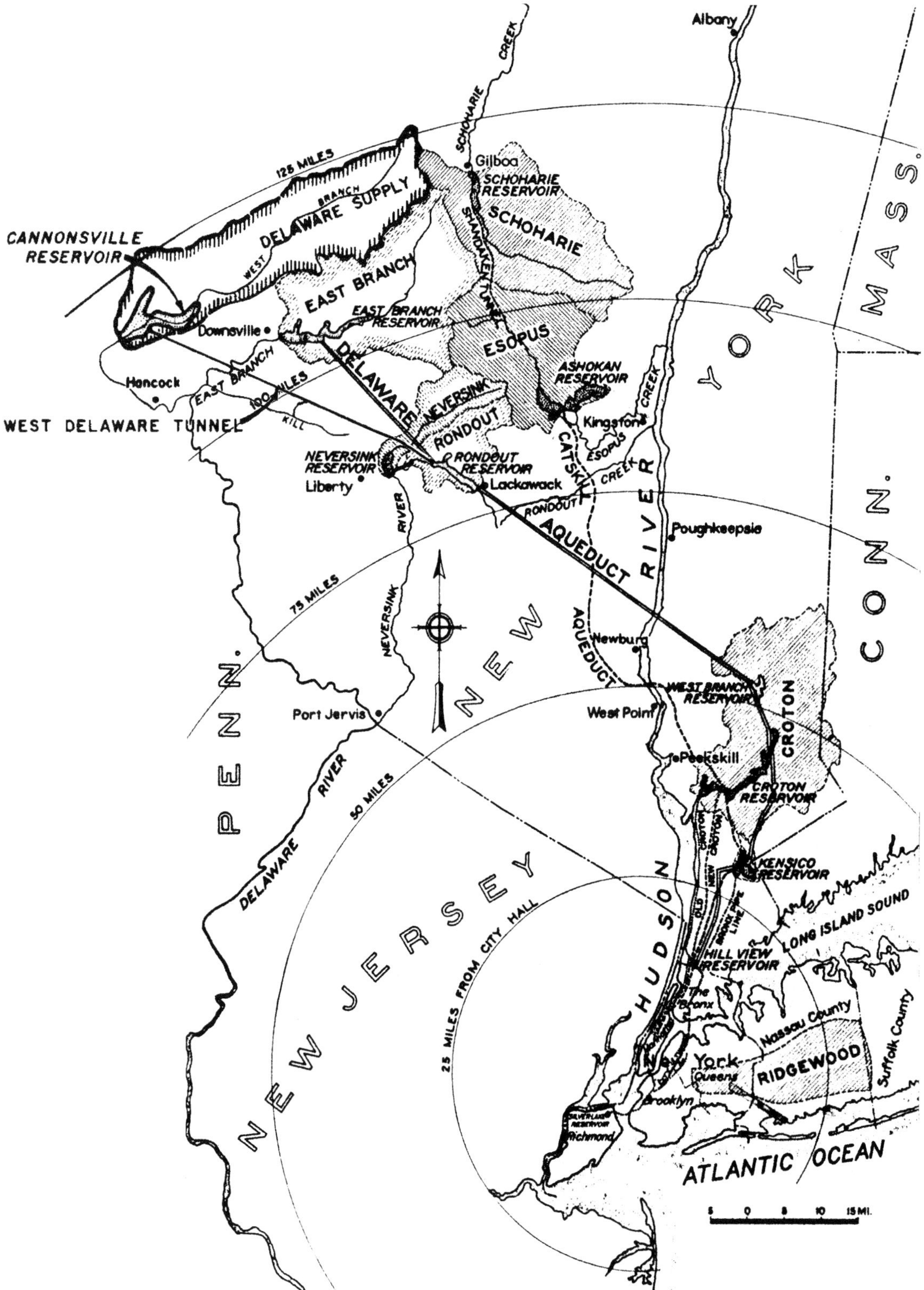

FIG. 2. *Map of the New York City Water Supply System showing Catskill, Delaware, and East-of-the-Hudson Divisions. 1977.*

The Neversink River and other western Catskill drainages considered by the BWS are tributaries of the Delaware River. The Delaware River forms the boundary between New York State and Pennsylvania from the town of Hancock, New York, down to Port Jervis, New Jersey. The river is also the boundary between New Jersey and Pennsylvania from Port Jervis south to the Delaware Bay. However, the 2,389 square miles of watershed that feed the headwaters of the Delaware River lie completely within New York State. This led New York State to believe it had jurisdiction over the water, but New Jersey and Pennsylvania, which also had access to the Delaware River water, disagreed. A bitter interstate conflict ensued, adding years of legal complications to the already monumental planning and building problems that the city faced in pursuing this water. Ultimately it took more than eleven years and a decision by the United States Supreme Court to settle the water rights by decree.

In March 1923, New York State established a commission to work with representatives of the State of New Jersey and the Commonwealth of Pennsylvania to draft a treaty for the use of Delaware River basin water. In 1925 and again in 1927, terms for the allocation of water were agreed upon by representatives of the three states, but in each case these agreements were ratified only by the legislature of New York.

New York City was undeterred by New Jersey and Pennsylvania's legislative impasses, and the relentless pursuit of and planning for water continued. The Board of Water Supply made recommendations as to the most available and best sources of additional water and submitted a three-phase plan for the development of the Delaware River system to the Board of Estimate and Apportionment on July 27, 1927.

The first phase called for the construction of reservoirs on the Neversink River and the Rondout Creek with a tunnel connecting the two basins. An aqueduct tunnel was planned to bring the waters across the southern Catskills, under the Hudson, and down to the city limits. This aqueduct would connect the new Rondout Reservoir, the West Branch Reservoir in the Croton group, the Kensico Reservoir in southern Westchester County and the Hill View Reservoir at the Bronx city limits. It ultimately became New York City's largest aqueduct and the longest continuous pressure tunnel in the world. The first phase would provide an additional 170 mgd [Fig. 2].

The second phase called for a dam on the East Branch of the Delaware River (a tributary north and west of the Neversink River) and an independent tunnel from this new reservoir through the mountains to the Rondout valley. There the tunnel would connect with the new reservoir and the main Delaware Aqueduct. The East Branch phase would add another 370 mgd, bringing the total of the new system to 540 mgd.

The third phase anticipated building other reservoirs in the upper Delaware basin, in locations on the Little Delaware River, the Beaver Kill, and the Willomemoc Creek. These sources could yield another 260 mgd, for a total of 800 mgd. Phases one and two were approved by the Board of Estimate and Apportionment on January 12, 1928, but before the State Water Power and Control Commission could complete its approval process (on May 25, 1929), the State of New Jersey commenced legal action to stop the project:

> On May 22, 1929, the State of New Jersey filed its original bill of complaint against the State of New York and the City of New York seeking to enjoin the state and city from diverting any water from the Delaware River or its tributaries. The Commonwealth of Pennsylvania requested and was permitted to intervene in the case. The court appointed a special master, the case was argued and a decision was rendered on May 4th 1931, and decree was entered on May 25th, 1931. Under the decree New York City was authorized to divert from that portion of the Delaware watershed located in New York State the equivalent of 440 million gallons per day for water supply purposes. As a condition of such diversion the decree required that releases be made from the City's reservoirs to the lower river during times of low flow in accordance with a formula, referred to as the decree formula.[4]

Justice Oliver Wendell Holmes wrote one of the great legal judgments dealing with conflicts over water resources, a reasoning that has been cited time and time again in the battles over water rights:

> A river is more than an amenity, it is a treasure. It offers a necessity of life that must be rationed among those who have power over it. New York has the physical power to cut off all the water within its jurisdiction. But clearly the exercise of such a power to the destruction of the interest of lower states could not be tolerated. And, on the other hand, equally little could New Jersey be permitted to require New York to give up its power altogether in order that the river might come down to it undiminished. Both states have real and substantial interests in the river that must be reconciled as best they may be. The different traditions and practices in different parts of the country may lead to varying results, but the effort always is to secure an equitable apportionment without quibbling over formulas.[5]

The agreement that enabled New York City to develop the Delaware projects is a critical milestone in the history of the water supply system. One of the great urban water supplies of the world was not only the product of heroic engineering and massive construction endeavors; it was also the result of negotiations, legal maneuvering, and court battles.

While the interstate negotiations and legal actions were crawling forward, the BWS, wasting no time, started to address the limitations of the existing inner-city delivery system. Any additional water had to reach a city system that was prepared for the increase, and many of the existing city conduits were already operating at full capacity. Additionally, any new water had to pass through a structure that could regulate and equalize its flow. It was no longer feasible to build another receiving reservoir within the city since large sites simply did not exist. The only reasonable option was to use the existing Hill View Receiving Reservoir. However, additional

FIG. 3. *City Tunnel #2. Concrete formwork for the transition between a 14-foot diameter shaft and two 48-inch risers. Circular formwork (in background) is for the placement of concrete tunnel lining. October 14, 1931.*

FIG. 4. *City Tunnel #2. View down shaft showing concrete plug in place around two steel riser pipes. Shaft caps not yet in position. Brooklyn, New York. September 9, 1931.*

water entering Hill View confronted a bottleneck because only so much water could be drawn out into the existing City Aqueduct (soon to be referred to as City Tunnel #1). Plans to upgrade the capacity of the distribution conduits within the city had to be developed before Hill View could be used as the receiving reservoir for the new water.

The board, with consulting engineers Freeman and Burr still advising on issues of additional water, argued that a second city tunnel was "urgently necessary."[6] They recommended that a new aqueduct line be built from Hill View through the city to serve existing conduits and the more remote areas of the expanding boroughs. The tunnel they proposed was designed to handle additional water, increase delivery capacity within the city, and provide a backup to City Tunnel #1. Built as part of the Catskill System, City Tunnel #1 had been in continuous use since coming on line January 22, 1917—a record for uninterrupted service of a large pressure aqueduct. The tunnel carried 60 percent of the city's water. Sooner or later the tunnel would have to be drained for service (or it could suffer a structural failure) and any resultant shutdown would cause severe shortages. In 1927, the Board of Estimate and Apportionment approved the plan for the second tunnel, which became known as City Tunnel #2. The new structure, a twenty-mile-long, concrete-lined pressure tunnel, extends from Hill View Receiving Reservoir through the Bronx, under the East River at Rikers Island, through Queens and Brooklyn, and connects with City Tunnel #1 and the distribution conduits at Fort Greene Park. The tunnel is seventeen feet in diameter and capable of transporting 700 mgd.

By April 1927 the task of subsurface exploratory drilling along the proposed route began. Thirty thousand vertical feet of drilling was done, a remarkable figure considering that ten linear feet per day was average progress for each drilling rig in the urban working environment.[7] Construction on City Tunnel #2 began in 1928, just as the nation moved toward the Depression. By 1930, nineteen vertical shafts used to build the deep rock tunnel (and later used as uptake risers for distributing the water) had been sunk, and the horizontal building of the actual tunnel had begun [Figs. 3, 4, 5, 6]. Techniques in tunnel construction had advanced to the point where 75 to 100 linear feet could be cut per week at each tunnel heading. The crews typically worked two headings, branching out horizontally in both directions from the shafts:

> During the past three years every operation on Tunnel No. 2 has been prosecuted with the utmost vigor and it is safe to say that no other work of similar kind and magnitude has ever been carried out with greater expedition or efficiency. The record, moreover, has been achieved in spite of the fact that just north of shaft 3A, geological conditions were encountered, which were as difficult as the worst that have ever been met in any tunnel. In July 1931, under the valley of the Bronx River a zone of soft ground carrying much water was met with. The conditions were so trying and difficult that a period of seven months was required to advance the tunnel 100 feet . . . Day and night through three years this work required both the combined resource and ingenuity of the supervising staff and the effort of over 2,500 men.[8]

City Tunnel #2 was built at a depth varying from 200 to 800 feet below ground and completed by 1936 [Figs. 7, 8]. This aqueduct has never been shut down, even for limited inspection. For a short time, it was the longest tunnel in the world. City Tunnel #2 was surpassed in length only by a tunnel that would bring New York City its next installment of water: the Delaware Aqueduct.

In the years following the Supreme Court battle, New York, with its newly won right to Delaware water, had to hobble along with limited construction funding. New York City's own financial mismanagement and the country's plunge into the Depression all but eliminated bond money and other funding resources for public works. In 1932, the Board of Estimate reduced its allocation for the Delaware Aqueduct Project from $50 million to $1 million.

FIG. 5. *City Tunnel #2. Bronze shaft caps ready for shipment from the contractor's plant. 1931.*

FIG. 6. *City Tunnel #2. Valve chamber under construction. Shaft caps on risers set in place with steel anchor bolts. January 12, 1932.*

FIG. 7. *City Tunnel #2. A series of six flashes (indicated by the rings of light) placed at 100-foot intervals were set off when the picture was taken. April 22, 1932.*

Between 1933 and 1936, despite severe funding limitations, the BWS, believing that a disastrous water famine was inevitable, continued to press the city to maintain board staffing so that research, design, and production of contract documents could continue. In a time when many governmental agencies were stripped of staff, the BWS was still able to employ 260 engineers. Throughout the Depression, the engineers continued to push forward with their preparations for the Rondout Reservoir and the Delaware Aqueduct.

The process of establishing the "taking area" (property that was condemned and purchased) and negotiations with landowners proceeded with the usual delays and heartbreak. Surveys defined land to be sacrificed for the good of a city that most upstate locals cared little for. "Four standard concrete monuments were set . . .one near each end of the reservoir above the elevation of the flow line and one at the foot of the slope [of the dam] on each side of the Rondout Creek on the valley floor."[9] With the simple act of setting benchmarks, residents were removed from the valley, their farms relocated, and the building of another dam begun.

The year 1933 saw the death of J. Waldo Smith, who had been employed by the board for more than thirty years. It also saw the resignation of Thaddeus Merriman, who had been with the board for twenty-eight years. Other changes shook up the quest for water as well. Fiorello LaGuardia had campaigned against excessive spending and the criminal misuse of city funds. LaGuardia attacked the Board of Water Supply for abuse of funds and sought to have it abolished. But, as had happened with his predecessor, Major Mitchell, who in 1916 also sought to abolish the BWS, the state stepped in to defend the best interests of the city and ensure the BWS's ability to complete its mission.

By 1937 contracts were established for the Delaware Aqueduct and the Rondout Reservoir. The aqueduct would differ substantially from its Catskill counterparts. The main section of the aqueduct was designed to be an 85-mile-long deep-rock pressure tunnel rather than the assortment of structure types that were employed on the Catskill Aqueduct. The terrain on the west and south side of the Catskill Mountains is irregular, and the new aqueduct had to confront a succession of ridges and valleys. The logical route ran across this ridge and valley topography. To get from the Rondout Reservoir site to the existing West Branch Reservoir in the Croton system, the aqueduct crossed the Rondout Creek, Vernooy Kill, Stoney Kill, Fly Brook, Palmaghatt Kill, Shawangunk Kill, Wallkill River, and the Hudson River more or less perpendicular to the line of the valleys before reaching the gentler terrain of Dutchess County. To build at or near the surface required a long and winding route along the hydraulic grade.

Although the least expensive method of construction was the cut-and-cover construction used on the Catskill line, that method was inappropriate for these geographic conditions. Between Rondout and the Croton West Branch Reservoir there are twenty-eight miles of valley too low to support a grade tunnel, and seventeen miles of ridges that would have to be tunneled through. The Hudson River, because of the extreme depth of its unstable geology, would have to be passed under using a deep-pressure tunnel (as was done with the Catskill Aqueduct). The lands east of the Hudson could support a tunnel built on grade, but the counties on the east side of the Hudson were changing rapidly. The population had increased and much of the land was developed with estates and high-end residential areas. The expense of acquiring surface rights and the disruption due to construction activity would have been intolerable to local residents and would add to the legal problems and settlement costs. Finally, the deep system offered the advantage of maximum protection against sabotage or accidental damage. This combination of factors—the geographical lay of the land west of the Hudson, the demands for pressure tunnels at key locations like the Hudson River, the difficulty of securing the right-of-way through the populated areas east of the Hudson, and the advantages to security and operation—led the BWS to make their decision. They chose the costly option of building the entire Delaware Aqueduct system as deep-rock tunnels that were blasted under the ridges and valleys and along alignments that would put the aqueduct lower than the hydraulic gradient. This allowed the entire tunnel system to be under pressure.

FIG. 8. *City Tunnel #2. View looking into the access drift from tunnel section, showing parts of the steel interlining in place after riveting. July 15, 1932.*

The final route of the aqueduct followed a straight line from the eastern end of the Rondout Reservoir to the West Branch Reservoir in the Croton system and then turned south to connect to the facilities at the Kensico Reservoir, terminating at Hill View. Routing the aqueduct this way offered several advantages:

> The aqueduct could pick up an additional 100 mgd that was being added to the storage capacity in the upper reaches of the Croton (but could not reach the city because of limits in the existing aqueducts) and could deliver that water at an elevation 160 feet higher than Croton aqueducts can deliver (because of increased hydrostatic head) thereby reducing the costs of pumping. Passing the water through the West Branch Reservoir allowed the water to stand in the open air, improving water quality by allowing settling time and by reducing bacterial content, it safeguarded a considerable portion of the supply in the event of a shutdown of either of the Croton Aqueducts (which at the time of Delaware construction were 60 and 110 years old respectively) and it made possible the replenishment of depleted Croton reservoirs with water from the Delaware system whenever rainfall is adequate on the Delaware but deficient on the Croton.[10]

New York City had been doing extensive tunneling for one hundred years, but the Delaware undertaking would include the longest tunnel built to date. It called for an enormous system of tunnels totaling over 170 miles in length. The methods of tunneling for water supply applications had changed significantly since New York City had built its first aqueduct. The Old Croton Aqueduct involved small tunnels about 8 feet in diameter with a maximum length of 1,263 feet, and most were shorter than that. Much of the New Croton Aqueduct tunnel was relatively near the surface and followed the hydrological gradient. The thirty miles of tunnel in the New Croton Aqueduct were built from forty shafts, the horizontal distance between shafts being from 400 to 7,500 feet apart. The logistical problems involved in moving materials, air supply, tunnel lighting, and the time involved to vacate the spoils between blasts all made the progress on the New Croton Aqueduct very slow. Construction rates averaged about 25 to 40 linear feet of tunnel per week. On the Catskill Aqueduct there were thirty-three miles of tunnels. Advances in mechanization allowed the construction shafts to be much farther apart and improved technology and labor skills increased tunneling rates to 55 to 70 linear feet per week. The Catskill system's Shandanken Tunnel was built at the rate of almost 100 feet per week, a rate three to four times that of New Croton. City Tunnel #2 progressed more slowly (about 75 feet per week) because the tunnel was significantly larger in diameter (17 feet) and the geological conditions were less favorable (harder rock, more ground water). On the basis of this experience, engineers planned on tunneling rates of 100 feet per week, after initial setup and mobilization. The actual pace would exceed plans: Deputy Chief Armstrong of the BWS described the process:

> In the old days it was pretty good performance if you got in one round per day, one shot per day. You will drill most of shift [to set the explosives and blast] and you would work two shifts to get the muck [loose rock] out by hand methods. On [the Delaware Aqueduct] the contractors set themselves a schedule of three rounds per day, making the shot and getting the muck out of the heading in an eight-hour shift. Before the work was over many of them were getting in more than three rounds per day . . . Average progress on this work, after the work was well organized, on all the 13½ foot diameter tunnels was about 200 feet per week. The average for the 15 ft tunnel was 140 feet per week and for the 19½ foot tunnel about 125 feet per week[11] [Fig. 10].

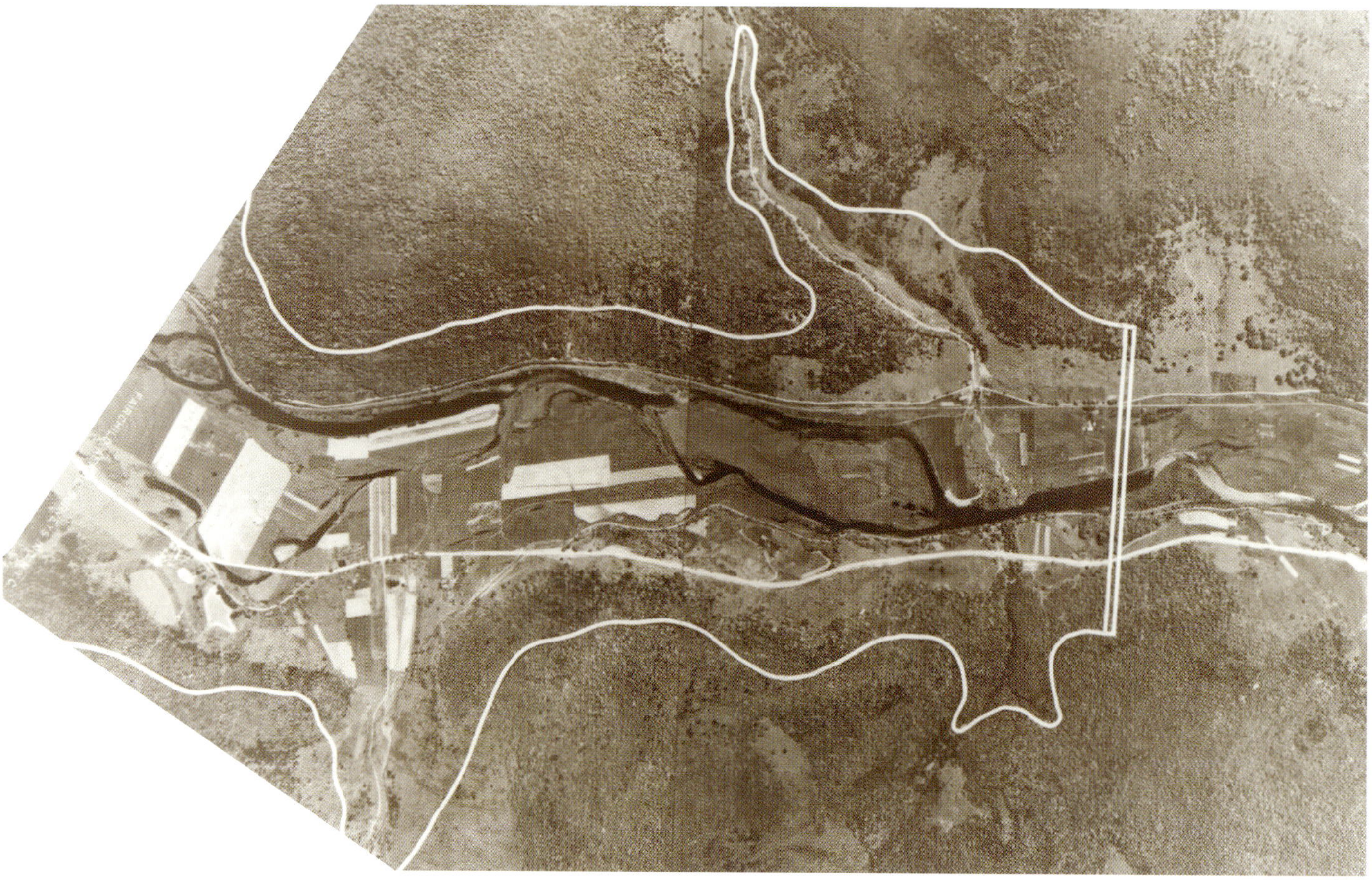

FIG. 9. *Neversink Reservoir. Aerial view showing extent of area to be flooded for reservoir. c. 1937.*

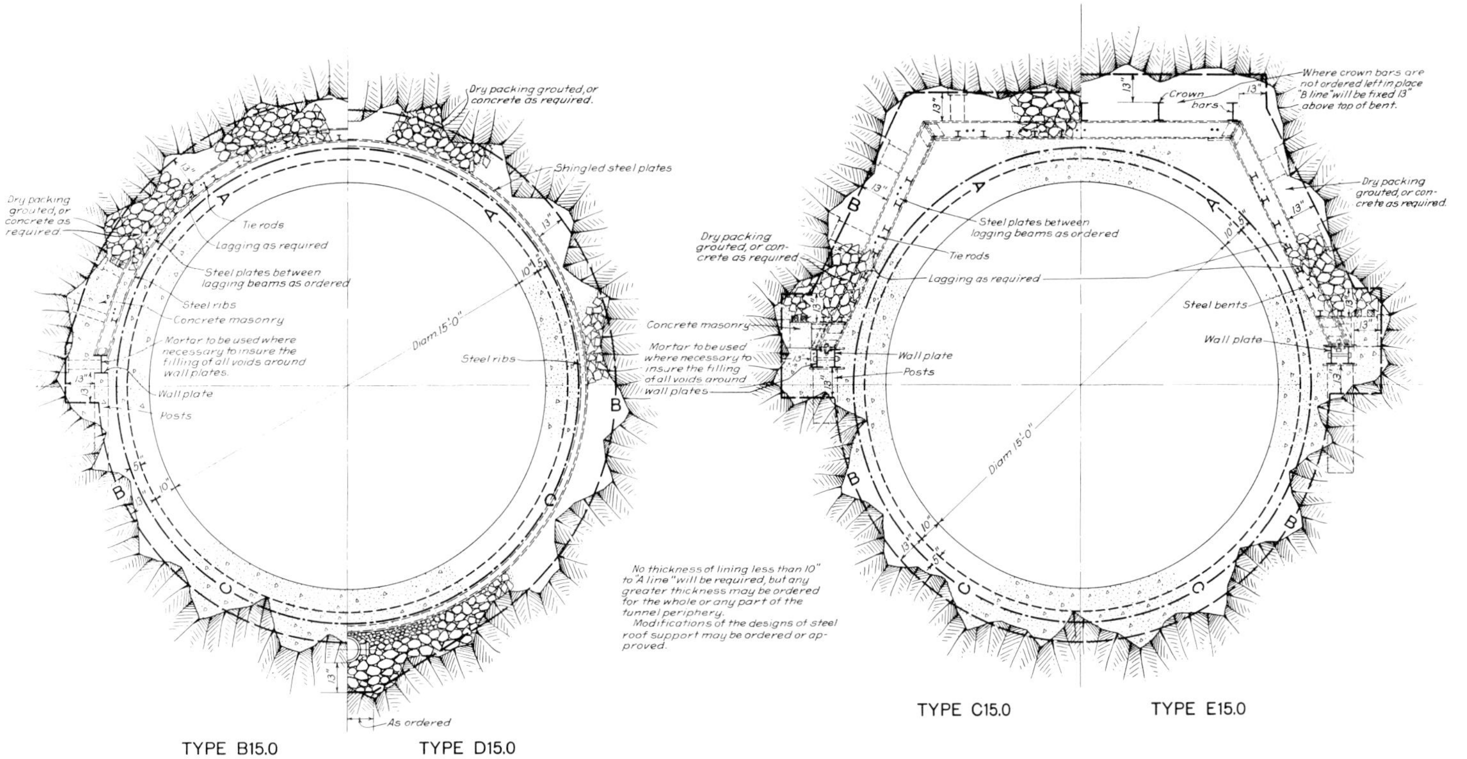

The main aqueduct was divided into three sections: the northerly section, from Rondout to the West Branch, is a 45-mile tunnel built at a finished dimension of 13½ feet. The middle section, from the West Branch Reservoir to Kensico, is 23 miles long, built to a 15-foot interior diameter. The southerly section, extending from the Kensico Reservoir to the Hill View Reservoir, is 13 miles long, and the interior tunnel dimension increased to a diameter of 19½ feet. Two bypass tunnels, each 2 miles long, were built at Kensico and at Hill View, allowing the Delaware water to continue downward without entering those reservoirs. These bypass tunnels are each 15 feet in diameter.

The route an aqueduct follows, as it appears on a map, is known as the horizontal alignment. The vertical alignment, known as the profile, is the relative position of the tunnel up or down in the earth. In the case of the Delaware Tunnel, the profile was determined not by hydrological grade, as with the gravity tunnels of the

FIG. 10. *Delaware Aqueduct. Cross sections of pressure tunnel showing different construction assemblies. November 21, 1938.*

FIG. 11. *Delaware Aqueduct. Kensico Bypass Tunnel. Effluent chamber at downtake shaft. Formwork of the baffle completed to final height. May 15, 1940.*

FIG. 12. *Delaware Aqueduct. Kensico East Influent Works above the West Branch and Kensico Bypass Tunnels. Steel frame of coagulating plant under construction. May 25, 1942.*

FIG. 13. *Delaware Aqueduct. Kensico East Influent Works above the West Branch and Kensico Bypass Tunnels. Superstructure of coagulating plant complete. September 11, 1942.*

New Croton and the Catskill Aqueducts, but by geologic conditions. Because it was built as a pressure tunnel that started at a relatively high elevation, the slope was less critical. The tunnel could, and did in some locations, run uphill. As long as the end point was lower than the starting point, and as long as the entire tunnel was a closed pressure system, the incline of individual sections could be adjusted to reflect the requirements of geologic conditions. Internal pressures in the deep tunnels could reach 500 pounds per square inch so a minimum safe cover of at least 200 vertical feet of sound rock was required. Zones of degenerated or faulted rock required the tunnel to dip deeper into the rock strata to bypass weak areas.

The building of the Delaware Aqueduct, which had been scheduled to start in 1932, would not gain real momentum until 1936 when the Board of Estimate and Apportionment was able to fund the project. Construction started on the section from Kensico to Hill View in February 1937 [Figs. 11, 12, 13, 14, 15]. By the end of 1939, six thousand men were employed in construction of the works. The tunnel was constructed from 32 shafts, with a maximum distance of five miles between shafts (compared to a maximum of just over one mile between construction shafts on the New Croton Aqueduct).

The shafts were first used for construction operations: men, machinery, and materials were lowered into the tunnel through these portals. The deepest shaft was almost 1,600 vertical feet below the surface. The shafts are generally circular in section, 14 feet in diameter, and contain valves and equipment for the operation and control of the water. After completion of the tunnels most of the vertical shafts were incorporated into the system: today, three shafts are used for drainage, eleven shafts supply water to local communities along the route, two shafts connect to the Croton System, three are left open as ventilators, nine were sealed completely, and one shaft—where the Delaware Aqueduct passes under the line of the Catskill Aqueduct—connects the two systems. This allows for a section

FIG. 14. *Delaware Aqueduct. West Branch and Kensico Bypass Tunnels. Influent chamber. January 21, 1942.*

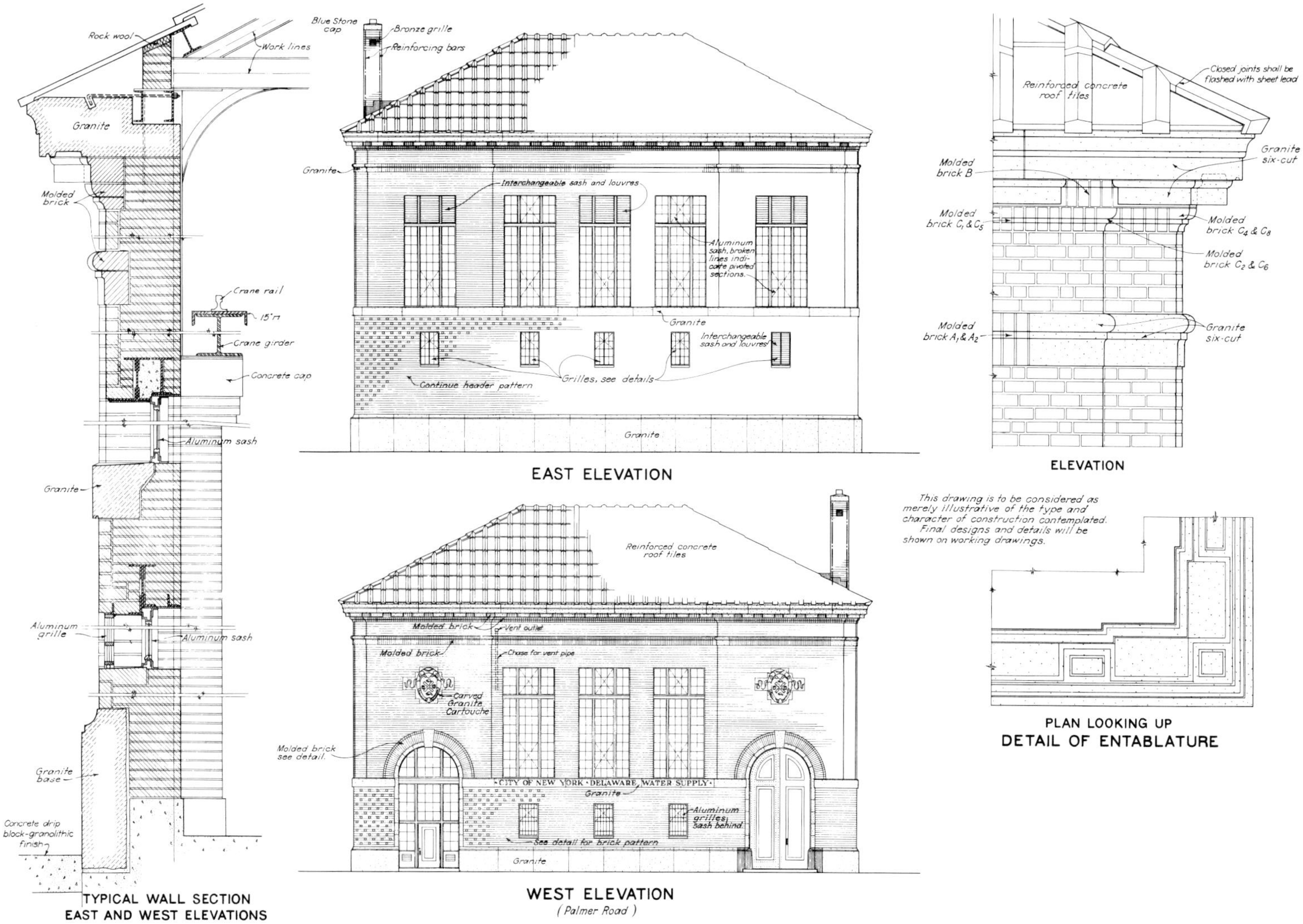

FIG. 15. *Delaware Aqueduct. Kensico–Hill View Tunnel. Headhouse at Shaft 23. Elevations, section, and construction details. December 1, 1937.*

of either of these conduits to be shut down and the water, at least in part, rerouted to the other aqueduct.

The tunnels, which are the actual conduit for the water, are blasted out of the rock, circular in cross section, and lined with concrete [Figs. 17, 18]. The concrete lining improves the hydraulic qualities of the aqueduct by reducing its resistance to flow. It prevents deterioration or collapse of the rock walls and serves to avert the outward leakage of water when the aqueduct is in service and under pressure. These tunnels are well below the water table, and they must withstand the pressure from the exterior groundwater when empty as well as the stress from the grouting that fills the seams between the concrete and rock walls [Fig. 19]. The cylindrical concrete lining is incredibly strong in compression and able to resist the huge external loads the structure is under before it is filled with water and under internal pressure. Once it is full, the tunnel is at equilibrium: the outward pressure is equal to the inward pressure.[12]

In calculating the required thickness of the lining, it was assumed that the entire lining would be exposed to the full hydrostatic head of the exterior ground water—a condition that is rarely, if ever, realized. In the primary Delaware delivery tunnels, the thickness of the concrete linings (that area into which no rock is allowed to encroach) ranges from 15 to 19 inches. The average thickness was 20 percent more than the minimum specified requirements:

> Every effort is made during its construction to bond the lining as closely as possible to the

rock against which it is placed. All rock debris is removed from the tunnel, and all dust and other coatings are removed from the rock surfaces before concreting is started. After the concrete has been placed, unfilled voids, shrinkage spaces, cracks and joints, as well as open seams in the rock behind the lining, are filled by pressure grouting through pipes left in the masonry or through holes drilled for that purpose.[13]

The Board of Water Supply went to great lengths to control the quality of work in the tunnels. The concrete, and its key ingredient Portland cement, were inspected and tested at every step of manufacture and placement to confirm the quality relative to standardized specifications, such as those of the American Society of Testing Materials (ASTM). But the BWS and its chief engineer, Thaddeus Merriman (a pioneer in advanced research into cement chemistry who wrote the BWS specification for concrete), took the

FIG. 16. *Merriman Dam. Inside the blow-off tunnel looking through the portal into the Rondout effluent chamber. August 26, 1942.*

FIG. 17. *Rondout Reservoir. Merriman Dam. Effluent chamber. East and west stilling basins prior to concreting. October 10, 1941.*

FIG. 18. *Rondout Reservoir. Merriman Dam. Effluent chamber. Twenty-foot diameter concrete-lined portals to east and west stilling basins. These portals mark the head of the Delaware Aqueduct. Opening to blow-off tunnel to left of portals. August 26, 1942.*

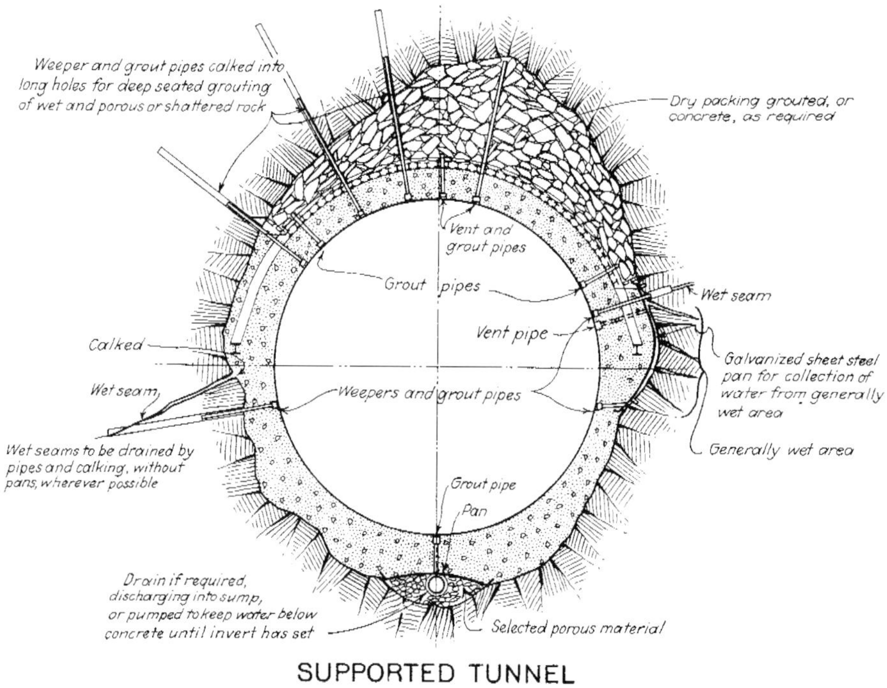

FIG. 19. *Delaware Aqueduct. Cross section of a typical pressure tunnel. c. 1937.*

FIG. 20. *Delaware Aqueduct. Rondout-West Branch Tunnel. Cleared rock tunnel looking toward completed concrete tunnel lining. May 7, 1942.*

additional step of placing inspectors at the mills where limestone for making cement was burned to verify the temperatures, the time in kilns, and the chemical composition of each mix. The mill inspectors maintained a history of each batch that detailed the differing compositions of base ingredients to ensure that the mixes were in full conformance with Merriman's specifications.[14] BWS specifications for concrete were far superior to the standard ASTM specifications and led to the rewriting of most of the current concrete standards in use.

BWS engineers tested every batch of concrete used in tunnel construction at the time it was placed. If the tests proved the concrete deficient, that section of work was removed. Formwork was inspected before any placement of concrete was allowed. In a difficult procedure, engineers crawled into the void between the formwork and the irregular rock wall to verify that no rocks had come loose, no seams or voids in the bedrock were overlooked in preparation for grouting, and all details of cleanup were complete and ready to receive the concrete. After the concrete was placed and the forms stripped, areas that showed signs of air pockets or other deficiencies were rejected and removed. It was considered a good rate of progress if 380 linear feet of tunnel could be concreted per month. The extensive testing, research, and records the BWS maintained on cement and concrete are critical to the body of knowledge associated with the modernization of concrete as a building material.[15]

The strongest tunnel sections were built of pure concrete bonded to the solid rock wall of an excavated space [Fig. 20]. This was the method used wherever possible. But where rock was unstable, reinforcement and steel interlinings were employed. Reinforcing the concrete with steel was limited to poor geologic conditions, since ferrous materials were both prone to corrosion and expensive. Techniques were developed for bending structural steel shapes, such as I beams, into semicircular sections. These curved beams were assembled into barrel

FIG. 21. *Delaware Aqueduct. West Branch effluent chamber. Waterway looking toward reservoir basin. June 19, 1942.*

FIG. 22. *Delaware Aqueduct. West Branch effluent chamber. Waterway looking toward closed sluice gate to aqueduct. June 19, 1942.*

vault–like cages and set against the loose rock walls of the tunnels [Fig. 23]. Steel braces and columns were added to further secure the assembly and stabilize the loose rock, which was then cast into and encapsulated by the concrete tunnel lining [Figs. 24, 25].

The Delaware Aqueduct was designed to carry up to one billion gallons per day, all the water that was potentially available to New York City from the Delaware watershed. The main tunnel, from Rondout to Hill View, is 84 miles long, and combined with City Tunnel #2 the aqueduct is, at 105 miles, the longest deep-rock pressure tunnel in the world. The tunnel was first put into use in April 1944: A temporary dam and tunnel diverted the waters of the Rondout Creek directly into the shaft at the head of the new aqueduct. This water (about 100 mgd) supplied New York City until 1949.

In addition to the main Delaware Aqueduct, the first phase of the Delaware system included two reservoirs: the Rondout Reservoir on the Rondout Creek and the Neversink Reservoir on the Neversink River [Fig. 9]. By 1937, as work on the shafts and tunnels of the aqueduct proceeded, work on the Rondout Reser-

FIG. 23. *Delaware Aqueduct. Rondout-West Branch Tunnel. Heavily reinforced section in loose rock (prior to concreting) with movable steel formwork in place. October 31, 1940.*

FIG. 24. *Rondout Reservoir. Inclined tunnel from the effluent chamber to Shaft #1 of the Delaware Aqueduct. The engineering team is measuring rock cross sections using a sunflower instrument. October 10, 1941.*

voir began. Though the Rondout Reservoir was built as an integral part of the Delaware system, the Rondout Creek, at the southwestern corner of the Catskills, is not part of the Delaware River drainage, but rather a tributary of the Hudson River. This creek drains south and east out of the Catskill Mountains before it bumps up against the Shawangunk Range and turns north and east to meet the Hudson River at Kingston. The Rondout Reservoir, at the head of the Delaware Aqueduct, was designed to collect and store the waters from the 95-square-mile watershed above the dam site. This reservoir also receives all the Delaware waters and acts as a control chamber for the entire Delaware system. By the time the system was complete, three separate tunnels from three Delaware River reservoirs discharged water into the Rondout Reservoir.

The dams built to create the Delaware system reservoirs were different from the others built for the New York City water-supply system. The locations of the dams of the Croton and Catskill systems were largely determined by subsurface conditions like accessibility to bedrock and the potential of the site to support the heavy loads of gravity-based, solid-masonry structures. But for the Delaware system, New York City engineers built dams where the maximum amount of water could be collected and stored. The locations in the Delaware basin that the BWS scouted were good for water collection, but the foundations presented difficulties for dam construction. The valleys of the western Catskill Mountains are filled with hundreds of feet of soft water-bearing soil with free-flowing subsurface streams. The excavation and removal of this overburden to expose the bedrock and prepare the site for a solid foundation would have required engineered and stabilized trenches in water-saturated ground up to 180 feet deep. Successfully excavating to a great depth in such difficult soil conditions was well beyond any previous construction endeavor. By comparison, the foundation trench for the New Croton Dam required a pit that was 65 feet deep in a location where the groundwater situation was relatively manageable, and that was a monumental endeavor. It was not an option to build directly upon the yielding soils of these valleys with any type of rigid monolithic structure because of the potential for structural failure due to settlement and displacement. The BWS had to adopt a new strategy.

In the years since the BWS built the Gilboa Dam on the Schoharie Creek, engineers throughout the world had gained experience with large earthen structures: "Prior to 1850, there was only one earth-fill dam more than 100 feet high, completed about the year 1500. By the year 1900, the number had risen to eight, by 1930 the total was 133."[16] An earthen structure offered the advantage of flexibility. The main body of the dam could be built directly upon the earth and was capable of deforming with the natural settlement of the soils. As a result, the body of the dam did not require a bedrock foundation to be stable. Earthen embankments had been used for dam building within the New York City water-supply system but only as appurtenant side structures and minor dikes, never on the grand scale being considered

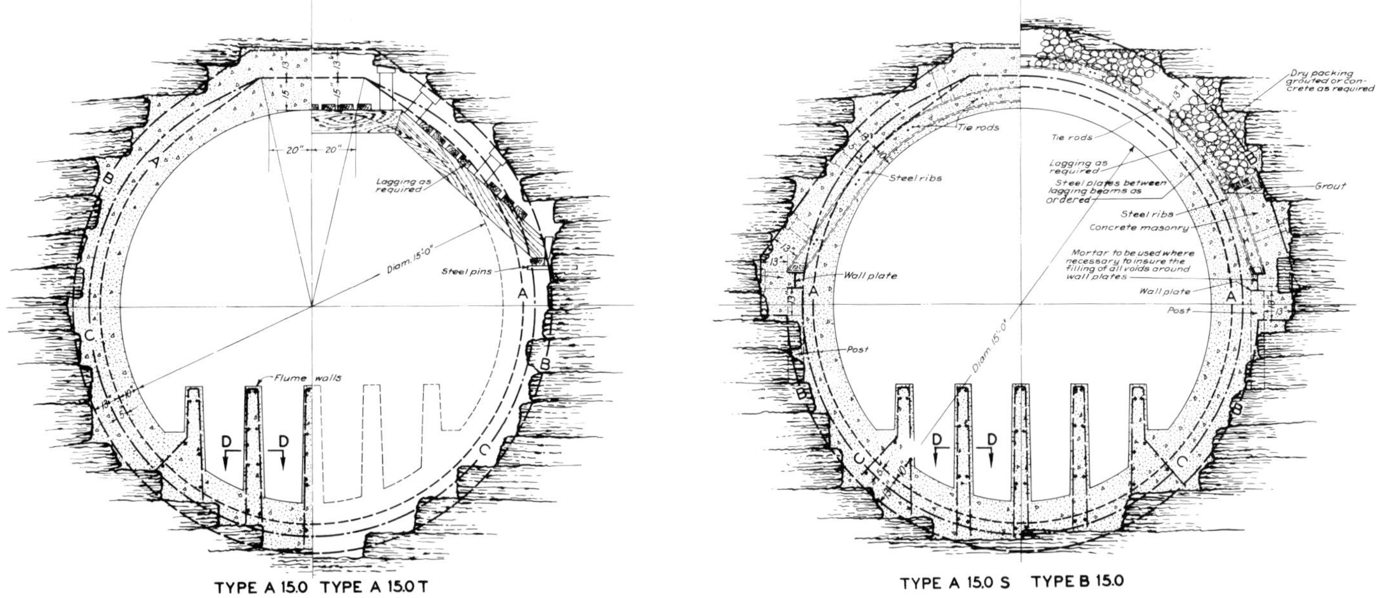

FIG. 25. *Delaware Aqueduct. Rondout-West Branch Tunnel. Alternate sections showing flow control flumes. April 1, 1939.*

FIG. 26. *Merriman Dam. Rondout effluent chamber. View from the west stilling basin to the inlet channel. August 26, 1942.*

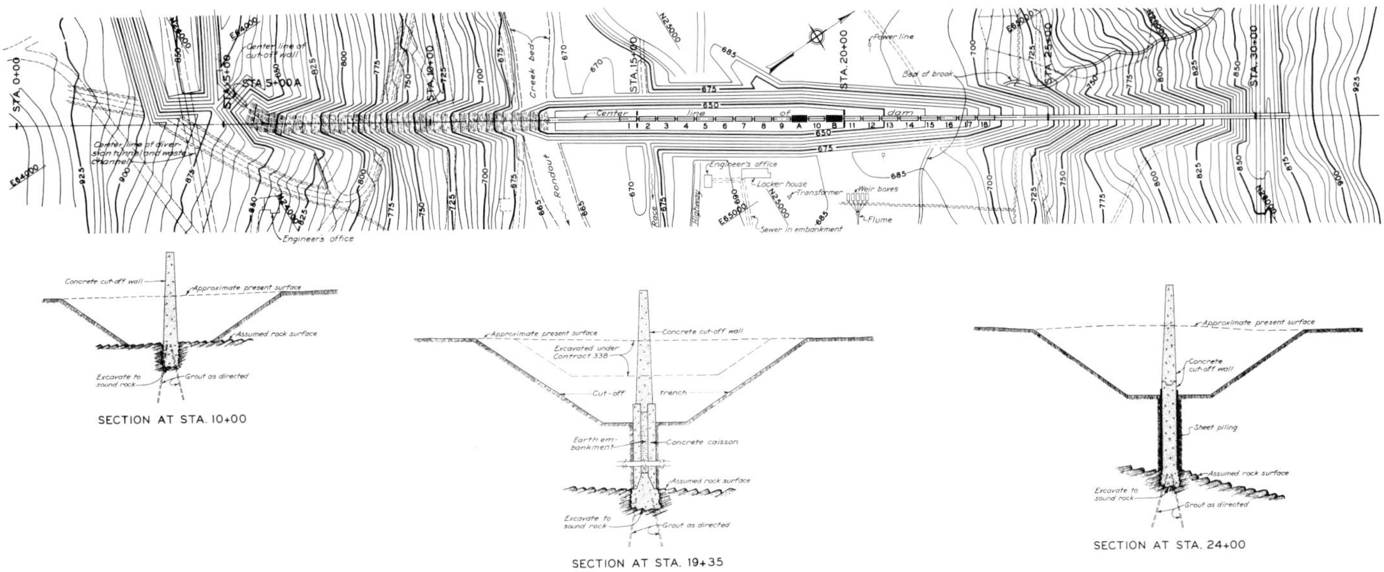

FIG. 27. *Merriman Dam. Plan and cross sections at caisson and core wall. April 1, 1939.*

for the Delaware system. However, methods were under development for controlling the selection, classification, placement, and compaction of earthen materials, and these methods dramatically improved the performance of the structures. This work was done scientifically and the behavior of the completed earth structures was mathematically predictable.

Core walls are a critical aspect of earth dam construction [Fig. 27]. The earthen embankments of a dam can be built upon the geologic overburden of valley floors, but to prevent leakage and subsurface hydrostatic action that could undermine the structure the dams require a core-wall water cutoff that acts as a diaphragm and eliminates subgrade water migrations. This prevents ground water and billions of gallons of impounded water from percolating under the dam. To accomplish this and eliminate the danger of undermining the structure, core walls have to be built down through all the layers of saturated soils and keyed solidly into the bedrock.

In earlier earthen structures, core walls typically extended from bedrock to the highest level the water was designed to reach. Improved earthwork engineering utilized compacted materials to make embankments impervious to the water pressures developed in dams. As a result, the practice of extending the core walls all the way to the flow line at the high water level of a reservoir was largely abandoned. This made building large-scale earth dams more practical. A new generation of heavy diesel-powered construction machinery made the huge earthmoving operations that were required more practical as well. These factors—the foundation issues involved in building a solid masonry dam on the soft valley floors; the advancements of engineering in the planning of large earth structures; and the improvement of construction machinery and methods, combined with the rising costs of labor and materials required for building solid masonry dams—led the BWS to build earthen dams with reinforced concrete core walls throughout the Delaware system.

The first of the new generation of earthen dams was built on the Rondout Creek in the southwestern Catskills near the small town of Lackawack. Preliminary core samples taken at the Rondout Reservoir site indicated that the loose, water-bearing materials in the valley floor extended 180 feet down before reaching bedrock. To learn more about the specifics of building on a site with such deep saturated soil cover, and to determine the feasibility of possible construction methods, two exploratory pneumatic caissons were driven into the earth along the axis of the dam at the point in the gorge where the bedrock was deepest.

Caissons are shaftlike structures that are driven down into saturated earth and used for foundation work in watery environments. Pneumatic caissons use pressurized air to keep groundwater out of the shaft work area. Caissons were historically used for building bridge piers in river bottoms, most famously in the Brooklyn Bridge, but they have also been used for foundations in buildings, particularly the skyscraper district of Chicago on the soggy banks of Lake Michigan.

The exploratory caissons developed for the Rondout site, and for the Delaware system in general, were solid reinforced concrete structures, typically 15 feet wide and ranging from 38 to 45 feet long [Figs. 28, 29]. The perimeter along the bottom of each caisson was covered with a continuous structural iron shoe configured as a sharp cutting edge, a design that reduced resistance from the earth and created a seal to keep surrounding groundwater out. The first forty feet or so of soil cover in the valley floor was scraped off to establish a

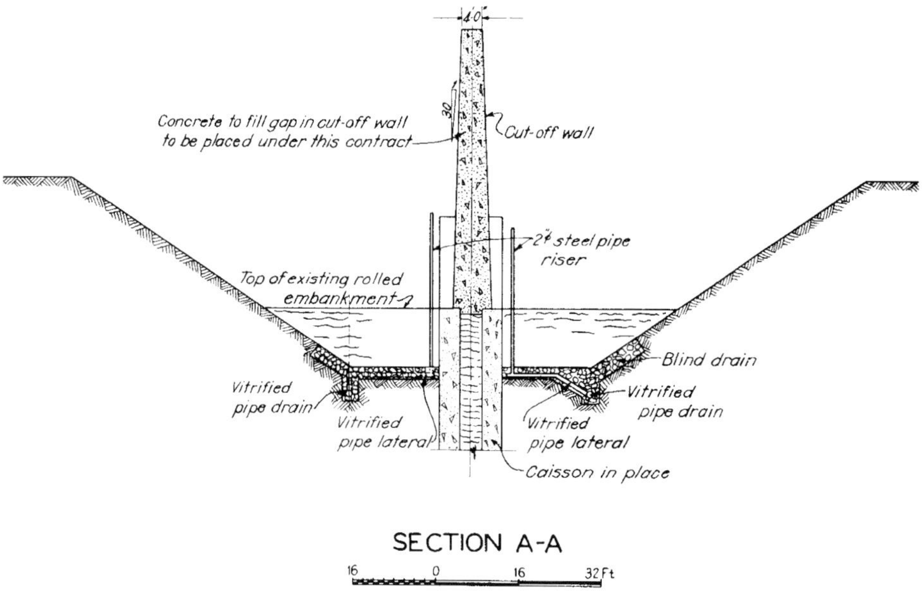

FIG. 28. *Neversink Reservoir. Interior of the working chamber of the west exploratory caisson. September 17, 1941.*

FIG. 29. *Merriman Dam. Typical cross section of a core wall and caisson. c. 1941.*

firm, level trench in which to build the caisson. The ironwork of the cutting edges was set and steel forms for concreting were erected around them. Wooden forms were built inside the area of the caisson to define the working chamber. Reusable collapsible forms were placed in the roof of the working chamber to provide the vertical passage for workers and the vertical muck shaft used to hoist out excavated materials. Ten to fifteen feet of concrete was placed into this formwork, and after the concrete set, the forms were removed and raised. As the resetting of the concrete forms proceeded on top of the caisson, crews of six men dug away the soils inside the caisson's working chamber with small pneumatic tools. The excavated muck—a sloppy combination of stones, sand, and mud—was hoisted up and out of the working chamber through the muck passage. In the reassembled formwork on the top of the caisson, another lift of concrete was placed and the process was repeated. Each successive addition of concrete increased the weight of the structure, and as the earth was slowly cut away from within the chamber, the weight pushed the giant shaft of concrete deeper into the ground. Boulders in the path of the cutting edge were broken up with small charges. As the caisson dropped farther into the earth and the exterior ground water increased, the working chamber was pumped full of pressurized air to limit water infiltration. Workers and materials would leave the chamber through pressure locks fitted onto the passages.

Other water-control methods around the caissons were used in addition to the pressurized air in the working chamber. At ground level, adjacent to the caissons, well holes were drilled to various depths, fitted with pipes and pumps, and operated continuously to drop water-table levels. Supplemental pumps were also set up in the working chamber and operated both during the sinking process and while the adjacent caissons were being set. The entire process was a constant battle to keep groundwater out of the work chamber.

If, in driving the caisson downward, the structure deviated from vertical, the crews were directed to carry out selective mining along a specified edge while the top of the caisson was asymmetrically loaded with pig iron. This process changed the balance of the loading, corrected the path, and kept the column of concrete within an acceptable range of true vertical. Every ten days another fifteen feet of reinforced concrete was added to the top of the structure, and every ten days the structure was driven fifteen feet deeper into the valley floor.

Working conditions in a caisson were severe. If workers were not properly reacclimated before leaving the caisson, they could suffer from "caisson disease," or decompression sickness, a painful, sometimes fatal condition caused by nitrogen bubbles in the tissues: "Men worked one three-hour shift per day, and wore badges that read 'COMPRESSD AIR WORKER.' If this man is stricken on the street, do not send him to the hospital, send him at once to the medical lock at Lackawack, Ulster County [Lackawack was the dam construction site]"[17]

After nineteen weeks, the two exploratory caissons hit "original rock" 185 feet below the river. Once the caissons were seated on the deep subterranean rock layers, the working chambers and passages were filled solid with concrete.[18] The data collected by driving the exploratory caissons on the Rondout Creek served to define the extent of the construction problem and provided an opportunity to modify the design specifications for the remainder of the caissons, both for the Rondout Dam and for those that followed. Information was provided to potential builders who were bidding on the work in order to minimize their uncertainties about the demanding and unfamiliar construction process. A total of twenty rectangular caissons, set side by side with the interstices keyed together with concrete

FIG. 30. *Merriman Dam. Concrete core wall with temporary gaps for transportation of construction material. October 15, 1941.*

FIG. 31. *Merriman Dam. Diagram showing vertical alignment of caissons, core wall, and top of dam. c. 1941.*

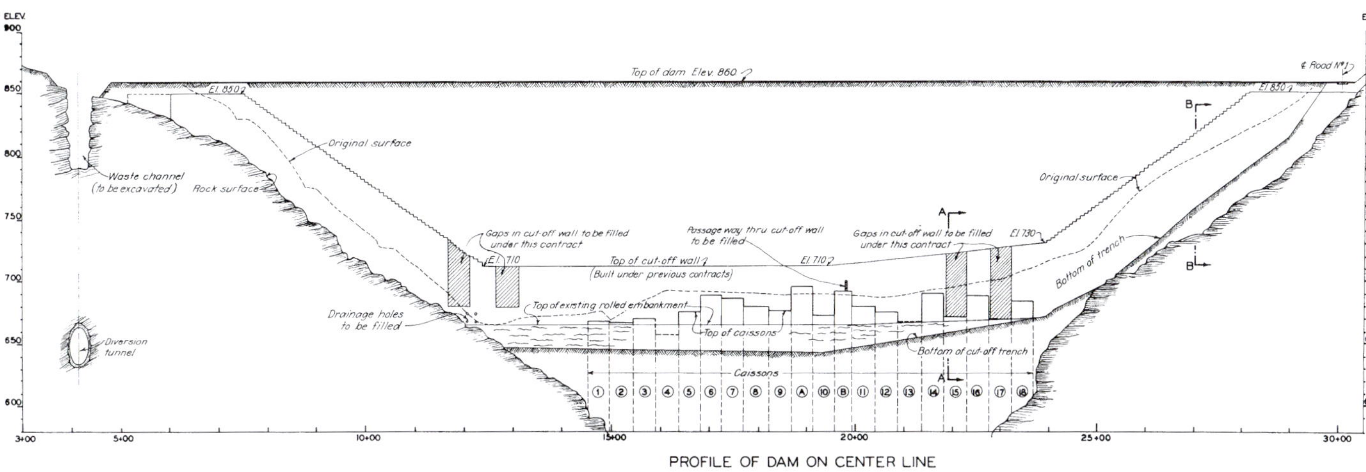

made up the foundation of the core wall through the deeper parts of the Rondout Valley. After the line caissons were set, a solid reinforced concrete wall averaging eighty feet above the original surface of the valley was built upon them, completing the core wall water cutoff assembly [Figs. 30, 31]. At the sides of the valley, where the bedrock rose up with the hills and the saturated soil cover was not so deep, caissons were not required and the cutoff wall was built directly on level benches blasted out of exposed bedrock. In these cases, a trench, typically 40 feet deep, was cut into the rock and the wall was built from the bottom of the trench.

In the summer of 1938, one year after work began on the exploratory caissons on Rondout Creek, the BWS let contracts on a cofferdam (a watertight embankment) and diversion tunnel to control the flow of Rondout Creek during the long construction process. The methods employed for stream control on the Rondout Creek (and at the other Delaware system dam sites to come) varied significantly from previous strategies used on New York City dams. The quantities of water released down the valleys during times of flood were potentially disastrous, and running the stream-control works through the heart of the dam, as had been done at Ashokan and Schoharie, was not a workable option with an earthen dam. To handle the water at Rondout, a huge diversion tunnel was cut through the ledge rock entering the hillside to the south and west about 400 feet upstream from the dam. The rough-cut tunnel is 35 feet in diameter and the finished tunnel is 30 feet in diameter; it was the largest diameter tunnel built to date by the BWS [Fig. 32]. The tunnel was so big that heavy construction equipment could operate freely within the space. Electric-powered shovels (electricity was used because of the exhaust problems pre-

FIG. 32 *Neversink Dam. Construction of the diversion tunnel for stream control. View from stilling basin, framed by outlet portal. May 27, 1942.*

FIG. 33. *Rondout Reservoir. Completed serrated steps of outlet channel stilling basin. November 18, 1939.*

sented by diesel equipment working in a tunnel environment) cleared the blasted rock within the cavernous space. A coffer-dam, constructed of earth, rocks, and steel-sheet piling, temporarily directed the flow of the creek away from the dam site and into the concrete-lined, 2,400-foot-long passage. The diversion tunnel would ultimately become part of the permanent spillway for the dam [Figs. 33, 35].

The body of the dam was built up along the core wall using carefully selected, processed, and graded earthen materials [Fig. 34]. Layers of each earth type were spread to a specified thickness in a specified sequence. These materials, distributed by trucks that opened on the bottom, were compacted to form a dense, impervious core. Compaction is done with a heavily weighted iron roller known as a sheepsfoot, a name derived from the compacting effect of sheep on the land. The roller has projecting steel teeth (or tamping feet) that mechanically compact the soil. The length and spacing of the feet and the weight of the roller are designed to meet the requirements of the particular soil and the desired density. Patterns and spacing were altered according to the soil density specified.[19] Permanent vertical measuring rods were placed within the layers of earth to monitor the compaction of material, the lateral movement, and the long-term settlement of the structure [Fig. 38]. Once the laying and compaction of the earthen materials were complete, the surfaces of the rolled embankments were covered with several feet of broken stone. These rock layers, known as riprap, form a stable shell covering that protects the compacted earthen materials from erosion [Fig. 39]. The upstream slope of the dam was covered in a rubble paving composed of large flat stones laid without mortar. The downstream slope was covered with an additional two feet of earth and one foot of topsoil and seeded with grass. The finished structure rises 200 feet above the original valley floor. It is 2,500 feet long and has a maximum width of 1,400 feet at the base [Fig. 36]. The total height from the base of the core wall to the top of the dam is 390 feet [Fig. 37].

FIG. 34. *Merriman Dam. View along centerline of dam showing the cleared foundation of the upstream cutoff trench and the concrete core wall. Earth compacting operations in progress. August 13, 1946.*

FIG. 35. *Neversink Reservoir. 30-foot high, 40-foot wide, horseshoe-shaped outlet portal of the diversion tunnel showing five completed steps of stilling basin. November 24, 1942.*

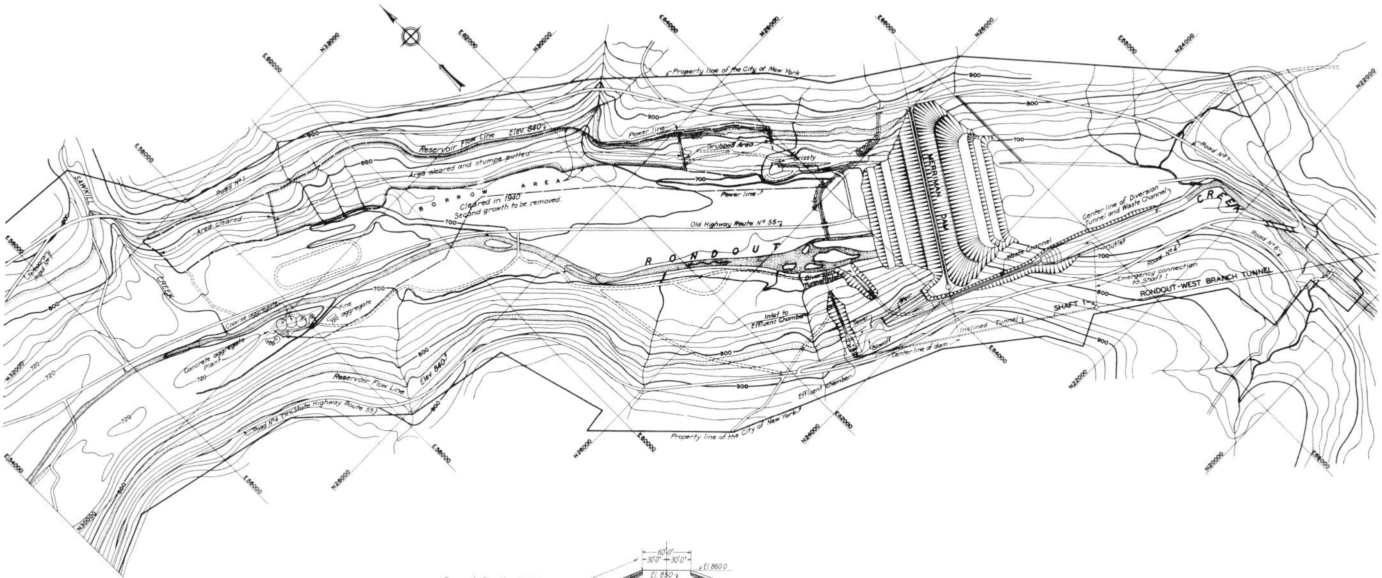

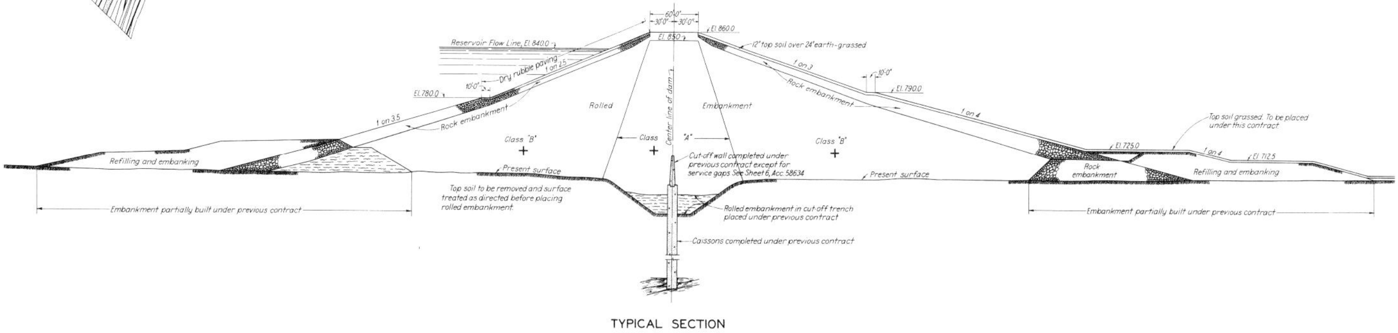

FIG. 36. *Merriman Dam. Contour plan showing completed structure. May 1, 1945.*

FIG. 37. *Merriman Dam. Cross section showing core wall, caissons, and embankment construction of completed structure. May 1, 1945.*

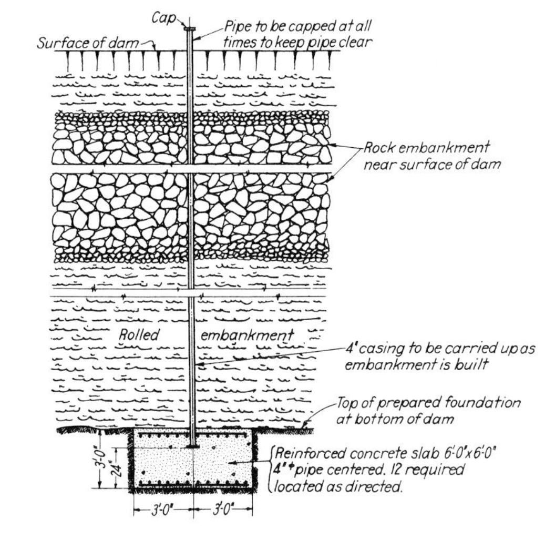

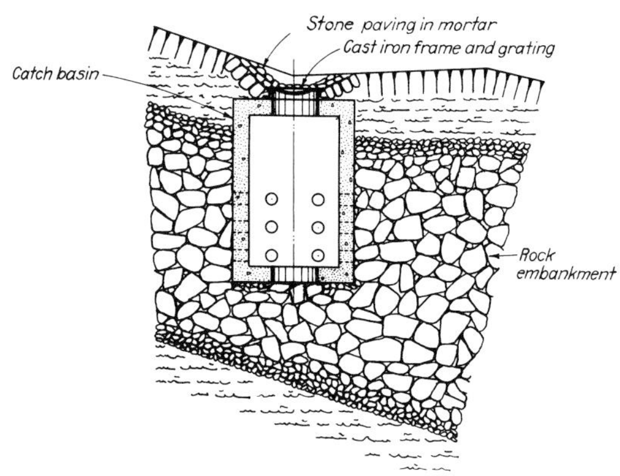

FIG. 38. *Merriman Dam. Typical method of measuring vertical settlement in earthen embankment. May 1, 1945.*

FIG. 39. *Merriman Dam. Typical detail of berm drain at base of embankment. May 1, 1945.*

The dam, started in 1937, took more than fifteen years to complete. Work proceeded well until 1941, when World War II intervened. The flow of workers and materials began to slow as federal restrictions limited access to construction materials and equipment that were deemed critical for national defense. By 1943 all but limited emergency work had been suspended, and contractors were released from their obligations. A small inlet chamber and tunnel were built, which allowed the Rondout Creek to be diverted directly into the first shaft of the Delaware Aqueduct and provided an interim supply of 100 mgd beginning in April 1944. The war shut down the project until the spring of 1946, when work resumed and the basin was cleared and prepared for the reservoir. In May 1951, the temporary connection to the aqueduct was sealed and water began to collect within the reservoir basin. By June water was flowing through the permanent effluent works into the Delaware Aqueduct [Figs. 40, 41]. The dam was completed in 1954, and water first overflowed the top in March 1955.

The Rondout Reservoir became the key to the Delaware chain of structures. It serves as the receiving basin for the waters from three separate Delaware reservoirs and houses the control works that regulate all water entering the Delaware Aqueduct. The dam that created the Rondout Reservoir was initially referred to as the Lackawack Dam, for the small village near the dam site. It was renamed the Merriman Dam in November 1939, in honor of Thaddeus Merriman.

Just a few miles west of the Rondout Creek, over the divide that separates the Hudson River drainage from that of the Delaware River lies the Neversink River. In 1935 preparatory fieldwork began to dam the Neversink River and build a tunnel to connect the valley to the Rondout Reservoir. The original BWS plans called for locating the dam near Curry, New York, to capture the water from about 70 square miles of watershed. But in the relentless quest for more water, the plans were revised and the site moved downstream to its present location at the village of

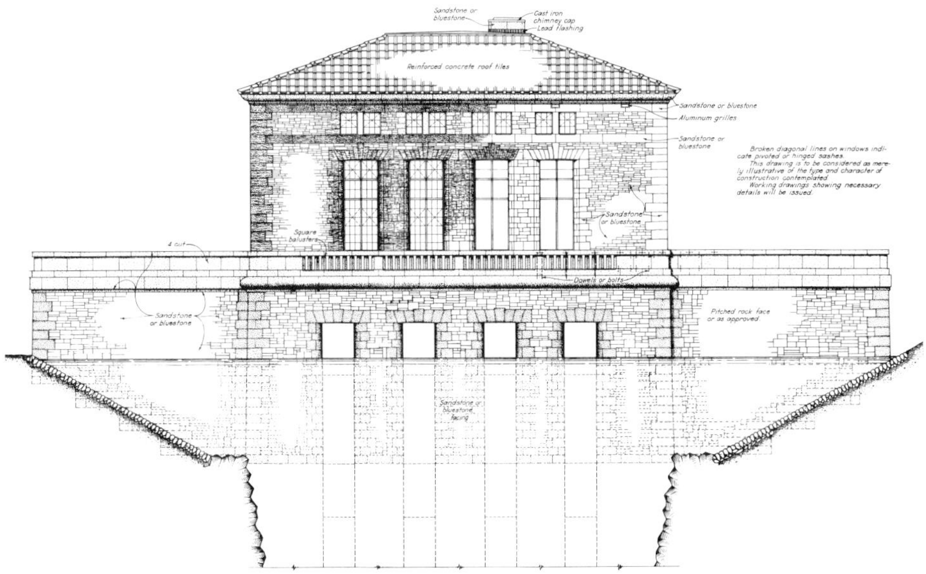

FIG. 40. *Delaware Aqueduct. Rondout effluent chamber. Elevation of superstructure. April 1, 1939.*

FIG. 41. *Delaware Aqueduct. Rondout effluent chamber. Four waterways under construction. July 21, 1949.*

FIG. 42. *Neversink Dam. Movable three-level work platform in diversion tunnel. July 14, 1942.*

FIG. 43. *Neversink Dam. Excavation of cutoff trench and construction of concrete core wall. Cutoff trench and caissons under construction on right. August 22, 1947.*

FIG. 44. *Neversink Dam. Contour plan showing completed structure. January 2, 1948.*

FIG. 45. *Downsville Dam. Formwork in outlet portal of diversion tunnel. May 13, 1949.*

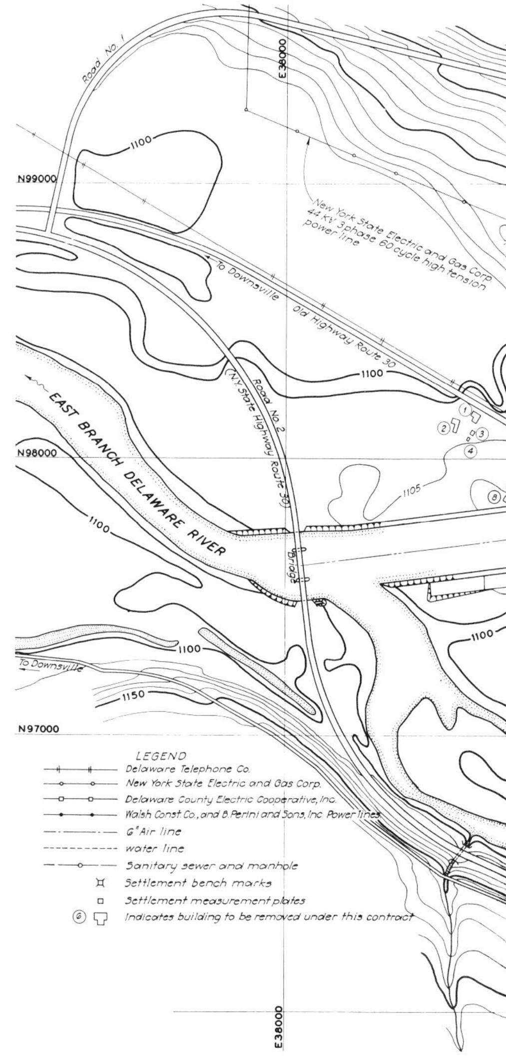

Neversink. The move increased the watershed to 93 square miles and, because of the unique shape of the valley, the new reservoir site could hold three times as much water as the Curry site.

The BWS started construction on the Neversink Reservoir in 1941. The first contract called for building the stream-control works—the diversion tunnel and the cofferdam—across the valley floor as well as two exploratory caissons [Fig. 42]. The design of the dam, stream control, and connecting tunnels is very similar to those at Rondout. The majority of the stream-control work was complete by the end of 1942.

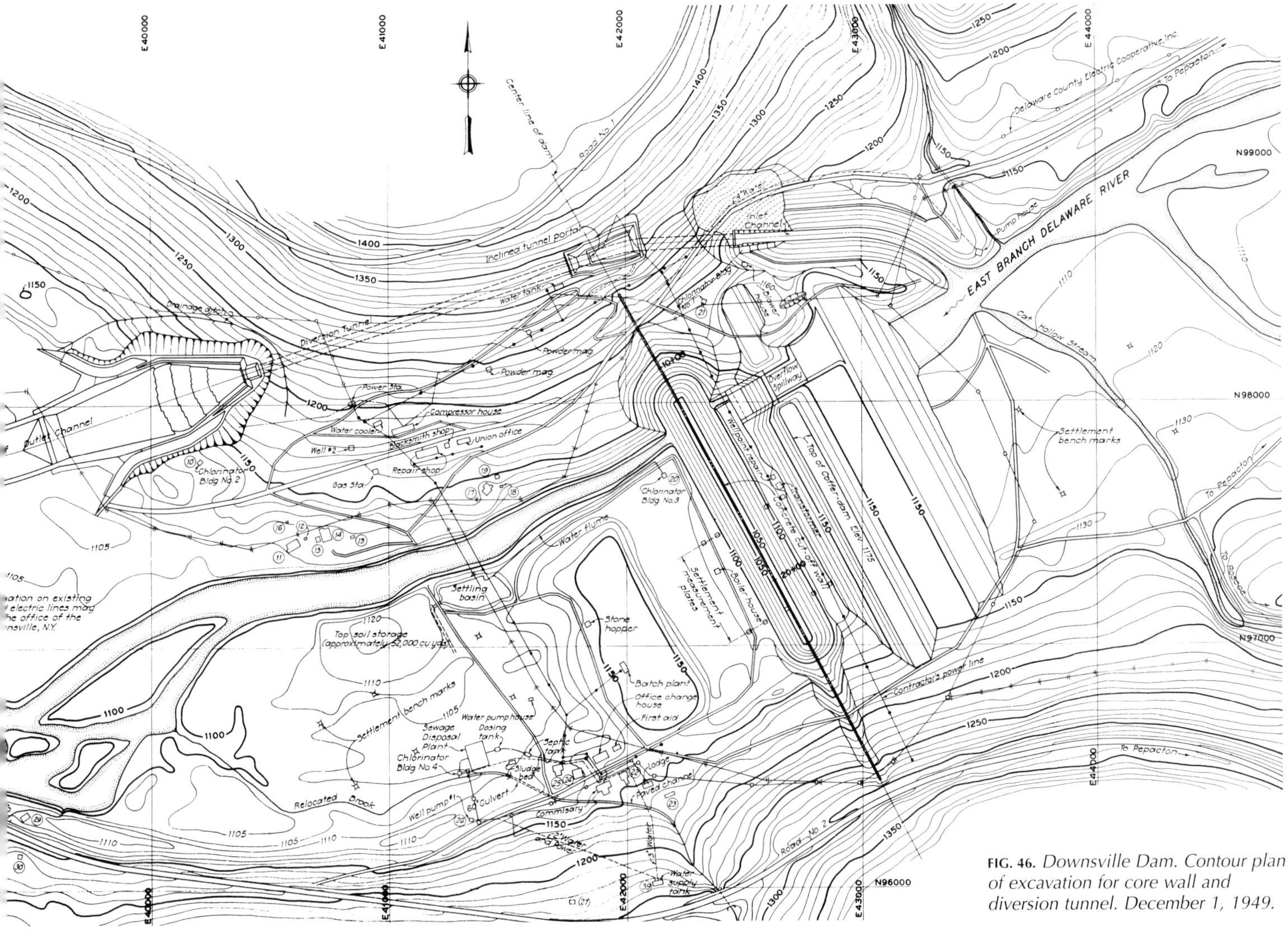

FIG. 46. *Downsville Dam. Contour plan of excavation for core wall and diversion tunnel. December 1, 1949.*

In November 1943 all work on the Neversink site was suspended because of restrictions on materials. Work on the 2,820-foot-long, 195-foot-high dam resumed in July 1946 when the construction of the cutoff trench and core wall began [Fig. 43]. The core wall was built on a line of twenty-one caissons sunk to a maximum depth of 200 feet. The impact of mechanization on dam building was evident by 1950, when work on the Neversink Dam was at its height. The contractor employed a maximum of 507 workers, compared to the thousands employed on the Catskill system dams.[20]

The Neversink River was dammed to create the second reservoir of the Delaware water system, but it was the first reservoir to actually tap Delaware River water [Fig. 44]. The diversion tunnel was closed and water began to accumulate in May 1951. The reservoir first overflowed the spillway in June 1953. The tunnel between the Neversink and the Rondout diverted water that once flowed three hundred miles south past Philadelphia and on to the Delaware Bay. Now it was redirected for use by New York City.

While the Rondout and Neversink Reservoirs were filling, the BWS pushed forward on the second stage of the Delaware watershed development. The East Branch River, a major tributary to the headwaters of the Delaware River, lies twenty-five miles to north and west of the Rondout Reservoir. The next reservoir to be built, the Pepacton, was created by building the 2,450-foot-long, 200-foot high Downsville Dam. Situated about one mile upriver from the village of Downsville, Pepacton was the largest reservoir built to date by the BWS. When filled to capacity it holds water equal to the combined volume of the Schoharie and Ashokan Reservoirs. The 18½-mile-long

Pepacton Reservoir holds 147 billion gallons of water collected from 372 square miles of the East Branch watershed. The flow line is 1,280 feet above sea level. The Pepacton Reservoir is connected to the Rondout Reservoir via a twenty-five-mile-long, concrete-lined pressure tunnel driven through rock, with a finished diameter of 11 feet, 4 inches. The Downsville Dam, like the two previous Delaware system dams, is a compacted earthen embankment with a concrete core wall.

Work on the stream control, diversion tunnels, and cutoff wall began in 1947. A diversion dike and cofferdam were built on the East Branch River upstream from the dam, along with a 2,155-foot-long, 40-foot diameter concrete-lined diversion tunnel [Figs. 45, 46]. The structures were completed and the river was turned off its natural course into the tunnel in 1949. Subgrade conditions on the East Branch site were less severe than those at Rondout and Neversink, and the elaborate caisson operations that were needed at those sites for the core wall foundation were not required. Instead, the foundation for the core wall was excavated after steel sheet piling was driven down to the bedrock along both sides of the trench. The sheet piling—a heavy gauge iron retaining wall system—preserved the position of the earth walls during the construction operations. One caisson was sunk at the low point along the line of the cutoff wall to facilitate the pumping of inflowing water into the trench during excavation. The deepest soil cover over the bedrock was about 110 feet thick. The 2,300-foot-long concrete core wall was set in a groove cut into the bedrock and rose out of the bedrock foundation. The surrounding bedrock was grouted solid by injecting concrete into holes drilled along the base of the wall. The cutoff wall was completed in March 1950, and the sheeted trench was backfilled and tamped to the proper density.[21] The same contract included provisions for removing 197 buildings within the taking area.

Work on the embankments began in February 1950. Two mechanized facilities for screening and processing raw earth were erected upstream from the dam site. Located adjacent to burrow areas where the materials for the embankments were mined, these heavy-duty plants, called "grizzlies," were capable of processing 1,000 cubic yards of material per hour. Construction of the rolled embankments began in July 1950. Soils for compaction were graded for use on two types of embankments. The more impervious mix contained a high level of clay and was used for the core of the dam structure, and a second classification, higher in rock content, was used for a protective shell layer that completely covered the core of the dam. The processed earth was taken by bottom-dumping Euclids (the earth movers that did much of the work on the Delaware dams) and placed in windrows (rows of flattened dirt) parallel to the centerline of the cutoff wall. These mounds of earth were then bulldozed to a uniform 7-inch thickness and 10-foot width. Each bed of earth was required to be passed over fourteen times by the 21,000-pound sheepsfoot roller for initial compaction. After the first compaction, the rollers were filled with water to increase the weight and rolled over the material for the balance of the compacting process until the specified density was reached. Successive layers of earth were built up until the desired height, density, and slope of the dam were achieved. At intervals along the embankments, drains were built into the earthwork to minimize erosion by intercepting and diverting rainwater runoff. As with other earthen dams in the system, vertical measuring rods, called settlement monuments, were placed within the embankments in order to measure settlement and lateral movement of the dam (by creep or thrust) that might result from the weight of the impounded water. Measuring and monitoring devices called piezometers determined the hydrostatic pressure and water percolation rates within the dam and foundation.[22]

The most notable architectural feature of the Downsville Dam is a long, curvilinear overflow spillway on the northern side of the valley [Fig. 47]. The overflow waste weir has a gentle S-shaped profile similar to the cross section of the Old Croton Dam. The curve of the profile directs the water and diminishes its scouring impact on the base of the spillway. The waters flow into a spillway channel cut into the natural rock of the valley wall. The walls of the channel are lined with monumental 3-foot-thick blocks of rough-cut granite quarried in Vermont, which protect the walls from the violent flow of floodwaters [Fig. 48]. The spillway leads into an inclined concrete-lined tunnel that connects to a portion of the original stream diversion tunnel (incorporated into the permanent spillway system). At the downstream outlet of the tunnel the water enters a huge stilling basin where 10-foot-high concrete steps absorb the energy of the discharging water, slowing down the flow before allowing it back into the natural channel of the river. The stilling basin is about 2,000 feet downstream from the dam, clearing the toe of the earthen structure, and protecting it from any potential flood danger.

Surveys for a tunnel to connect the East Branch to the Rondout Reservoir and the main Delaware aqueduct began in 1937, but the World War II–related interruptions that slowed the completion of the Rondout and Neversink Reservoirs also affected this project, and construction did not commence until August 1949. Once underway, the 25-mile-long, 11-foot-diameter concrete-lined tunnel took six years to complete. The structure has a capacity of 700 mgd. Local geology presented a special construction problem: "The rock that was encountered varied from grey and red shale to a thinly laminated sandstone. Joints ran in different directions with bedding horizontal. Rock was structurally weak and had a tendency to spall off in flat slabs."[23] The roof of the tunnel was supported either by gunite (a sprayable concrete mixture that secured smaller loose pieces), arch rib supports, suspension roof supports (where expansion bolts are set into the rock), or a combination of these methods. Wherever possible, the preferred technique was to spray the rock of the upper portion of the tunnel with gunite. In cases where the rock was

FIG. 47. *Pepacton Reservoir. Downsville Dam. Construction of the overflow waste weir and the spillway channel. July 31, 1952.*

severely degenerated and the gunite application was not sufficient to secure the tunnel, a steel frame was used. The tunnels were also grouted by pumping a loose slurry of concrete into weakened portions of the rock walls. Work on the East Delaware Tunnel went on 24 hours a day. The concrete lining was completed in November 1954, when water was first directed down to the Rondout Reservoir and on to New York City.

The third stage of the Delaware project was realized with the building of one more earthen dam. This dam is located twenty miles west of the Pepacton Reservoir near the town of Cannonsville, in the valley of the West Branch of the Delaware River. Because the water taken from the Neversink and the East Branch was approaching the limits of the Delaware water allocation, New York City had to return to the Supreme Court to seek

FIG. 48. *Pepacton Reservoir. Downsville Dam. Overflow Spillway. Granite stone facing on the north wall of the weir channel under construction. September 7, 1951.*

a modification of the original decree. This battle was protracted; the project was bitterly opposed by those who argued that the western Catskills had given up too much already for the city and that they should not have to forfeit yet another valley to its thirst. But New York again prevailed, and the 1931 court decision was amended. In 1954 the city won the right to take up to 800 mgd from the Delaware River system as long as it maintained specified minimum stream flows downriver. The new reservoir at Cannonsville had the storage capacity to meet those requirements and bring the city a new source of water.

Like the Rondout, Neversink, and Pepacton Dams, the Cannonsville Dam is an earthen embankment, but just as construction advances twenty years earlier led to the revision of the core wall concept (allowing a drastic reduction in overall height), later advances allowed the solid core wall to be eliminated altogether.

Engineers determined that soils alone, if properly processed, could be compacted into an impervious core. The dam still required a cutoff trench cut into the bedrock to prevent the impounded water from passing under the structure, but instead of building a wall out of the ground to render the dam watertight, the trench was packed with layers of highly impervious soils. Other features of the Cannonsville Dam were similar to its Delaware cousins. During construction the water was diverted through a diversion tunnel cut through the valley side. The spillways were lined with granite, and the embankments were covered with stone on the reservoir side and seeded with grass on the downstream slope.

The BWS built one more aqueduct in the western Catskills: the 44-mile long, 11-foot-4-inch diameter West Branch Tunnel. Unlike the Catskill system, where the most remote reservoir was at the highest elevation and water flowed downhill into the chain of structures, the Delaware reservoirs are not located at decreasing elevations and so cannot be connected by a continuous tunnel system. The Rondout Reservoir is at elevation +860 and the Neversink Reservoirs at elevation +1,440, but the Pepacton Reservoir is at elevation +1,279. Cannonsville was built so far down the West Branch drainage (at elevation +1,151) that it was downhill from both reservoirs. Because of this elevational relationship, each reservoir required its own aqueduct to get back to Rondout, where it joined up to the main Delaware Aqueduct. The West Branch Tunnel runs roughly parallel to the East Branch Tunnel, and they both lead to the Rondout Reservoir. Multiple parallel aqueducts may seem inefficient, but the reason was simple: more water. The farther down a watershed a dam is placed, the greater the area from which water can be collected. New York was not going to give up any of the Delaware water it could possibly capture.

The Cannonsville Dam was the last structure built to impound water for New York City's use. Completed in 1965, the 204-foot-high structure holds up to 95 billion gallons of water and first filled to capacity in 1968. By the time Cannonsville was complete, New York City had been building waterworks for 131 years. During that period the city constructed 315 miles of aqueduct and primary tunnel, 22 dams and storage reservoirs covering 31,949 acres, five distributing and balancing reservoirs adding 282 acres, numerous headworks, control chambers, siphons, aerators, chlorination facilities, and 6,181 miles of water mains to distribute the water within the city. The Delaware collection, storage, and primary delivery system supplies the majority of the water used by the region's nine million people.

In 1951 New York City's water usage (including suburban customers) averaged 1.07 billion gallons per day. By 1960 that figure had risen to 1.26 billion gallons per day.[24] In ten years consumption increased by a volume of water greater than all the water supplied annually by the Croton system. During the mid-1960s a sustained drought in the northeast depleted reservoir levels to all-time lows, and the disparity between supply and consumption became more evident. Despite unrelenting engineering and construction efforts and despite the huge influx of water from the final phase of the Delaware system, the historical pattern was repeating. The system was barely able to supply the water the city needed.

But the era of aqueduct and reservoir building was over, and the focus of the water supply community had to shift. In its 1965 Annual Report, the Board of Water Supply outlined options for increasing the supply:

> Considering the unprecedented drought during the past five years and the uncertainty as to how much longer this condition will remain, it is advisable to seek new sources. Consideration is being given to the practicability of increasing the capacity of the Hudson River Pumping Plant: and skimming off the flood water flows of the Beaver Kill and Willowemoc Creek and directing the water into the East Delaware and West Delaware Tunnels for storage in Rondout, Pepacton, and Cannonsville Reservoirs . . . other studies are underway on . . . obtaining a water supply from the Hudson River in the vicinity of Hyde Park, and the feasibility of the production of water in connection with the operation of Consolidated Edison's atomic plant at Indian Point.[25]

After the drought ended and planning for water became less of a priority, these schemes were set aside. Work turned to the practical business of addressing weaknesses within the existing system. The Hudson River Pumping Plant, built in 1950 and in disrepair, was reconstructed so that an emergency supply of 100 mgd could be drawn directly from the river and added to the Delaware Aqueduct at the Hudson crossing. Work was also initiated to improve system connections to Staten Island. A deep-rock cross-harbor tunnel was designed and constructed between 1962 and 1970, and the 100-million-gallon Silver Lake Park Storage Tanks, built to replace the original reservoir, were also completed by 1970.

The 1965 Annual Report also detailed the potential vulnerability of the local distribution aqueducts attributable to their age and limited capacity. The BWS embarked on studies and cost estimates for a complex of city tunnels, valve chambers, and controlling works known as City Tunnel #3, a project that is still underway and has become one of the largest and most expensive construction efforts in the world.

The thirty years following the completion of the primary system have been dedicated to protecting the quality of the water at the source, eliminating waste, and reducing consumption. This era of conservation and environmental protection recognized that water is a finite resource and that good management could be a more efficient and cost effective approach to increasing supply than construction of new infrastructure. Future needs will bring future projects. The great structures will age and require rebuilding, and civil and structural engineering ingenuity will be tested once again. But the far greater challenge will be engineering of a different sort: environmental engineering and ecological servicing that will protect and improve this most precious resource.

THE HOGHOUSE MAN

It's John, John O'Flaherty. A lot of people will tell you Flaherty, but no, that's not it. It's O - Flaherty. That would be O apostrophe F.L.A.H.E.R.T.Y. - O'Flaherty. I've been a miner for over some 40 years. Started up in Boston, working on their harbor tunnel. My brother was killed standing almost right next to me. I guess that would be a high point or a low point . . . depending on how you're speaking of it. That's when I came down to New York City, hooked up with Local 147. Been in this racket ever since. It's a good union. A good union. Well, it's been good to me. We do all the work in construction that's under 50 feet. Anything below 50 feet that's where we come in. Tunnel workers. Miners. Sandhogs . . . that's the nickname for us. If it was ever derogatory, we've lost track of that a long time ago. The Brooklyn Bridge—she's our job. The caissons, I mean. All the tunnels—Holland, Brooklyn, Battery. The subways, most of the sewers. That's all our work. It's very noisy down there. Your joints are aching from the damp. If you make it to sixty, you're doing all right.

Excerpt from *City Water Tunnel #3*,
written and performed by Marty Pottenger

THE POST-DELAWARE ERA

ALBERT F. APPLETON

As construction of the Delaware water system drew to completion in the mid-1960s, the Board of Water Supply (BWS) proposed construction of a third deep-rock water tunnel to link the Kensico Reservoir to the Hill View Reservoir, and Hill View to the city. This proposal had two justifications: the tunnel was needed as a backup water supply and as a supplement to the two existing city water tunnels. The board had projected that the population would reach 10 million by the year 2000. Only a new water tunnel could serve the estimated demand. Equally important, both existing tunnels had been in continuous use since they had been opened and the shutoff valves had never been operated. City Tunnel #1, completed in 1917, was nearly fifty years old, and City Tunnel #2, completed in 1936, was thirty years old. The Board of Water Supply feared that if the valves were closed in an emergency, they might not reopen, and water supply to vast areas of the city would be cut off.

Over the next several years, a series of studies concluded that despite some opposing views, a third water tunnel was needed, but that the proposed diameter, an unprecedented 27 feet, was significantly oversized. The maximum diameter was reduced to 24 feet, and the project was broken into four construction phases. When completed, the concrete-lined pressure tunnel would extend nearly 60 miles and reach depths of 800 feet. Phase one, a 13-mile stretch from the Hill View Reservoir in Yonkers down through the Bronx and into Manhattan, would roughly parallel the upper route of City Tunnel #1 but would turn left at Central Park and cross under the East River into Queens. Phase one would also include the construction of subsurface valve chambers below Van Cortlandt Park, Central Park, and Roosevelt Island [Figs. 1, 2]. Phase two would consist of two distinct sections: a 5.5-mile-segment extending from Maspeth, Queens, to Red Hook, Brooklyn, and connecting with the Richmond Tunnel, which brings water to Staten Island, and a 9-mile loop around the perimeter of lower Manhattan. Phase three would connect the Kensico Reservoir in Westchester to the new Van Cortlandt Valve Chamber in the Bronx via a 16-mile-long tunnel. Phase four would be a tunnel from the Van Cortlandt Park Valve Chamber to the eastern parts of the Bronx and Queens, required on the basis of projected population growth in those parts of the city, their remoteness from the existing tunnels, and to some extent, the expectation that the groundwater in eastern Queens would eventually become too polluted to use. In 1970 construction of the legendary City Tunnel #3—the largest capital improvement project in the history of New York City—began [Fig. 3].

One significant design departure from past tunnel projects was the addition of subsurface chambers to house the valves and meters that would measure, direct, and control water flow. The valves for City Tunnels #1 and #2 had been placed in line with the tunnels, hundreds of feet underground. For City Tunnel #3, valve chambers were incorporated into the design. These would be accessible for maintenance and repair, and segments of the tunnel could be connected as they were completed without having to drain the tunnel or take parts of it out of service [Figs. 14, 15, 19].

Technologically, the tasks for the team that began to dig the City Tunnel #3 differed only slightly from those involved in digging Tunnels #1 and #2 since conventional blasting methods are still used today [Figs. 4, 5, 6]. It was not until the late 1980s and early 1990s that the transition to tunnel boring machines (TBMs) took place. TBMs (also known as "moles") utilize rotating steel cutting heads rather than explosives to remove rock and they are safer and less costly since they carve out a smooth tunnel, which requires less concrete to finish [Fig. 10]. Socially, however, the context could not have been more different. Wages, working

FIG. 1. *City Tunnel #3. Van Cortlandt valve chamber. View of valves during assembly. Bronx, New York. May 8, 1986.*

FIG. 2. *City Tunnel #3. Central Park valve chamber. Formwork in place for lining the chamber arch with concrete. July 17, 1984.*

conditions, and safety levels had advanced enormously. The work forces on City Tunnels #1 and #2 had been largely immigrant laborers who were cheap, docile, desperate for work, and unassertive regarding their safety and rights. On City Tunnel #3, the workforce was (and is) the descendants of those workers. Known as "sandhogs," they are organized and protected by economically and politically powerful unions and safety standards. Two hundred eighty-nine men died in the seven years it took to construct both City Tunnel #1 and the Catskill Aqueduct. By contrast, twenty-four men have died in the more than thirty years of active construction on City Tunnel #3.

The financial environment had changed as well. City Tunnels #1 and #2 were driven forward by specific fears of immediate water shortages that only their construction could solve, and cost was not questioned. There is no such immediate need for the City Tunnel #3. It is being built in anticipation of future need and for backup in case of a catastrophic and completely unpredictable service disruption of City Tunnels #1 and #2. When such threats do not materialize, the urgency of preparing for them dissipates.

Beginning with the construction of the Croton Aqueduct system in the 1830s, the city financed its water projects with money raised by issuing municipal bonds. By the 1970s, virtually all of the city's infrastructure improvements were financed by this method, which had evolved into a sophisticated capital budget structure that used municipal bonds backed by a pledge on general city revenues. Water supply projects enjoyed certain privileged positions within that system, but essentially water infrastructure was paid for out of general city revenues and had to compete with other capital projects.

FIG. 3. *City Tunnel #3. Survey car with feeler gauges for measuring dimension of excavated rock. September 18, 1980.*

The 1975 fiscal crisis left New York City bankrupt. It was unable to issue any new bonds for infrastructure projects because it could not afford any increases in its debt service obligations. By the end of 1975, virtually all capital expenditures for infrastructure, including City Tunnel #3, were frozen. Lean years followed, but the city slowly rebuilt its finances. In the late 1970s, inflation and steep interest rates (which peaked at 21 percent in 1980) created brutal financial obstacles that prevented the use of municipal bonds. Thus in 1978, when Francis X. McArdle, Mayor Ed Koch's newly appointed commissioner of the Department of Environmental Protection, boldly proposed to commit $100 million to restart the City Tunnel # 3 project, a bitter internal debate followed. Many in city government argued that the proposed funds would be better spent on sewer-system repairs rather than on a project with no immediate benefits. McArdle prevailed and the tunnel work was restarted, but its long-term claim on scarce city resources remained fragile and uncertain. Consequently, McArdle, Mark Page (then the Budget Bureau's municipal finance coordinator), and others realized they needed a new way to fund not only the tunnel, but also the capital programs to modernize the water mains and sewers and to build new sewage treatment facilities required by the Clean Water Act.

Careful study of water and sewer utilities across the United States, and of the finance infrastructure of New York State, suggested that a self-funding authority was the most promising approach. In 1984 the city's water institutions underwent their most significant restructuring since the creation of the Board of Water Supply in 1905. The board was abolished and an interconnected three-part entity was created. It consisted of two new agencies, the

FIG. 4. *City Tunnel #3. Placing and wiring explosives at tunnel heading. January 27, 1984.*

FIG. 5. *City Tunnel #3. Tight spots loaded with dynamite to complete rough excavation of rock tunnel. January 22, 1981.*

New York City Water Finance Authority and the New York City Water Board, and a third, the Department of Environmental Protection, which was restructured to formally consolidate the management of the water and sewer systems under a single organization. The DEP's mission is to "protect and maintain the City's water supply for nine million city and upstate residents, and treat waste water and sewage at 23 treatment plants." The financial reforms associated with this new structure had an enormous impact on the future evolution of the water system, and the city was relieved of any further financial responsibility for both capital and operating costs. Henceforth, water and sewer rates were legally required to be set by the Water Board, and at a level that would pay all of the system costs with annual adjustments to cover increases. The City of New York would no longer issue municipal bonds to pay for City Tunnel #3 and other water and sewer infrastructure. Nor would the city be responsible for the repayment of Water Finance Authority bonds out of general city revenues. Instead, the Water Authority would pay the debt service from revenue generated from water and sewer rates.

This sensible transfer of funding from general municipal revenues to user payments immediately ended the financial crisis for water and sewer infrastructure and solidified the financial future of the City Tunnel #3. It gave the DEP the luxury—as long as it maintained reasonable cost controls—to plan a long-term capital investment and system management program with reasonable certainty that it could be funded.

The benefits were equally dramatic for the city. It diversified its borrowing structure, gave the water and sewer system economically rational cost-based financing, and removed a major competitor for the city's scarce capital budget resources. By keeping the DEP as the system operator, while having a separate financing authority to issue water and sewer bonds, the city managed to obtain the advantages of the authority model for funding without losing political accountability over water and sewer system operations.

FIG. 6. *City Tunnel #3. Dynamite loading operations at tunnel heading. February 29, 1984.*

Construction on the City Tunnel #3 moved steadily forward, and the DEP churned out a flood of new capital programs, both to meet the mandates of the Federal Clean Water Act and from a commendable desire to make up for years of underinvestment in water mains and trunk sewers. But by the mid-1980s, the city was again facing a potential water supply crisis. The city was using on average over 1.5 billion gallons of water daily. Safe yield for the water system (defined as the average daily yield during the driest year on record—in the city's case, 1964) was calculated at 1.29 billion gallons daily. Extended droughts in 1985 and 1988–89 (in February 1989 reservoirs dropped to an all-time low of 27 percent of capacity) drove this vulnerability home.

In response, the city's water supply planners recommended building a new facility on the Hudson River at Peekskill, the site of the Chelsea Pump Station, that would utilize water from the river. Capacity recommendations ranged from 300 million gallons to one billion gallons per day. The estimated cost of this new waterworks, including a filtration plant to purify the river water and new underground conveyance facilities, approached $2 billion.

This recommendation was a definitive break from 150 years of New York City water supply tradition. The city had always been proud of its history of going far afield to tap pristine water sources that enabled

FIG. 7. *City Tunnel #3. Van Cortlandt valve chamber. Grouting operations inside clad steel pipe. October 30, 1984.*

it to provide superb drinking water from surface water sources without filtration. Instead of seeking out a new clean source of water, the city proposed using polluted water from the Hudson and spending money to purify it, rather than investing in the long distance transmission of pure water. Had this plan gone forward, it would have profoundly altered the ethos of the water system.

The Hudson River waterworks proposal provoked a firestorm of opposition from Hudson Valley communities and environmentalists. They attacked the facility as unnecessary and denounced the city as a wasteful and profligate water user that did not even meter its residential buildings. These opponents demanded that the DEP consider water conservation as an alternative. They mobilized widespread Hudson Valley support and voiced their opposition to the state government, which would have to authorize any new waterworks.

In 1988, under pressure from the New York State Department of Environmental Conservation (DEC), the city initiated a ten-year program to meter all residential buildings, which had always paid a "frontage charge" for water—a fixed annual amount based on building characteristics that was supposed to approximate water usage. Now building owners had to pay for actual water usage and at a time when water and sewer rates were skyrocketing. The stock market crash of 1987 and the subsequent recession had dealt a major blow to real estate values and income. Now there was neither a ceiling nor a predictability regarding water and sewer costs. Moreover, analysis soon showed that the buildings that would have the highest cost increases were moderate- and lower-income multifamily housing. At the end of 1989, the real estate industry concluded that out-of-control rate increases, mandatory cost recovery, and water metering were a potentially fatal combination that had to be avoided at any price.

But another crisis was in the making, and this one was even worse. New York City had always been proud of its position as one of the largest unfiltered urban water suppliers in the world. Due to the generally pristine character of its remote watersheds and the ability of its enormous reservoir system to naturally purify the water, the city had avoided the need for filtration. By the 1980s, that no longer appeared good enough.

Filtration is a purifying process in which source water is passed through a filter medium, most often fine sand, to remove contaminants and bacterial agents. By the early twentieth century, filtration had been adopted as one of the foundation stones of modern American public health and, combined with chlorination, it had played the central role in largely eliminating waterborne enteric disease from American urban life. In the 1980s, a series of giardia-related incidents occurred in unfiltered American water systems, and new, troubling research about the extent of cryptosporidium led public health experts to doubt that the safety of any unfiltered water system could be ensured. Meanwhile, the growth of exurban development in the Catskills, combined with the industrialization of Catskill dairy farming, began to change pollution patterns in the watershed. The city, historically fearful that an active program of watershed protection would trigger local antagonisms, chose to ignore these warnings, relying instead on the size of its watersheds and the natural purifying abilities of its reservoirs to harmlessly dilute any pollution increases. More and more, this course appeared to public health experts to be dangerously outdated and culminated in a massive national debate on safe drinking water (in which New York City was prominently featured) that took place in Congress in 1985 and 1986. Convinced by the public health lobby that only filtration could guarantee drinking water safety, the Environmental Protection Agency (EPA) introduced legislation that required all water systems to be filtered by 1993. New York City and other major cities with unfiltered water systems (principally Boston, Syracuse, Seattle, Portland, and San Francisco) fought this legislation, arguing that filtration should be based on performance criteria. The EPA countered that filtration was not only remedial but

FIG. 8. *City Tunnel #3. Sections of clad steel pipe in Hill View Tunnel prior to being moved into chamber. September 14, 1984.*

FIG. 9. *City Tunnel #3. Sandhog in fully concreted tunnel before removal of temporary rail tracks and utilities. September 15, 1978.*

FIG. 10. *City Tunnel #3. Steel cutting heads on Tunnel Boring Machine (also referred to as the "Mole" or "TBM") in maintenance tunnel of the Van Cortlandt valve chamber. May 2, 1994.*

FIG. 11. *City Tunnel #3. View around the only curve in the 521B section of the tunnel. February 28, 1986.*

also preventive, and given the lead times necessary to construct filtration works, performance criteria would not trigger a filtration requirement until it was too late.

The result was the 1986 Safe Drinking Water Act, which declared filtration the standard for all water systems but allowed exceptions if a city could demonstrate that its water had always met objective quality standards and that there was a program of watershed protection that would effectively prevent any future deterioration of water quality. Surface water systems were given five years, until November 1991, to decide whether to filter or apply for an exemption to the new regulations. All filtration facilities had to be completed by 1993.

Prompted by this legislation, New York City added more than two hundred people to its water-sampling and laboratory operations between 1986 and 1989, but did little else to prepare for an exemption application. Environmentalists, who were concerned about both drinking water quality and about the environmental impacts of filtration on the Croton and Catskill watersheds, grew increasingly frustrated. Led by Robert F. Kennedy Jr., they began publicly attacking the city for a passive approach to watershed protection, a lack of environmental enforcement in the watersheds, a failure to update the obsolete 1953 watershed regulations, and an unwillingness to purchase environmentally critical land parcels.

Public health advocates feared that the financial costs of filtration would lead New York City to avoid its implementation. A filtration plant large enough to handle the water from the Catskill reservoirs with suitable backup would need a design capacity of over two billion gallons of water a day. An estimate of the minimum cost in 1990 of such a plant was $4 billion; maximum estimates went as high as $8 billion. In addition, the amount of filter material that needed to be used and replaced was so enormous that the annual operating cost for the plant would be at least $200 million. Filtration could easily double New York's water and sewer rates, and a revolt against rate increases would have dire consequences for the

FIG. 12. *City Tunnel #3. Collapsible steel formwork in position on the completed concrete invert. August 30, 1985.*

financing and management of the water system as well as for New York City's economy.

As the 1990s and the new mayoralty of David Dinkins began, the DEP faced a number of major problems: runaway water and sewer rates; the potential requirement to filter all its upstate water sources; escalating water usage that could lead to future water shortages; limited sewage treatment capacity that could curtail new construction; an impossibly short deadline for ending ocean dumping of sewage sludge; demands to control nitrogen discharges into Long Island Sound; and demands to control combined sewage overflows (or CSOs). The city sewer system, which dates from the nineteenth century, is a combined sewer that carries both human wastewater and storm water runoff. When the amount of liquid exceeds sewer capacity during heavy rainfall, the excess is discharged directly into the rivers. Traditional solutions for these problems, such as construction of new facilities, would cost well over $20 billion, and operating costs could total another $400 million annually. These were financially impossible numbers. Strategies were needed that would solve these problems in an environmentally acceptable but affordable manner.

Surprisingly, given the urgency of the watershed protection issue, the first decisive move came through water conservation. In the spring of 1990, the DEC publicly questioned whether, without flow

FIG. 13. *City Tunnel #3. Hill View Reservoir. View up Shaft #1B at concrete lining and deck at top of shaft. March 20, 1985.*

controls, it would be able to renew the discharge permit of the Newton Creek sewage-treatment plant, which served the business districts of Manhattan. Until then, the real estate industry had not made the connection between growing water use and the increasing crunch in sewage treatment capacity. The more water a city uses, the more sewage treatment capacity it needs. During the spring and early summer of 1990, the DEP met with members of the real estate industry to explain the differences between the costs of conserving 300 million gallons of water versus the cost of creating 300 million gallons of new supply. The difficulty and unpredictability of new facility construction, especially when it involved siting new sewage treatment plants in New York City neighborhoods, were compared with the relative ease and reliability of a water conservation program.

The decision of the real estate industry to support a water conservation strategy proved to be a historic turning point. After extensive negotiations and refinements, the DEC accepted the DEP's comprehensive water conservation program and ended the threat of a sewer connection moratorium. Two aspects of the conservation program were particularly noteworthy. First, it had a dual focus: to reduce consumer demand and to reduce system losses. System water loss is the water that disappears during operations. Reduction of that loss is pure profit. Unlike reductions in consumer water use, which also reduce system revenue, saving system water costs virtually nothing. Second, it was agreed that water conservation measures that positively impacted low-income housing (in other words, a social criteria) would be factored in.

Between 1990 and 1993 the DEP reduced system water losses significantly. Street pipes were notoriously leaky, so water main replacement was dramatically accelerated. In 1992 it reached an annual total of over one hundred miles. Sonar leak detection programs for street pipes were greatly expanded and, backed up by new staff for immediate remedial action, the loss from leaks was further controlled.

FIG. 14. *City Tunnel #3. Van Cortlandt valve chamber. Installation of lateral pipe. March 1, 1985.*

FIG. 15. *City Tunnel #3. Van Cortlandt valve chamber. Looking from Hill View-Van Cortlandt Tunnel toward 42-inch diameter clad steel nozzle. March 20, 1985.*

Maintenance crews were enhanced and valve settings were checked and rechecked to avoid water loss. The city finally succeeded in developing an unbreakable hydrant lock, which ended the summer spectacle of hydrants spraying tens of millions of gallons of water an hour out onto the pavement. Instead, controlled-flow sprinkler caps were made available at precinct houses and community centers. These measures accounted for an estimated 25 percent of the nearly 300 million gallons a day saved since 1990.

The conservation program for individual water consumers was equally comprehensive. Free water-saving aerating shower and faucet fixtures that use 30 percent less water were distributed. Water audits were offered free of charge to building owners, and a flood of consumer material on water conservation was produced and distributed. The DEP offered consumers educational programs instructing them on ways to save money through wise management of water use.

The ultimate goal was to get the city's average daily water use down to roughly 1,200 mgd, well below the safe yield. Thus, in 1993, the DEP established two new conservation programs that enabled it to reach that target. First, it created an ambitious rebate program as an incentive for buildings to convert to low-flow toilets. This program, at $300 million the city's single largest expenditure for water conservation, was an unqualified success. Between 1993 and 1995, 1.5 million of the city's 4 million toilets were changed to low-flow models. Since the difference between a regular toilet and a low-flow one is more than three gallons per flush, this program alone saved 100 million gallons of water a day and reduced daily use well below the safe yield.

Second, price incentives such as rate-cap programs were established for the affordable-housing industry in exchange for full participation in the city's water conservation programs. In 2005, roughly 20 percent less water per person was used than in 1990, with virtually no impact on the water consumer. Moreover, for an expenditure of about $500 million on water conservation, the city saved about $5 billion in capital costs. A water supply reservoir had been created in the city itself through water conservation. The era of great and heroic waterworks engineering had ended, to be replaced by the wise stewardship of the resource.

With these conservation programs in place, the DEP could turn its attention back to the problem of watershed protection. Effective watershed management requires effective land-use control, and much of the EPA's support for filtration was driven by the assumption that it was difficult, if not impossible, to control American land use.

First, the DEP needed to restore its credibility as a good watershed steward. During 1990 the agency developed a comprehensive program to upgrade the sewage treatment facilities it operated in the watershed areas, reversing decades of non-action with a surge of enforcement, and began attending reviews of development proposals that affected areas in the watershed whenever it saw potentially adverse environmental implications. It also began a long and contentious dialogue with the EPA, whose professional staff was virtually unanimous in favor of ordering the city to filter. Fortunately, the U.S. EPA regional administrator for New York at the time, Constantine Sidamon-Eristoff, took a much more holistic approach to the environment. Once he was convinced of the city's commitment to watershed protection, he became a stalwart champion of the cause in the EPA's internal debates.

By the end of 1990, the DEP's watershed protection effort had been energized, and it appeared that the city was on course to develop a meaningful and credible program. One constant criticism had been the failure to update the 1953 watershed regulations, so in November 1990 the DEP released a first draft of new, modernized regulations. The city was determined to show it was serious about regulating non-point source pollution, particularly agricultural pollution. The draft outlined a series of generic practices for rural farms and homes that caused a firestorm of protest in the agricultural community. Bitter allegations raged that the regulations were designed to drive farmers out of business and depopulate the Catskills so the city could purchase the land cheaply [Fig. 16].

At this critical point, Dave Taylor, a local farmer, staked out his land to show that compliance would deprive him of 45 percent of his productive acreage. He then issued a public challenge to the DEP to see the potential impact of their regulations. More than fifty Catskill residents, local reporters, elected officials, DEP staff members, and a scattering of others converged on Taylor's farm. After a tour of the farm and much discussion, it was clear to the DEP that the regulatory approach should be abandoned.

Immediately thereafter, Dennis Rapp, deputy commissioner of the New York State Department of Agriculture, contacted the DEP. Rapp and his boss, Richard McGuire, shared the perception that the city's commitment to watershed protection might in fact be a major opportunity for Catskill agriculture. They wanted to bring Catskill farmers and the city together, but Rapp wisely recommended that before any negotiations took place there should first be a program of mutual education. The DEP water quality experts briefed the farmers on the city's regulatory obligations, non-point source pollution control, the standards the city had to meet to obtain a filtration avoidance, and what was at stake for the city. In turn, the farmers gave DEP staff an education on farming in the watershed, its practices, problems, economic imperatives, landscape constraints, and past experiences with environmental regulation. Each side realized that the other had legitimate concerns that had to be addressed. Together they concluded that agriculture was inherently watershed friendly, even a preferred land use. Whatever pollution management arrangements were needed, agriculture and the environment had fundamentally common interests, and this recognition gave both sides a mutually beneficial strategic target.

The agricultural community proposed a voluntary farmer-run program of pollution

FIG. 16. *Farmland in New York City watershed. Catskill Mountains.*

control, and over the next year the farmers, in ongoing discussions with the DEP, created such a program. Instead of a series of fixed solutions, each farm would have a customized pollution-control plan developed by the farmer and local farm experts. Moreover, it would be integrated into the farmer's business plan to ensure direct economic benefits as well as relief from environmental compliance. A Watershed Agriculture Council, run by the farmers with DEP and state agency participation, was established, and Cornell University provided outside technical review. In November 1991, the Whole Farm Agreement was signed and the program recruited its first ten model farms. Ultimately, 93 percent of the watershed's farmers joined the program.

Over the years, the Whole Farm program expanded to include a forest management program for farm woodlots, a conservation easement program, local farm marketing, and other farm-friendly features. Much of the program's success was due to Richard Coombe, an Orange County dairy farmer and political leader who became chairman of the Watershed Agricultural Council and who guided it successfully for nearly a decade. Coombe believed that the long-term future of farming depended upon recognition of the role of farmers and farm communities as environmental stewards and that maintaining the vitality of farm landscapes through programs such as Whole Farm was the best way to control exurban sprawl.

Today, the Catskill farm program is internationally regarded as the model for successful urban/rural partnerships and how to effectively address rural non-point source pollution. Its contributions to the maintenance of a viable agricultural landscape in the Catskills are widely envied at a time when agricultural communities are in decline throughout the United States and the world. Its strategy of rewarding farmers for successful environmental stewardship is seen as an essential economic foundation for future farm policy everywhere, and it has significantly shaped what is now called ecosystem services.

Perhaps most important, the creation of the Whole Farm urban/rural partnership was the foundation upon which other Catskill community cooperative programs were built. It established the credibility of the city's goal-oriented partnership approach and completely changed the dynamics of the debate over watershed protection.

In the same month that the Whole Farm Agreement was signed, the New York State Department of Health, in accordance with the requirements of the 1986 Safe Drinking Water Act, formally granted New York City its first filtration avoidance, with a term of two years, suggesting that the city's successes might overcome the deep skepticism about watershed protection.

In 1992 the city moved to consolidate its gains and prepare for the 1993 exemption decision, which it knew would be the decisive one for the future. The era of non-enforcement against watershed polluters was over, and the city built a new staff of inspectors and natural resource experts to review new developments and identify targets for enforcement. Under the leadership of Carolyn Summers, who became the DEP's first assistant commissioner for Natural Resources, a natural resources management strategy for the watershed was developed. Fifty-nine watershed towns in the Catskills banded together to form the Coalition of Towns and began productive negotiations with the city over what was called Whole Community Planning.

These measures were not without opponents; it seemed many had completely missed the point of watershed protection. The ultimate goal was not to avoid filtration, but to guarantee the purity of the water that nine million people—one out of every thirty Americans—drank, and the best guarantee of good water is good environment. Thus, the DEP's watershed program focused both on the reduction of existing pollution loadings and on the creation of safeguards to ensure that pollution did not climb back over time. If the city could accomplish this by the preservation and restoration of the Catskill environment, supported by enhanced water quality research and laboratory testing of Catskill waters—from the tiniest upland rivulet to the giant reservoirs—then it might have such a margin of safety that it would never have to filter. To combat existing pollution, the DEP enacted whole-farm planning, comprehensive sewage treatment facility upgrades, remediation of faulty individual septic systems, aggressive enforcement against pollution hot spots, stream corridor remediation, and the cleanup of the Kensico Reservoir's seasonal pollution problems.

The prevention program included new watershed regulations, purchase of environmentally sensitive lands to keep them from development, comprehensive review of new development and city-provided technical assistance, partnerships with local Catskill communities to identify watershed-compatible programs for local economic development, and natural resource management to enhance the ecological vitality of the watershed.

By the end of 1992, implementation was in full swing, with the important exception of land acquisition. This fueled the doubts of local environmentalists and national water-quality experts about the city's long-term commitment to watershed protection and spurred on many who still felt filtration should be required. In January 1993 the issue came up for final resolution. The EPA decided to assert jurisdiction over the upcoming filtration decision and resolve whether to order the city to filter its water or to allow it to file for exemption. Virtually everyone urged EPA Administrator William Reilly to deny New York the avoidance except Eristoff. His strong support of the city's efforts and the environmental philosophy behind them (as exemplified by the farm program), helped convince Reilly that the city was creating an effective watershed protection program. Reilly confirmed the exemption.

That year, 1993, saw the climax of the campaign to establish an effective program of watershed protection. The city finally began its land acquisition program and by the fall, it had reached a conceptual agreement with the Watershed Coalition of Towns on the core elements of a Whole Community Planning program, based upon watershed-compatible economic development. It also mounted a

FIG. 17. *City Tunnel #3. Tunnel interior with finished concrete lining. November 29, 1985.*

successful, non-lethal, goose-and-gull control campaign that ended the seasonal pollution problems in Kensico Reservoir, the last open issue for filtration advocates. On December 31, 1993, the EPA granted New York a new three-year exemption.

In the fall of 1995, a draft upstate/downstate agreement was signed by the city and the watershed communities. Conceptually, little in the agreement had changed. The basic consensus about watershed protection and its scope that had earned the critical first two avoidances was not altered, and the exemption was renewed again at the end of 1996. The last documents that memorialized the upstate/downstate agreement and the new watershed regulations were finalized in the spring of 1997.

The city's watershed program is now mostly a settled and mature policy, with the normal ups and downs of everyday life. The land acquisition still disappoints some advocates, since only about 45,000 acres have been purchased, many of which are not high priority. In the 2002 avoidance, the EPA imposed on the city an obligation to build a $250 million ultraviolet water treatment plant, and this requirement can be viewed as an attempt to placate those who still believe all water systems should be filtered. But basically New York prevailed in creating a successful watershed protection program, which ultimately represents a significant accomplishment in water management.

Though the early 1990s were a pivotal and transformative time for the DEP, with struggles over water conservation and watershed protection, for the most part, the period from 1994 to 2001 was a much less dramatic one for the water system. With the exception of an accelerated push of Phase two to dig two sections of the City Tunnel #3 simultaneously, the construction of the tunnel continued to plod forward during this time, encumbered by contract disputes, issues of appropriate methods for excavation, and stabilization of overburden as well as controversies over shaft sites. Phase one was finally put into service in 1998, twenty-eight years after the project started [Figs. 9, 17].

FIG. 18. *City Tunnel #3. High car work platform on invert rail system. January 17, 1984.*

Two new developments were particularly notable during the years 1994 to 2001. The first was the proposal by Mayor Rudolph Giuliani to privatize the water system. This was not a true privatization, but rather a financial scheme developed by the Budget Bureau to carry out a paper transfer of assets to justify selling $2 billion in Water Authority bonds that would produce an immediate $800 million cash subsidy for the city's capital budget program. The idea of adding $2 billion in water debt that would not produce $1 of benefit for the water system, combined with the institutional ambiguity it would produce around a treasured and vital city asset, produced outrage throughout the civic and financial communities. City Controller Alan Hevesi used the contract-approval powers of his office to first review and then squash the proposal.

The city then reversed gears and bought out the Jamaica Water Supply Company, the last of the privately owned water companies that furnished groundwater to communities in southeastern Queens. This was a long overdue step that completed the transfer of all water service in the city to the DEP and greatly benefited southeastern Queens.

But the more significant development—and controversy—of the late 1990s concerned the decision whether or not to filter the Croton water system. In the early 1990s, when the city first organized its watershed protection program, it decided to apply the program uniformly to the three water systems. It planned to avoid filtering Catskill and Delaware but had to make a final decision whether or not to filter Croton, a course the DEP had long assumed it would have to follow because of suburbanization in the Croton watersheds. In the review that followed, the DEP reluctantly concluded that the Croton watershed was too far gone to avoid filtration. Residential densities were far too high and continuing to grow. Six out of twelve reservoirs were eutrophic in the summer, and water quality standards were being maintained only by careful manipulation of withdrawal regimes. To avoid filtering Croton would require not only watershed protection but also a program of watershed restoration of unknown cost, dimension, and duration. Such a program would probably cost more than filtration. The New York State Department of Health and the EPA made it clear that failure to filter Croton would be viewed as irresponsible and would compromise efforts to avoid filtration in the Catskill watersheds. In April 1992, the DEP reluctantly signed a consent order to build a filtration plant at the Jerome Reservoir but made it clear that the Croton watershed would continue to receive the same protection as the Catskills in order to minimize source-water contamination and stabilize water quality.

Over the next several years, residents who had enjoyed the reservoir as a quiet, tranquil neighborhood resource mounted a growing campaign against the filtration plant. Their goal was to move the plant, not necessarily to reverse the decision to filter Croton. The DEP had concluded the reservoir was the preferred location based on engineering grounds, but the community made a powerful and appealing argument that an established middle-class neighborhood should not be grievously affected by the construction of such a facility in its midst unless there were no other alternative.

In response to these concerns, the site was moved to a location in Van Cortlandt Park. More controversy resulted. Now it was Van Cortlandt Park interests and adjacent neighborhoods that objected. Ultimately, they sued the DEP and blocked the use of the site on the grounds that the project represented an alienation of parkland that only the legislature could authorize.

As the debate continued throughout the 1990s, it began to transmute into an attack on the decision to filter Croton. The problem, as the move to Van Cortlandt Park had shown, was that opposition at one site merely opened up controversy at the next. Local Westchester interests in the Croton watershed also decided to oppose filtration to force the city to be more aggressive in its pursuit of land acquisition and its protection of the Croton watershed.

Filtration advocates and jurisdictions elsewhere assailed the EPA for allowing New York to stall Croton filtration. Public-health advocates were appalled at the idea of not filtering Croton. Given the development densities in the Croton watershed and the character of its suburban landscape, not filtering Croton was simply unacceptable.

Finally, in 2002, the EPA threatened to sue the city unless it settled the issue and finalized a site for the plant. Mayor Michael Bloomberg had appointed Christopher Ward as DEP commissioner. Ward realized that the position of not filtering Croton was indefensible, and he moved forcefully to resolve the issue. Unfortunately, by that time most of the downstate environmental community had begun to see the issue as a choice between watershed protection and filtration; they opposed the state legislation that the DEP and the city had submitted to obtain authorization to place the filter plant in Van Cortlandt Park. The DEP finally prevailed, but only after agreeing to make significant payments for the improvement of Van Cortlandt and other Bronx parks.

In the meantime, ongoing system surveys of the Delaware Aqueduct had identified a probable leak, near Newburgh, New York. Initial information suggested that the leak was not large—less than 10 mgd out of a daily transmission volume that could be as high as 800 million gallons. The leak appeared stable, and the general conclusion was that it had probably occurred relatively early in the life of the aqueduct and then had stabilized, a common occurrence in water tunnels. But over the remainder of the decade, studies of the leak gradually painted a more troubling picture. Estimates of the volume of water lost rose to more than 30 mgd. Fears were expressed that the stability of the leak could not be presumed forever.

In 2002 the DEP inserted a submersible camera vehicle into the tunnel to locate and photograph the leak in order to determine a repair procedure. The project was successful, but the results were inconclusive. Ward initiated a major review of potential repairs. One critical question was whether an attempt to repair the leak would be more dangerous than leaving it undisturbed. Many issues were involved, including how long the city could function without the Delaware Aqueduct and whether shutting it down and eliminating the water pressure against the sides of the tunnel would cause the overlying rock to collapse.

For Ward, the defining concern of his tenure became insuring the reliability of water delivery. He took what steps he could to accelerate the City Tunnel #3, but pre-existing contracts and decisions had largely determined that schedule. More important, he accelerated planning for Phase three of the project—the new delivery tunnel between Kensico and the city—so that construction could start as soon as Phase two was finished. Ward's mission was to educate the public that reliability of water delivery had to be the next great charge of the department. As the cost of meeting the Clean Water Act's mandates began to increase again after being under control for a decade, he sought creative ways to address them, lest they preempt resources that the city needs for this task.

In most of its key aspects, the water system has reached a point of comfortable stability. The success of the watershed-protection and water-conservation programs have ensured that, barring gross mismanagement or political weakness, the city will not need to build any new major water sources or filtration facilities for the foreseeable future as long as it maintains its level of effort and continues to refine its programs. The one major problem that remains, one that planners first confronted forty years ago and proposed to solve with City Tunnel #3, is reliable backup for the delivery infrastructure. The Delaware Aqueduct leak made the need for backup very apparent and, perhaps more important, reminded the city that the water system is aging. Water systems age well, but they are not ageless. Moreover, the need for security against potential terrorist attack has brought the issue of backup transmission capacity to the forefront again.

The question for the future is how much and where to invest next. Should the city proceed to Phase three of City Tunnel #3 or, given the leak in the Delaware Aqueduct, should the first priority be a backup system to the Delaware Aqueduct, including a new crossing under the Hudson River? Or is there another strategy? How much and how fast should the city invest in backup transmission capability when its water consumption is essentially stable? The challenge for the future is to give these questions well-considered answers because the financial, social, and cultural implications are profound.

FIG.19. *City Tunnel #3. Van Cortlandt valve chamber. Primary valve assembly. Valve caps at right used when valve is removed for servicing. Bronx, New York. May 18, 1989.*

In the mid-1960s, the Board of Water Supply looked forward and saw a future very much like its past—more engineering, more new facilities, and more refinements to a system that was an engineering jewel. Forty years later, those expectations have been radically transformed. Now the challenge is the wise management of a great water infrastructure, management that requires not only the most sophisticated use of water engineering, but also the integration of those skills with many new and non-traditional ones. Even the enormous transmission facilities that the DEP needs to consider as system backup and to maintain system reliability will have a far different function than in the past. No longer will they simply bring water to a thirsty city; rather, they will ensure that water service to a thirsty city remains uninterrupted.

Former Mayor Ed Koch once said that New York City has the "champagne" of drinking waters. As the last forty years have shown, preserving that heritage is a never ending, ever more challenging task. But 170 years of unparalleled success has proved the city capable of meeting those challenges. And hopefully it will continue to do so in the future.

ACCIDENT STORY

The walking boss heard it coming—but he was only halfway to the heading about 2 miles out. He started running, muck sucking his boots off, he kept running, barefoot as the motorcar passed him by, picking up speed, coming our way. I heard it first.

A train had come uncoupled at the bell out. We didn't have direct communication back then between the bottom bellhouse and the heading, just these phones along the rockwall every mile or so. Walkie-talkies were considered a luxury.

On that crew, I was considered an old-timer. Something didn't sound right. I can't explain it. You develop a sixth sense down in the hole. I dropped the wires I was working on, knelt down and put my hand on the rail, listening. The slow roll of a motorcar and no hum of a locomotive. That's when I knew for sure we had a runaway.

I started to get up, turned and fell, yelling "Runaway" but at the heading, another quarter mile up, the big jumbo was working full out, six bits at the same time getting ready for the next blast and nobody would've heard me if I'd been yelling in their ear.

14 men working at the heading, 14 men working right on the track. I got up on my knees, and looked behind me, that's when I saw her, but I already knew what she was gonna look like from the stories my dad told me about building Lincoln Tunnel—two hundred tons of steel and no sign of a heart.

Before she could get by me, I started grabbing anything I could get my hands on, throwing it on the tracks, trying to make her derail before she got to the men. Bits of wood, busted rocks, broken scraps of metal from the rock bolts, my flashlight, my work gloves, my hardhat.

And as I was crawling, stumbling, shoving off the side of the tunnel wall to keep from falling—this horrible metal scream cut me in half as she came off the tracks, riding on her outside wheels, bucking over, riding up the side of the tunnel, sparks flying from metal against rock, tearing out the pipes as she came straight at me.

I threw myself behind an empty wood blasting carton and put my hands over my head. The noise was hurting me to my bones, but inside my head, everything was dead quiet cause I knew for certain that she wasn't going to make it to the men up ahead.

Excerpt from *City Water Tunnel #3*,
written and performed by Marty Pottenger

THE FUTURE OF WATER

PETER H. GLEICK

How much water is there? A clear understanding of the natural hydrological cycle requires reliable estimates of the stocks and flow of freshwater resources. Despite vast improvements in the technology of monitoring, data storage, and computer modeling, information on the amount of fresh water on earth is still neither reliable nor accurate—any estimate of water stocks and flow thus remains an approximation. This is partly the result of inadequate efforts to actually monitor water stocks, and partly the result of the impossibility of accurately measuring—and integrating multiple measurements of—factors such as soil moisture, water in wetlands and groundwater aquifers, and the vastly complex and chaotic flow of water vapor and liquids. Even the quantity of water in lakes, glaciers, polar ice, and other better-defined stocks can only be approximated. That said, the total volume of water on earth is around 1.4 billion cubic kilometers (km^3) and only 2.5 percent of it, or about 35 million km^3, is fresh water. Even this small fraction of fresh water is largely inaccessible to humans since it is locked up in permanent ice or snow in the inaccessible regions of Antarctica and Greenland or buried in deep groundwater aquifers. The principal sources of water for human use are lakes, rivers, soil moisture, and relatively shallow groundwater basins. The usable portion of these sources is estimated to be only about 200,000 km3 of water—less than one percent of all fresh water on earth and only one-hundredth of a percent (0.01%) of all water on the planet. And much of this water is located far from human populations.[1]

Water is fundamental to all human activities. Clean and adequate water grows the food we eat, generates the energy that sustains a modern economy, which, in turn, produces the goods and services we desire, and indeed, ensures human and ecosystem survival. The development of basic water services has proven to be a key element in the advancement of social development and in the elimination of a host of debilitating and costly diseases. Modern cities began to flourish only when humans figured out how to deliver reliable water resources to large numbers of people and to remove the concentrations of wastes that cities produce, thus reducing the prevalence of water-based diseases.

As the twenty-first century unfolds, meeting the water needs of rapidly expanding megacities remains a daunting challenge. While some historical approaches will continue to serve us well, we will need new ideas, technologies, and tools in order to provide universal access to safe, reliable, and affordable water and wastewater services. Despite the progress made in satisfying the water needs of urban concentrations, global needs remain unmet. The United Nations World Health Organization estimated that at the beginning of the twenty-first century, more than a billion people lacked access to reliable safe drinking water and more than 2.6 billion people lacked access to adequate sanitation services. Hundreds of millions of these people live in sprawling peri-urban areas that lack reliable services for energy, transportation, schools, health care, or drinking water. Extending urban standards of service to these areas is an enormous challenge. Indeed, the failure to provide these services to all people is perhaps the greatest development failure of the twentieth century.[2] The most egregious consequence of this failure is the high rate of sickness and mortality from essentially preventable water-related diseases such as cholera, dysentery, and various parasitic diseases. As many as five million people die annually from a combination of different water-related diseases, and more than two million people, mostly young children, die from diarrheal diseases each year.

What is the most effective and efficient form of water management? Where will the money come from to maintain and upgrade existing water infrastructure? Should public or private agencies

FIG. 1. *Schoharie Creek near site of future Gilboa Dam. July 10, 1920.*

manage that water infrastructure? Unless municipalities in at-risk countries ask these questions—and answer them—sustainable human development will not be possible.

The engineering and social systems that provide water services to cities emerged millennia ago, in the ancient civilizations of the Incas, the Romans, the Greeks, and the cultures of the Indus Valley. As demands for water outstripped local resources, innovations in the form of small dams, open-air gravity-fed aqueducts and irrigation canals—and ultimately closed aqueducts, pipelines, and simple pumping systems—were invented, refined, and built.

In Asia, excavations at Harappa and Mohenjo Daro in the Indus Valley uncovered ceramic pipes for water distribution. Brick conduits found for drainage under the streets are thought to have been in operation around 3,000 B.C.[3] In the second millennium B.C., the Minoan civilization on Crete used running water and flush latrines. The towns of Athens, Pompeii, and Morgantina, like most Greco-Roman towns of their time, maintained elaborate systems for the supply, use, and drainage of water and sewage.[4] The first of the Roman aqueducts was completed around 312 B.C. By the height of the Roman empire, nine major systems supplied the occupants of Rome with as much water per person as is provided in many parts of the industrialized world today. This water was distributed through an extensive system of lead pipes in the streets, and well-built sewers provided drainage for the city.[5] Even in Thomas More's *Utopia*, written in 1516, the ideal cities of the perfect republic are all provided with clean, protected water sources as part of their basic design. Dora Crouch, in her groundbreaking book on water management in ancient Greek cities (1993), notes: "Attention to water supply and drainage is the sine qua non for . . . that human condition we call civilization. In fact, development of water supply, waste removal, and drainage made dense settlement possible."[6]

In the past century, advances in pumping technology, chemical and biological disinfection, and more sophisticated purification processes involving filtration, advanced membranes, and electrochemical processes, have improved water availability and quality faster than ever before. Yet despite these advances in providing urban water, more than half the population of our "civilized" world still suffers today with water services inferior to those of the ancient Greeks and Romans. As a result, water-related diseases continue to be a serious problem in many parts of the world. Water-related diseases include four types or classes: the waterborne, the water-washed, the water-based, and water-related insect vectors. The first three are most clearly associated with a lack of improved domestic water supply and adequate sanitation.

Waterborne diseases include those in which transmission occurs by drinking contaminated water, particularly contamination by pathogens transmitted from human excreta into water. Among these are most of the prevalent enteric and diarrheal diseases caused by bacteria, parasites, and viruses. Waterborne diseases also include typhoid and over thirty species of parasite that infect the human intestines. Evidence also suggests that these diseases contribute to other diseases not detected or reported explicitly as outbreaks.

Water-washed diseases occur when there is insufficient water for washing and personal hygiene or when contact with contaminated water occurs. Some of these are trachoma, typhus, and diarrheal diseases that can be passed from person to person. The World Health Organization estimates that there are seventy to eighty thousand deaths, and as many as 2.5 million illnesses from these diseases every year.

Water-based diseases come from hosts that live in water or require water for part of their life cycle. They are passed on to humans when contaminated water is ingested or comes into contact with skin. The most widespread examples in this category are schistosomiasis and dracunculiasis. Schistosomiasis is a parasitic worm that currently infects two hundred million people in seventy countries each year, causing seizures, paralysis, and organ damage. Safe drinking-water systems can eliminate exposure to these hosts.

The final category—water-related vector-borne diseases—includes those spread by insects that breed or feed near contaminated water, such as malaria, onchocerciasis, and dengue fever, rather than diseases associated with lack of access to clean drinking water or sanitation services. It must be noted, however, that the spread of these illnesses may be facilitated by the construction of large-scale water systems that create conditions favorable to their hosts.

Perhaps the clearest example of a water-related disease resulting from unsafe drinking water is cholera, which is an acute diarrheal illness caused by intestine infection from the bacterium *Vibrio cholerae*. Although cholera can be life-threatening, it is preventable and treatable. This disease was prevalent in the United States and many other countries in the 1800s, but was virtually eliminated in wealthier countries by modern sewage and water-treatment systems. As a result, cholera has become very rare in industrialized nations for the last one hundred years. It remains epidemic in many developing countries.

Until the 1800s cholera was found almost exclusively in the Ganges/Brahmaputra River basins in Asia.[7] From 1817 to 1923 cholera spread along trading routes, causing six major pandemics throughout Europe and in the United States. By the early 1830s, cholera was found mostly in major urban centers in the United States, and severe cholera epidemics were prevalent in many parts of the world. In the late 1840s John Snow, a London physician, proposed that cholera was a contagious disease caused by a poison that reproduces itself in the human body. He believed that the principal means of transmission was water contaminated by human wastes. Snow was able to prove his theory in 1854 by documenting and mapping cholera cases among subscribers to London's two main water companies. His work showed that cholera occurred much more frequently among the customers of a water company that drew its water from the lower Thames, where it had become contaminated by London sewage. The other company had obtained cleaner water from the upper Thames.

Even more compelling evidence soon appeared. In one neighborhood the concentration of cholera cases was so great that the number of deaths reached over five hundred in ten days. Snow carefully mapped the cases and observed concentrations around a particular water pump. He advised officials to remove the pump handle, forcing people to use other uncontaminated water sources, successfully halting the epidemic. Snow's fame has led to the creation of a John Snow Society, whose membership consists of people who have visited the John Snow Pub in London, which claims ownership of the "original" pump handle.[8]

As a result of such research, cities throughout the industrializing world began to dramatically improve public health by protecting water sources and treating drinking water supplies through filtration, chemical treatment, and ozonation. Water filtration was introduced in the late 1800s in the United States, building on early efforts in Scotland. The city of Nice, in France, introduced ozone as a disinfectant in 1906, and Jersey City, New Jersey, pioneered the use of chlorination of public water supplies in the U.S. in 1908. These approaches to disinfection spread quickly to most large towns and cities in Europe and America, and cholera and typhoid rates began to plummet.[9] Today, filtration and various chemical processes remain the primary approaches to drinking-water treatment worldwide, though new processes using sophisticated membranes and ultraviolet light, among other approaches, continue to be explored.

As a result of these efforts, water-related diseases associated with the lack of access to clean and safe water and sanitation have largely been eliminated in wealthier nations. Unfortunately, they remain a major concern in much of the developing world. According to the World Health Organization, which regularly publishes water- and health-related information in the Human Development Report series and the annual World Health Report,[10] there are on the order of 250 million cases of water-related diseases annually, excluding common diarrheal diseases. In the 2000 World Health Report, the UN reported 4.37 billion cases of diarrhea annually. Even this horrifically large number may be too low: the true extent of these diseases is unknown, and many cases of water-related illnesses go undiagnosed and unreported.

The most severe consequence of water-related diseases is death. However, statistics are inadequately monitored and reported. A wide range of estimates is available in the public literature, ranging from 2 million to 12 million deaths per year.[11] Current best estimates appear to be between 2 and 5 million deaths per year. Of these, the vast majority consist of small children struck by virulent but preventable diarrheal diseases. Water-related deaths are tragic and unnecessary. And there are other significant health consequences of the failure to provide adequate water services as well. These include morbidity, lost workdays, lost educational opportunities, official and unofficial health-care costs, and the draining of family resources. These consequences are poorly understood and even more poorly measured and assessed.

In order to address overall poverty, including the vast unmet human need for basic water supply and sanitation, the United Nations and the international community announced explicit goals in 2000—the Millennium Development Goals—for human development over the next several decades.[12] Two of these goals call for reducing by half the proportion of people unable to reach or afford safe drinking water and sanitation by the year 2015.

The majority of people lacking access to basic water services live in small rural communities, largely in Africa and Asia. Yet substantial numbers of urban dwellers also continue to lack adequate water and sanitation. As many as 115 million people in major cities lack access to an adequate supply of water, while over 400 million people lack adequate sanitation, and these numbers are growing as populations seeking work move from rural areas to

FIG. 2. *Esopus Creek at the site of Olive Bridge Dam. March 27, 1908.*

FIG. 3. *Ashokan Reservoir. View from the Stone Church Bridge showing water flowing down the natural spillway (at the rate of 224 million gallons per day). May 1, 1928.*

cities.[13] As population densities rise in cities, meeting these needs offers great potential benefits for reducing water-related-disease rates.

Achieving these goals is a laudable objective, but even if they can be achieved, hundreds of millions of people will still lack basic water services twenty years from now. Moreover, given current levels of financial and institutional commitment, there now seems to be little chance that the Millennium Development Goals will be met. This problem must be considered one of the most serious public health crises facing us and deserves far more attention and resources than it has so far received.

Unfortunately, the UN's water goals and solutions to other water problems are also unlikely to be achieved. Despite growing awareness of water-related issues, international economic support for water projects of all kinds is marginal and declining. Official development assistance for water supply and sanitation projects from countries of the Organisation for Economic Co-operation and Development (OECD) and the major international financial institutions has actually declined over the past few years, down from $3.4 billion per year (the average from 1996 to 1998) to $3 billion per year (the average from 1999 to 2001).[14] Moreover, those most in need receive the smallest amount of aid. For example, ten countries received about half of all water-related aid, while countries where less than 60 percent of the population has access to an improved water source received only 12 percent of the

FIG. 4. *Ashokan Reservoir. View from Stone Church Bridge looking up the natural spillway channel. October 27, 1928.*

money.[15] In particular, all of Africa, with its enormous unmet needs, received less than $20 million dollars in total aid from the United States for water supply and sanitation projects in 2002.

Solving these problems will require a new focus on urban water systems. The old approaches to addressing urban water problems—building large-scale dams, pipelines, and irrigation systems—brought great benefits to hundreds of millions of people, but they also had great costs. As described above, by the end of the twentieth century, billions of people still struggled without access to the most basic of water services. In an effort to meet these needs, new ideas—both good and bad— must be considered, evaluated, and tested.

Among these ideas are advances in technology, new economic tools, and changes in the focus and priorities of water institutions. Water must be considered both a social and an economic good, subject to the rules and power of regulation, markets, prices, and social constraints. Innovative financing mechanisms are being developed to help local communities find the funds to build and operate water systems, and new technologies can be developed and built to work at scales appropriate to peri-urban areas, or that use new methods for purifying and distributing water.

The question of whether water should be managed by public or private agencies is highly controversial and complicated. Until recently the question of whether individuals or groups have a legal right to a minimum set of resources—specifically

FIG. 5. *Coxing Kill Falls on Esopus Creek. December 11, 1906.*

FIG. 6. *Buttermilk Falls on the Wallkill River. September 27, 1918.*

water—and whether there is an obligation for states and other parties to provide those resources when they are lacking, has not been adequately addressed. However, in an assessment written for the United Nations, Professor Stephen McCaffrey, one of the world's leading international water lawyers, concluded that there is at least a right to sufficient water to sustain life, and that a state has the "due diligence obligations to safeguard those rights."[16] What is the value of explicitly acknowledging a human right to water, in the way that the international community has explicitly acknowledged a human right to food and life? After all, despite the declaration of a formal right to food, nearly 850 million people remain undernourished. One reason for this "right" is to encourage the international community and individual governments to renew their efforts to meet the basic water needs of their populations. Secondly, acknowledging such a right adds pressure to translate that right into specific national and international legal obligations and responsibilities. A third reason to articulate this right is to spotlight the deplorable state of water management in many parts of the world. Even where progress is being made in meeting basic water needs, such as in South Africa, where the human right to water is enshrined within their new constitution, systems are not always being put in place to ensure their long-term sustainability. A fourth reason is to help focus attention on the need to more widely address international watershed disputes and to help resolve conflicts over the shared use of water by identifying minimum water requirements and allocations for all basin parties, such as the effort under way in the Nile Basin, where ten countries share access to the river. Finally, explicitly acknowledging a human right to water can help set specific priorities for water policy: meeting a basic water requirement for all humans to satisfy this right should take precedence over other water allocations and investment decisions.[17]

Urban water needs have been satisfied by various combinations of public and private involvement over the past several

FIG. 7. *High Falls on Esopus Creek. April 19, 1910.*

centuries, and by those of technologies and policies. Because water supply tends to be a monopoly and is considered a public good, in most cities provision of water services is undertaken by public agencies, with government oversight.

The privatization of the water sector involves transferring some or all of the assets or operations of public water systems into private hands. There are numerous ways to privatize water, such as the transfer of the responsibility to operate a water-delivery or treatment system, a more complete transfer of system ownership and operation responsibilities, or even the sale of publicly owned water rights to private companies.

There is a long history of forms of private participation in water supply. Indeed, in some places, private ownership and provision of water was the norm, until governments began to assume those responsibilities. In the United States, municipal services were often provided by private organizations through the early- and mid-1800s. Toward the latter half of that century, municipalities started to confront problems with access and service, and began a transition toward public control and management. In particular, private companies were failing to provide access to all citizens in an equitable manner. Private water companies provided 94 percent of the U.S. market in the nineteenth century, dropping to only 15 percent by the year 2000.[18] As Nelson Blake points out:

> Private companies supplied water to Boston from 1796 to 1848, and to Baltimore from 1807 to 1854. As late as 1860, 79 out of 156 water works in the United States were privately owned. But eventually most cities turned to municipal ownership. The profit motive was ill suited to the business of supplying water to city dwellers. Private companies were reluctant to invest enough capital; they preferred to lay their distributing pipes through the wealthier sections of the city and to hold back from carrying water into the poorer districts.[19]

This was New York's experience with the private Manhattan Company in the early 1800s. Christopher Anderson notes similar experiences in Chicago and other cities in the United States.

Private companies were notorious for choosing a water source that would minimize the initial investment outlay and for ignoring the concomitant shortcomings in water quantity and quality. Only municipal governments, so the argument goes, had the foresight and the latitude to invest large sums now in order to gain a future payoff in the form of years of excellent water.[20]

In nineteenth-century France, the trend moved in the opposite direction. Municipalities that previously had the responsibility for providing water services began to contract services to private operators. Over the years, these operators expanded beyond the borders of France and as a result, they now have a dominant position in much of the world in providing private water services.

The push to shift to private water systems and markets is a relatively recent phenomenon, with major transfers taking place only over the past ten to fifteen years. As part of this effort, prices have been set for water that was previously provided for free. Private companies have been invited to take over the management, operation, and sometimes even the ownership of previously public water systems. International development agencies that used to work with governments to improve water services are now pushing privatization efforts.

By the end of 2000, at least ninety-three countries had partially privatized water or wastewater services,[21] including Argentina, Chile, China, Colombia, the Philippines, South Africa, Australia, the United Kingdom, and Central Europe, though less than 10 percent of all water is currently managed by the private sector.[22]

Alternatively, various combinations of private and public financing are possible, such as soliciting private investment in the development of new facilities, with the transfer of those facilities to public ownership after investors have been repaid, or hiring private contractors to operate and maintain parts of a water system, while retaining public ownership and control. Increasingly, offers to privatize water services come from large multinational corporations, such as the French companies Suez and Veolia, and RWE in Germany. As these efforts intensify, so does opposition at the local, regional, and international levels.

When the service being privatized has "public good" characteristics, as does water, government regulation or oversight has traditionally been required. Goods and services previously provided by public officials or agencies may become less vulnerable to political manipulation when privatized, but private entities, with their responsibility to the "bottom line," may also become less responsive to public interests, including protection of water quality, a commitment to efficiency improvements that reduce the volume of water used, the maintenance of basic service levels, transparent prices and billing practices, and investments in water reclamation or additional sources of water supply.

These ideas and trends have generated enormous controversy. On the positive side, we are seeing lively and productive debates and arguments among water managers, activists, and policymakers. On the negative side, privatization is provoking social and political conflict and discord, such as was seen in the violence that accompanied protests over privatization of water in Cochabamba, Bolivia, and it may prove to be ineffectual in solving unmet urban water needs.[23]

In South America, public water systems were most common through the mid-1990s, but, as in other world regions, public water systems consistently failed to provide universal coverage, to treat most wastewater, and to find and reduce extreme water losses. Because of these failures, governments in South America have increasingly sought private-sector involvement. In some cases, such as Buenos Aires, governments have sold or leased water facilities, allowing private

FIG. 8. *Titicus Dam. Water spilling over the stepped weir. c. 1896.*

operators to sell services directly to the public, with government regulation. Mexico City took another approach to privatization, contracting the rights to operate parts of the city water system to different parties. Chile went beyond offering operating contracts and combined the granting of concessions with private ownership of water resources. By 2000, almost all countries in the region began to commit themselves to long-term private concessions.

Major cities in Asia also suffer from inadequate infrastructure, huge water losses, inadequate sewage treatment, and the lack of service to large numbers of peri-urban residents. Australia, New Zealand, Malaysia, and the Philippines have all explored various forms of privatization, and the water and wastewater utilities in almost every major city in Oceania have contracted some important services to private companies. Recent efforts in Manila and Malaysia have run into political or economic controversy, however, causing private companies and governments to rethink the writing of contracts, with their conditions for concessions, but privatization efforts continue.

Nations in Europe have also explored a variety of models recently, and the UK, Germany, France, and Italy all now encourage forms of water privatization. In France and the UK, the process is far advanced. The British, under Prime Minister Margaret Thatcher, for example, sold state-owned water operations to private investors more than a decade ago. Some of those newly privatized companies, such as Thames Water, have become multinational players in privatization markets. Service providers in all four countries initially tried to keep prices low, but have recently imposed large price increases in order to upgrade their plants and distribution systems to accepted European standards. These price increases have led to growing consumer distrust, though some argue that governments would have similarly had to raise taxes or increase borrowing to make comparable improvements, or worse, would have failed to make them at all.

The United States and Canada have moved more slowly toward privatization.

FIG. 9. *Coxing Kill Falls on Esopus Creek in the Catskill Mountains. April 24, 1907.*

The United States has long had a mix of privately owned and publicly regulated water and wastewater utilities, though an estimated 85 percent of residences still receive water from public agencies. Indeed, the ability of U.S. cities to sell public bonds to fund water projects has been a major reason for continued public management.

Moreover, several well-publicized privatization efforts in the U.S. have failed, putting a damper on enthusiasm for turning public water agencies over to private management. The city of Atlanta, for example, signed a long-term contract with a multinational water company only to cancel that contract a few years later because of growing customer dissatisfaction, concerns about the quality of service, and increasing costs. The city of Stockton, California, went through a long and contentious debate over privatization and finally signed a contract, only to have a state superior court throw out the contract because of various concerns, including inadequate public review and oversight.

Many unanswered questions remain about the true implications and consequences of privatizing water and whether these new approaches can effectively, equitably, and adequately serve human and environmental needs. However, there is little doubt that the headlong rush toward private markets has failed to address some of the most important issues and concerns about water. In particular, water has vital social, cultural, and ecological roles to play that cannot be protected purely by market forces—especially because water services are, as mentioned earlier, a natural monopoly. Certain management goals and social values require direct and strong government support and protection. Some of the consequences of privatization may be irreversible; hence they deserve special scrutiny and control. As a result, proposals to privatize or commodify water must be evaluated far more carefully than they have been and must be accompanied by

FIG. 10. *Esopus Creek at site of Ashokan Reservoir. c. 1905.*

guarantees to respect certain civil principles and support specific social objectives.

New technologies are available to purify water, distribute it to users, and collect and treat wastewater. The focus of water managers in the twentieth century was on building traditional centralized water systems, primarily because that is what the engineers and agencies already working to meet urban water needs in industrialized nations knew how to do. While these systems often brought substantial benefits to many urban users, their high costs often led planners to design systems with insufficient capacity to satisfy all users and needs, and their technical complexity called for trained personnel to maintain operations—a workforce not always readily, or affordably, available. In Asia, the Near East, and Africa many expensive systems were built but never operated, or operated briefly before failing because of insufficiently trained personnel.

In recent years, municipalities have begun to explore alternative approaches for obtaining, cleaning, and distributing water. These range from small-scale methods for non-water-based sanitation systems to legislative efforts to protect watersheds in order to ensure high-quality supplies as an alternative to expensive treatment.

Over the past decade, New York City has led a highly successful effort to manage watersheds in order to protect ecosystems and water quality, rather than build a large new water-treatment infrastructure. In lieu of constructing a filtration plant that was estimated to cost city residents $6 to $8 billion, New York City elected to design and implement an innovative environmental protection strategy to protect water quality in the entire watershed.

On a much smaller scale, water agencies in Durban, South Africa, developed a public-private partnership that used innovative computerized water delivery management systems combined with small-scale infrastructure and community management to help meet the water needs of a rapidly growing population. Using geographic information systems to map water distribution, Durban was able to provide maps of service areas, facilities, and customers, which have served to help policymakers and corporations better understand the water usage and needs of the people in the region. It also improved billing and collection, which helps fund system improvements, which in turn encourage more efficient use. This improvement was then coupled with small-scale technology in the form of low-pressure tanks and electronic valves that connect twenty households at a time to improve reliability and service, creating, in essence, a network of neighborhood water systems. These low-pressure tanks provide a maximum of two hundred liters per day per household, or six thousand liters per month, which is provided free of charge. Water use above these levels is charged at rates that increase with the volume of use. This effort has improved both overall access to water and the reliability and operation of the water system as a whole, though time will tell whether the approach proves acceptable and sustainable.

Providing adequate sanitation systems is even more of a challenge for cities undergoing sudden and significant expansions. Like many other cities in developing countries, Tegucigalpa, the capital city of Honduras, has experienced rapid growth over the last twenty years. The population in the year 2000 was around one million, more than half of which lives in settlements on the outskirts of the city. Provision of water and sanitary services by the city has been especially difficult because of the local geography: new settlements tend to be located on steep hillsides where very little surface water is available and groundwater is often polluted or runs too deeply. In the late 1980s the United Nations under UNICEF, together with local water and sewage authorities, began an innovative program to supply water services to more than two hundred peri-urban communities growing around Tegucigalpa. This program developed alternative technologies that can provide small-bore and simplified sewage systems at a per-connection cost that is only slightly higher than latrines.

In a small-bore or solids-free system the sewage is first settled in a septic tank at each house, and only liquid waste is transported by the system. A simplified sewage system uses shallower pipes and simple inspection boxes, which are far less costly than conventional sewage systems. The Tegucigalpa program uses appropriate low-cost technology, local labor and materials, and revolving loans, and has had considerable success in extending water and sewerage to the peri-urban settlements.

Another example of innovative urban sewer projects is the development of "condominial" sewers in Brazil, and more recently the La Paz-El Alto region of Bolivia. Condominial sewers are a particular form of the lateral sewer used to connect homes to main trunk sewers. In traditional approaches, one lateral sewer is provided for each home. In the condominial scheme, groups of homes share a single lateral sewer that runs from the home farthest away from the trunk sewer, passing under or alongside homes, until finally reaching the trunk sewer. The group of homes shares maintenance of the lateral through a users' group. Small-bore and simplified sewage systems and condominial sewers are examples of systems that can reduce the costs of providing sewerage by as much as 70 percent, compared to traditional methods.

However these new technologies are financed, emerging economies face significant hurdles such as finding the capital to expand basic water supply in rapidly growing urban areas, maintaining existing infrastructure, and treating wastewater to even minimal quality standards. Water supply projects can be extremely capital intensive, though estimates of future needs vary widely. The World Bank estimates that new investment required for water infrastructure over the next decade exceeds $60 billion per year. In mid-2001, the American Water Works Association released a study suggesting that $250 billion may be needed over the next three decades just to upgrade and maintain the existing drinking water system in the United States.[24] The Framework for Action that emerged from the Second World Water Forum in the Hague in March 2000 called for an additional $100 billion annually from private sources for the next twenty-five years to meet basic water needs worldwide.[25]

FIG. 11. *Hanging Rock Falls on Bear Kill Creek. September 24, 1918.*

Construction costs for water supply and sanitation vary widely, depending on location and type of service. Simple onsite systems tend to be far less costly than centralized sewer services, which can run as high as $500 per person in industrialized countries. Many options are available at costs below those of centralized systems, some for as little as 25 percent of the costs incurred in industrialized countries.[26]

The failure to meet basic human needs for water is widely acknowledged as a major development failure of the twentieth century. Especially in rapidly growing urban centers, meeting water supply and sanitation needs is an expensive and complex challenge. Yet failing to meet these needs causes vast human and ecological suffering. Despite a new international focus emerging from the adoption of the Millennium Development Goals, practical actions and commitments to provide universal coverage for water and sanitation continue to be inadequate. The price for this will be paid by the poorest populations of the world in sickness, lost educational and employment opportunities, and, for a staggeringly large number of people, early death. Even if the official Millennium Goals set for water are met—which is unlikely, given the current level of commitments by national governments and international aid agencies—as many as 32 million people, and perhaps many more, will die by 2020 from preventable water-related diseases. This is morally unacceptable in a world that values equity and decency. At present, it appears unavoidable.

WEST INDIAN SANDHOG

This is the accomplishment. This is what men died for. You could fit three St. Patrick's Cathedrals in this space. I worked on this Van Cortlandt Valve Chamber for seven years. Part of my heart is in this place. I never dreamed it could be this beautiful. It's like heaven down here. The men who died would be proud to see this. Believe me. I feel proud of myself coming back down here.

Excerpt from *City Water Tunnel #3,*
written and performed by Marty Pottenger

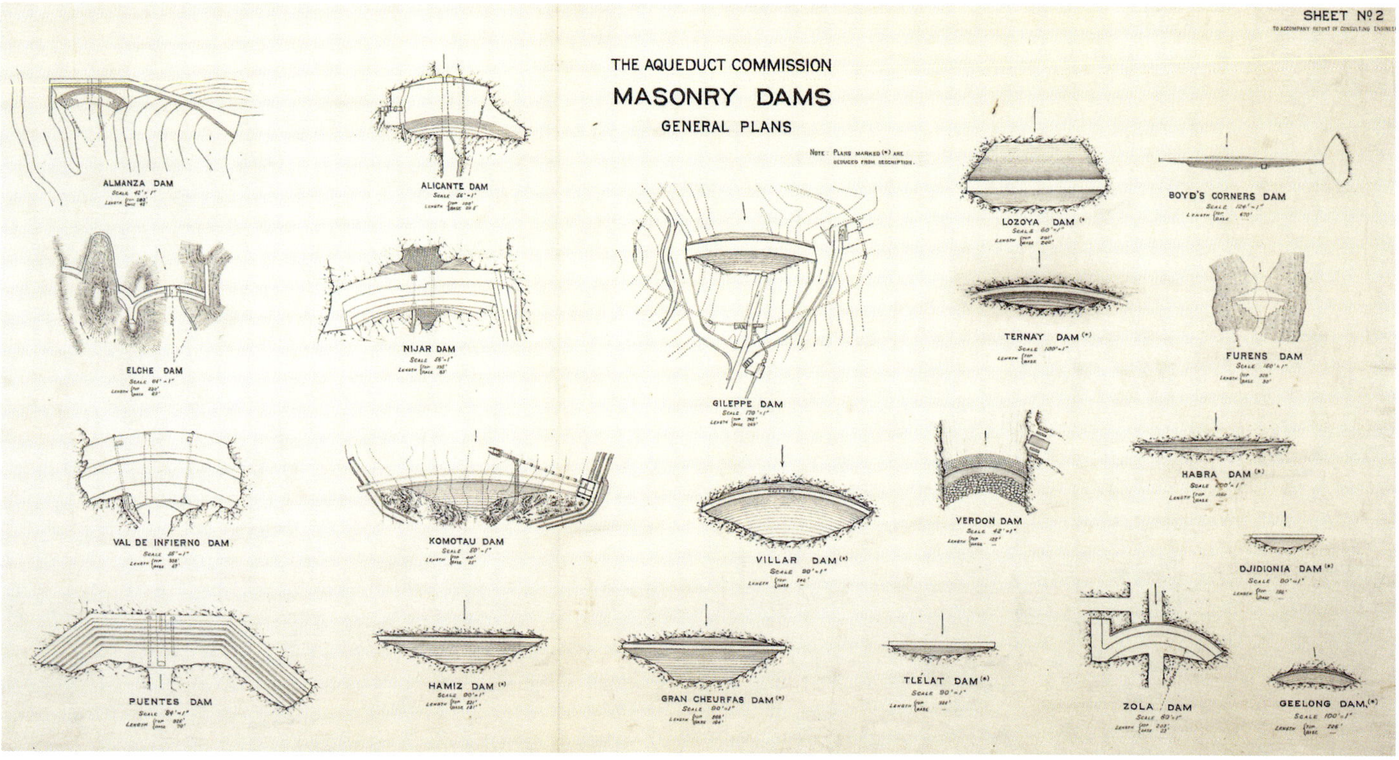
SHEET No 2
THE AQUEDUCT COMMISSION
MASONRY DAMS
GENERAL PLANS
NOTE. PLANS MARKED (*) ARE DEDUCED FROM DESCRIPTION.
ALMANZA DAM
ALICANTE DAM
LOZOYA DAM (*)
BOYD'S CORNERS DAM
ELCHE DAM
NIJAR DAM
GILEPPE DAM
TERNAY DAM (*)
FURENS DAM
VAL DE INFIERNO DAM
KOMOTAU DAM
VILLAR DAM (*)
VERDON DAM
HABRA DAM (*)
DJIDIONIA DAM (*)
PUENTES DAM
HAMIZ DAM (*)
GRAN CHEURFAS DAM (*)
TLELAT DAM (*)
ZOLA DAM
GEELONG DAM (*)

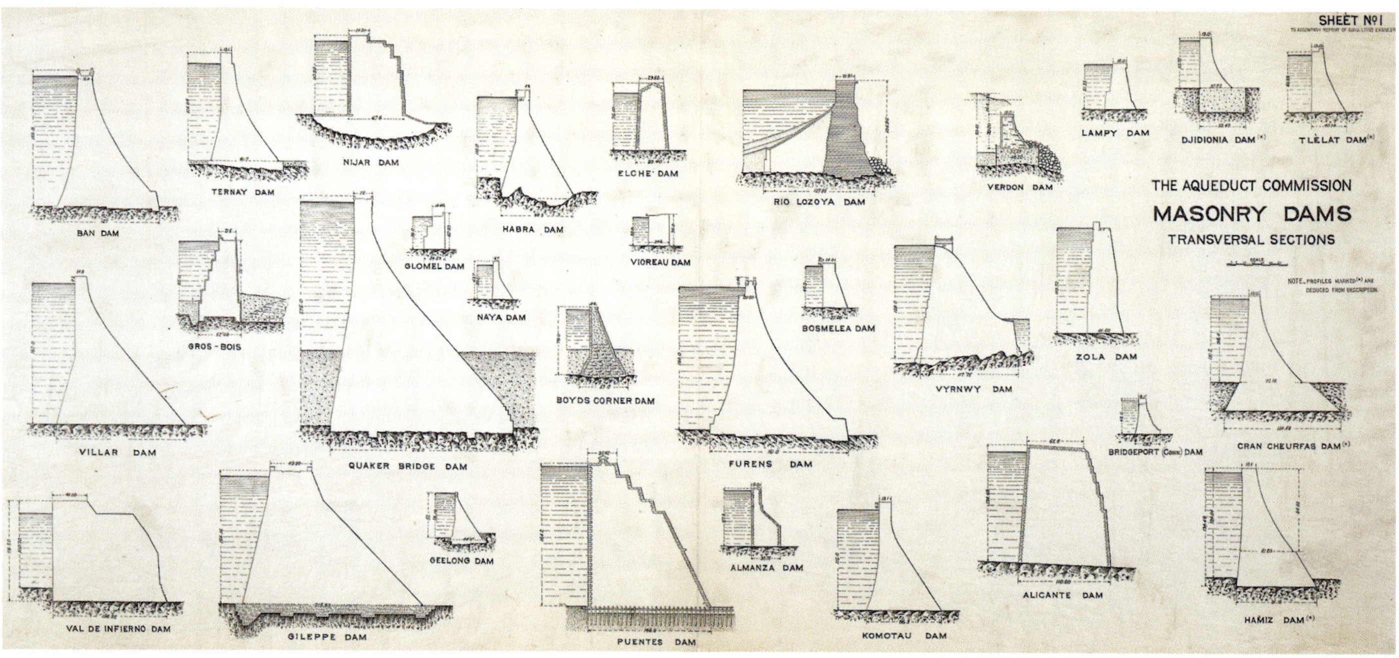
SHEET No 1
BAN DAM
TERNAY DAM
NIJAR DAM
HABRA DAM
ELCHE' DAM
RIO LOZOYA DAM
VERDON DAM
LAMPY DAM
DJIDIONIA DAM (*)
TLELAT DAM (*)
THE AQUEDUCT COMMISSION
MASONRY DAMS
TRANSVERSAL SECTIONS
NOTE. PROFILES MARKED (*) ARE DEDUCED FROM DESCRIPTION.
GROS-BOIS
GLOMEL DAM
NAYA DAM
VIOREAU DAM
BOSMELEA DAM
VYRNWY DAM
ZOLA DAM
VILLAR DAM
QUAKER BRIDGE DAM
BOYDS CORNER DAM
FURENS DAM
BRIDGEPORT (CONN) DAM
CRAN CHEURFAS DAM (*)
VAL DE INFIERNO DAM
GILEPPE DAM
GEELONG DAM
PUENTES DAM
ALMANZA DAM
KOMOTAU DAM
ALICANTE DAM
HAMIZ DAM (*)

NOTES

FOREWORD

1. Kenneth T. Jackson, quoted in Glenn Collins, "In City's History, a Glass Half Full: Unearthed City Archives Reveal Artistry of Water Supply," *New York Times* (January 8, 2001).

THE RISE TO CROTON

1. William Bridges, *Map of the City of New-York and Island of Manhattan; with Explanatory Remarks and References* (New York: T&J Swords, 1811), 24.
2. Wayne Andrews, "A Glance at New York in 1697: The Travel Diary of Dr. Benjamin Bullivant," *New-York Historical Society Quarterly* 40, no. 1 (January 1956), 65.
3. Alexander Hamilton, *Itinerarium* (New York: Arno Press and *New York Times*, 1971 [1907]), 107.
4. Peter Kalm, *Travels into North America*, trans. John Reinhold Forster, 2 vols., 2nd ed. (London: Printed for T. Lowndes, 1773 [1753]), 196–197.
5. Andrew Burnaby, *Travels Through the Middle Settlements in North America in the Years 1759 and 1760* (New York: Augustus M. Kelley, 1970 [1775]), 112.
6. Henry Brockholst Livingston to Henry Remsen, April 11, 1810, in Gregory S. Hunter, *The Manhattan Company: Managing a Multi-unit Corporation in New York, 1799–1842* (New York: Garland Publishing, 1989), 104–105.
7. Irving to Charles King, August 9, 1832, in *Works of Washington Irving*, 10 vols. (New York: G. P. Putnam & Co., 1910), 24:713.
8. Edmund B. O'Callaghan, ed., *Documents Relative to the Colonial History of the State of New-York*, 15 vols. (Albany: Weed, Parsons, and Co., 1856–87), 2:500.
9. I. N. Phelps Stokes, *The Iconography of Manhattan Island*, 6 vols. (New York: Arno Press, 1967 [1915]), 4:262.
10. John W. Francis, "Reminiscences of Christopher Colles," *The Knickerbocker Gallery: A Testimonial to the Editor of the Knickerbocker Magazine from Its Contributors* (New York: S. Hueston, 1855), 197–208.
11. Minutes of the Common Council of the City of New York, 1675–1776, 8 vols. (New York: Dodd, Mead, 1905), 12:203.
12. *The Spectator* 22 (September 29, 1798).
13. *Commercial Advertiser* 1 (September 5, 1798).
14. Nicholas Low to Rufus King, May 24, 1799, in Beatrice G. Reubens, "Burr, Hamilton and the Manhattan Company, Part II: Launching the Bank," *Political Science Quarterly* 73, no. 1 (March 1958), 109.
15. *Proceedings of the Corporation of New-York, on Supplying the City with Pure and Wholesome Water, with a Memoir of Joseph Browne, M.D., on the Same Subject* (New York: John Furman, 1799), 9–29.
16. Certificates, Relative to Manhattan Water in N.Y. (New York[?], 1824[?]), 1.
17. *New York Evening Journal*, undated clipping in Stephen Allen Scrap Book, New-York Historical Society. The paper was published from 1829 to 1832.
18. New York City Board of Aldermen, "Semi-Annual Report of the Water Commissioners," Document No. 9 (August 8, 1842), 86–87.
19. MCC, 1784–1831, 19:718.
20. *The Family Magazine*, or, *Monthly Abstract of General Knowledge* 6 (1839), 117.
21. Edward Wegmann, *The Water Supply of the City of New York 1658–1895* (New York: John Wiley & Sons, 1896), 24.
22. *New-York Evening Post*, March 18, 1845.
23. *NYEP*, March 18, 1845.
24. *Morning Courier* and *New-York Enquirer*, November 12, 1840.
25. Renwick v Morris, Court for the Correction of Errors of New York, 7 Hill 575 (1844).
26. John B. Jervis, *The Reminiscences of John B. Jervis, Engineer of the Old Croton*, ed. Neal FitzSimons (Syracuse, N.Y.: Syracuse University Press, 1971), 143.
27. Fayette B. Tower to (brother) Julius Tower, June 12, 1839, Tower letters, private collection of Helen Tower Wilson.
28. Junius Henri Browne, *The Great Metropolis: A Mirror of New York* (Hartford: American Publishing Company, 1869), 642, 644.

29. *Hudson River Chronicle*, January 16, 1838.
30. *New York Morning Herald*, April 13, 1840.
31. Norval White and Elliot Willensky, eds., *AIA Guide to New York City*, 4th ed. (New York: Three Rivers Press, 2000), 367.
32. Fayette Tower to "brother," June 28, 1842, Tower letters.
33. Fayette Tower to Helen M. Phelps, July 5, 1842, John Wolcott Phelps Papers, New York Public Library.
34. *The Diary of Philip Hone, 1828–1851*, ed. Allan Nevins, 2 vols., (New York: Dodd, Mead and Company, 1927), 2:624.
35. The ode was commissioned by the city from minor Knickerbocker Group poet George Pope Morris.
36. Lydia Maria Francis Child, *Letters from New-York* (New York: Charles S. Francis, 1843), 200.

THE EXTENSION OF THE CROTON SYSTEM

1. Edward Wegmann, *The Water Supply of the City of New York 1658–1895* (New York: John Wiley & Sons, 1896), 48.
2. Wegmann: 65.
3. Elizabeth Barlow Rodgers, *Rebuilding Central Park* (Cambridge, Mass.: MIT Press, 1987), 106.
4. Robert J. Kornfeld Jr., "Jerome Park Conservancy Preservation Report," 1998: 106.
5. Annual Report of the Croton Aqueduct Department Made to the Common Council of the City of New York for the Year 1858: 8.
6. Annual Report 1858: 10.
7. Annual Report of the Croton Aqueduct Department Made to the Common Council of the City of New York for the Year 1860: 7–11.
8. Annual Report 1860: 1–14.
9. Kenneth A. Jackson, *The Encyclopedia of New York City* (New Haven: Yale University Press, 1995), 1149.
10. Charles Weidner, *Water for a City* (New Brunswick, N.J.: Rutgers University Press, 1974), 62.
11. Kornfeld: 7.
12. Jackson: 1206.
13. Weidner: 49.
14. Annual Report 1858: 111.
15. Wegmann: 90.
16. Kornfeld: 10, quoting Union League, "Union League Club Report," (January 1882), 9.
17. Wegmann: 110.
18. New York City Aqueduct Commission, Report I (1883–87), 7.
19. Isaac Newton, *New York Water Supply: Report on Storage Reservoirs in the Croton* (New York: Martin B. Brown Co., 1883), 8–11.
20. Weidner: 74.
21. Weidner: 81.
22. Edward Wegmann, *The Design and Construction of Dams* (New York: John Wiley & Sons, 1911), 159–60.
23. Wegmann, Dams: 170.
24. Weidner: 119.

THE CATSKILL SYSTEM

1. New York City Board of Water Supply, Office of the Chief Engineer, History of the Catskill Aqueduct and Related Structures 1905–1917, 12 vols., 1:3. This document is an unpublished history of the first phase of the Catskill system. Attribution of specific writings is not given in the text. Dates in the manuscript suggest that the documents were produced in 1917 and 1918.
2. Charles Weidner, *Water for a City* (New Brunswick, N.J.: Rutgers University Press, 1974), 141.
3. *New York Times*, April 20, 1899.
4. Weidner: 147–48.
5. Weidner: 150–51.
6. HCARS 1:7–8.
7. HCARS 1:9–10.
8. HCARS 1:11.
9. HCARS 1:13–14.
10. Weidner: 180–81.
11. HCARS 2:5–6.
12. New York City Board of Water Supply, First Annual Report, December 31, 1906: 21.
13. HCARS 2:18.
14. HCARS 2:60–62.
15. HCARS 2:58.
16. HCARS 2:23–25.
17. HCARS 2:24–25.
18. HCARS 2:28–30.
19. HCARS 2:82.
20. HCARS 5:27–29.
21. HCARS 5:31–32.
22. HCARS 5:35.
23. Weidner: 215.
24. William H. Burr, Rudolf Hering, and John R. Freeman, Report of the Commission on Additional Water Supply for the City of New York Made to Robert Grier Monroe, commissioner of Water Supply, Gas and Electricity, November 30, 1903 (New York: Martin B. Brown Co., 1904), 3.
25. HCARS 6:8–12.
26. HCARS 7:74–76.
27. HCARS 7:79–83.
28. HCARS 3:27.
29. HCARS 3:23–24.
30. HCARS 3:43–45.
31. HCARS 4:1–5.
32. HCARS 10:131–42.
33. HCARS 9:13–15.
34. HCARS 9:19–23.
35. HCARS 9:28–31.
36. HCARS 9:36.
37. HCARS 4:23–27.
38. Weidner: 274–75.
39. New York City Board of Water Supply, Catskill Water System: A General Description, September 1915.
40. New York City Board of Water Supply, The Water Supply of the City of New York, A Volume Descriptive of its Sources, Storage Reservoirs and Transportation . . . , September 1950: 10–11.
41. BWS, Water Supply: 12–13.
42. BWS, Catskill, December 20, 1913.

THE DELAWARE SYSTEM

1. New York City Board of Water Supply, Fifteenth Annual Report, January 1, 1921, 42.
2. BWS, Twenty-first Annual Report, January 1, 1927: 103.
3. BWS, Twenty-first Annual Report, January 1, 1927: 6.
4. Delaware River Case 1952–1954, Supreme Court of the United States, II: 1–4.
5. State of New Jersey v State of New York and City of New York, Supreme Court of the United States, (decision May 4, 1931).
6. BWS, Twentieth Annual Report, January 1, 1926: 7–8.
7. BWS, Twenty-second Annual Report, January 1, 1928: 64.
8. BWS, Holing Through City Tunnel No. 2, March 30, 1932: 2.
9. BWS, Annual Report, December 31, 1935: 43.
10. BWS, Origin and Achievements of the Board of Water Supply, City of New York, September, 1950: 43.
11. Roger W. Armstrong, "Construction of the Delaware Aqueduct," *Journal of New England Water Works Association* LV, no. 2 (1941): 161.
12. Walter E. Spear, "Civil Engineering, Deep Tunnels for Delivery of Water Supply," March, 1933.
13. Armstrong: 140.

14. *Delaware Water Supply News* 69 (June 1, 1941).
15. DWSN 27 (September 1, 1939), 3.
16. DWSN 30, "Development of Deep Cut-Off Walls for Earth Dams," (October 15, 1939), 132.
17. DWSN 52 (September 15, 1940), 216–220.
18. Diane Galusha, *Liquid Assets: A History of New York City's Water System* (Fleischmanns, New York: Purple Mountain Press, 1999), 199, quoting *Ellenville Journal*, March 24, 1938.
19. DWSN 80 (Nov. 15, 1941), 365.
20. BWS, Annual Report 1950: 131.
21. BWS, Annual Report 1950: 151.
22. BWS, Annual Report 1950: 157–63.
23. BWS, Annual Report 1950: 172.
24. BWS, Annual Report 1960: 95.
25. BWS, Annual Report 1965: 34–35.

THE FUTURE OF WATER

1. Peter H. Gleick, *The World's Water 1998-1999: The Biennial Report on Fresh Water Resources* (Washington, DC: Island Press, 1998), 39.
2. H. Rouse and S. Ince, *History of Hydraulics* (Ames, Iowa: Iowa State University, Iowa Institute of Hydraulic Research, 1957).
3. Dora Crouch, *Water Management in Ancient Greek Cities* (New York: Oxford University Press, 1993), 24–28.
4. Rouse and Ince.
5. Crouch: 19.
6. P. R. Epstein, "Cholera and the Environment: An Introduction to Climate Change," *Physicians for Social Responsibility Quarterly* 2, No. 3 (1992), 146–160.
7. C. E. Rosenberg, *The Cholera Years* (Chicago: The University of Chicago Press, 1962).
8. J. Snow, *Snow on Cholera* (London: Oxford University Press, 1936).
9. L. Nash, "Water Quality and Health," *Water in Crisis: A Guide to the World's Fresh Water Resources*, ed. P. H. Gleick, (New York: Oxford University Press, 1993), 25–39.
10. World Health Organization (WHO), "World Health Report 2003—Shaping the Future," (January 2004), http://www.who.int/whr/en/.
11. P. H. Gleick, N. Cain, D. Haasz, C. Henges-Jeck, C. Hunt, M. Kiparsky, M. Moench, M. Palaniappan, V. Srinivasan, and G. Wolff, *The World's Water 2004–2005. The Biennial Report on Fresh Water Resources* (Washington, D.C.: Island Press, 2004).
12. United Nations, The Millennium Goals, United Nations Millennium Declaration A/RES/55/2, September 8, 2000.
13. WHO, "Global Water Supply and Sanitation Assessment 2000 Report," (January 2004), http://www.who.int/water_sanitation_health/Globassessment/GlobalTOC.htm.
14. Gleick et al.
15. Organisation for Economic Co-operation and Development (OECD), Creditor Reporting System, Aid Activities in the Water Sector: 1997–2002, Report prepared by the Development Assistance Committee (Paris).
16. Steven C. McCaffrey, "A Human Right to Water: Domestic and International Implication," *Georgetown International Environmental Law Review* 5, no. 1 (1992), 1–24.
17. Gleick.
18. J. A. Beecher, "Water Utility Privitization and Regulation: lessons From the Global Experiment," *Water International* 22, no. 1 (1997), 54–63.
19. Nelson P. Blake, "Water and the City: Lessons from History," *Water and the City*, (Chicago: Public Works Historical Society, 1991).
20. Christopher Anderson, "Cholera Epidemic Traced to Risk Miscalculation," *Nature* 354 (1991), 255.
21. E. Brubaker, "The Promise of Privatization," Energy Probe Research Foundation (April 2001), http://www.environmentprobe.org/enviroprobe/pubs/Ev548.htm.
22. G. LeClerc and T. Raes, *Water: A World Financial Issue*, Sustainable Development Series (Paris: PriceWaterhouseCoopers, 2001).
23. M. Palaniappan, P.H. Gleick, V. Srinivasan, C. Hunt, "Principles and Practices for the New Economy of Water: Protecting the Public Interest in Water Privatization," Pacific Institute working paper, Oakland, California, March 2003.
24. American Water Works Association, Reinvesting in Drinking Water Infrastructure, (May 2000), http://www.awwa.org/govtaff/infrastructure.pdf.
25. Global Water Partnership, "Toward Water Security: A Framework for Action to Achieve the Vision for Water in the 21st Century," Stockholm, Sweden, 2000.
26. http://www.who.int/water_sanitation_health/monitoring/en/Glassessment3.pdf.

BIBLIOGRAPHY

Adams, Arthur G. *The Catskills*. New York: Fordham University Press, 1990.

Aicher, Peter J. *Guide to the Aqueducts of Ancient Rome*. Wauconda, Illinois: Bolchazy-Carducci Publishers, 1995.

Baker, Ira O. *A Treatise on Masonry Construction*. New York: John Wiley & Sons, 1900.

Bone, Kevin, ed. *The New York Waterfront: Evolution and Building Culture of the Port and Harbor*, rev. 2d ed. New York: The Monacelli Press, 2004.

Capossela, Jim. *Good Fishing in the Catskills*. Woodstock, Vermont: BackCountry Guides, 1989.

Carpenter, Rolla C., and Herman Diederichs. *Experimental Engineering and Manual for Testing: For Engineers and for Students in Engineering Laboratories*. New York: John Wiley & Sons, 1911.

Evans, Harry B. *Water Distribution in Ancient Rome: The Evidence of Frontinus*. Ann Arbor: University of Michigan Press, 1994.

Fitchen, John. *Building Construction Before Mechanization*. Cambridge, Mass.: MIT Press, 1986.

Frisbie, Richard. *Water for New York City*. Saugerties, N.Y.: Hope Farm Press, 1993.

Galusha, Diane. L*iquid Assets: A History of New York City's Water System*. Fleischmanns, N.Y.: Purple Mountain Press, 1999.

Giedion, Sigfried. *Building in France, Building in Iron, Building in Ferro-Concrete*. Santa Monica: Getty Center for the History of Art and the Humanities, 1995.

Gleick, Peter H. *The World's Water, 2000–2001*. Washington D.C.: Island Press, 2000.

Greenberg, Stanley. *Waterworks: A Photographic Journey Through New York's Hidden Water System*. New York: Princeton Architectural Press, 2003.

Harris, Cyril M. *Dictionary of Architecture and Construction*. New York: McGraw-Hill, 1975.

Hodge, A. Trevor. *Future Currents in Aqueduct Studies*. Leeds: Francis Cairns Publications, 1991.

Hufeland, Otto. *A Check List of Books, Maps, Pictures, and Other Printed Matter Relating to the Counties of Westchester and Bronx*. White Plains, N.Y.: Westchester County Historical Society, 1929.

Jackson, Donald C. *Great American Bridges and Dams*. Washington D.C.: Preservation Press, 1988.

Jackson, Kenneth, ed. *The Encyclopedia of New York City*. New Haven: Yale University Press, 1995.

Kidder, Frank E. *Building Construction and Superintendence*. New York: William T. Omstock, 1905.

Kidder, Frank E., and Thomas Nolan. *The Architects' and Builders' Handbook*. New York: John Wiley & Sons, 1884.

Koeppel, Gerard T. *Water for Gotham: A History*. Princeton: Princeton University Press, 2000.

Lomax, E., and T. Gunyon, eds. *Nicholson's Dictionary of the Science and Practice of Architecture, Building, and Carpentry from the Earliest Ages to the Present Time*. New York: London Printing and Publishing Company, c. 1861–64.

Mock, Elizabeth B. *The Architecture of Bridges*. New York: Museum of Modern Art, 1949.

Morgan, Morris Hicky. *Vitruvius: The Ten Books on Architecture*. New York: Dover, 1960.

Newton, Isaac. *New York Water Supply: Report on Storage Reservoirs in the Croton*. New York: Martin Brown, 1883.

Rodgers, Elizabeth B. *Rebuilding Central Park: A Management and Restoration Plan*. Cambridge, Mass.: MIT Press, 1987.

Silver, Nathan. *Lost New York*. New York: Schocken Books, 1971.

Steuding, Bob. *The Last of the Handmade Dams: The Story of the Ashokan Reservoir*. Fleischmanns, N.Y.: Purple Mountain Press, 1989.

Sturgis, Russell, et al. Sturgis' *Illustrated Dictionary of Architecture and Building*. 3 vols. Mineola, N.Y.: Dover, 1989.

Museum of Modern Art. *Twentieth Century Engineering*. New York: Museum of Modern Art, 1964.

Urell, Catherine, and Rosmunda Nonnenmacher. *Big City Water Supply*. Chicago: Follett, 1953.

Ward, Diane R. *Water Wars: Drought, Flood,*

and the Politics of Thirst. New York: Riverhead Books, 2002.

Wegmann, Edward. *The Water Supply of the City of New York, 1658–1895.* New York: John Wiley & Sons, 1896.

———. *The Design and Construction of Dams.* New York: John Wiley & Sons, 1911.

Weidner, Charles H. *Water for a City: A History of New York City's Problem from the Beginning to the Delaware River System.* New Brunswick, N.J.: Rutgers University Press, 1974.

White, Lazarus. *The Catskill Water Supply of New York City.* Vol 1. Brooklyn: The Scientific Press, 1913.

White, Lazarus, and Edmund Astely Prentis Jr. *Modern Underpinning: Development, Methods, and Typical Examples.* New York: John Wiley & Sons, 1917.

ARTICLES

Armstrong, Roger W. "Construction of the Delaware Aqueduct." *Journal of the New England Water Works Association*, 55, no. 2 (1941).

Breen, Virginia, Austin Fenner, and Bill Hutchinson. "A Water Main Spews Chaos." *Daily News*, July 16, 1998.

Fan, Maureen. "Ready To Open 3rd Tunnel Tap." Daily News, August 13, 1998.

Galusha, Diane. "Liquid Assets." *Ellenville Journal*, March 24, 1938.

Grann, David. "City of Water." *The New Yorker*, September 1, 2003.

Martin, Douglas. "Decades After Its Conception, 3d Water Tunnel Set to Open." *New York Times*, August 6, 1998.

Revkin, Andrew C. "Shift in Rules at Reservoirs Draws a Suit to Protect City's Water." *New York Times*, October 14, 1998.

Sugarman, Raphael. "Showdown Over Plant." *Daily News*, May 1, 1998.

Wadler, Joyce. "Master at the Noble Art of Supplying Water." *New York Times*, January 9, 1998.

OFFICIAL RECORDS, MUNICIPAL PUBLICATIONS AND TECHNICAL REPORTS

Board of Water Supply of the City of New York. *Annual Reports of the Board of Water Supply of the City of New York.* [New York], 1935–38; 1940–47; 1950; 1954; 1959–60; 1965–66; 1969; 1971; 1973.

———. *Catskill Water Supply. A General Description.* New York, January 1913; September 1915; 1916; June 1919; October 1923; December 1928; August 1929.

———. *Catskill Water System News.* New York, December 20, 1913.

———. Commission on Additional Water Supply. *Report of the Commission on Additional Water Supply for the City of New York Made to Robert Grier Monroe, Commissioner of Water Supply, Gas, and Electricity,* November 30, 1903. By William H. Burr, Rodolf Hering, and John R. Freeman. New York: Martin B. Brown, 1904.

———. *First Annual Report.* New York, December 31, 1906.

———. *History of the Catskill Aqueduct and Related Structures.* Vol 1. 1905–1917. New York, Does Vol. 1 cover history only from '05-'17? If so, give date of publication for entire series.

———. "Holing Through City Tunnel No. 2." [New York], March 1932.

———. *Origin and Achievements of the Board of Water Supply City of New York.* [New York], September, 1950.

———. *Report of the Board of Water Supply of the City of New York to the Board of Estimate on the Third City Tunnel, First Stage, in the Boroughs of The Bronx, Manhattan and Queens.* [New York], July 1966.

———. *Report of the Board of Water Supply of the City of New York to the Board of Estimate on the Third City Tunnel, Second Stage, in the Boroughs of Manhattan, Queens and Brooklyn.* [New York], May 1973.

———. *Testimonial to the Chief Engineers, 1905–1965.* [New York], 1965.

———. *The Water Supply of the City of New York.* [New York], September 1950.

———. *Work of the Board of Water Supply, A General Description of the Catskill Water Supply and of the Project for an Additional Supply from the Delaware River Watershed and the Rondout Creek.* New York: Beacon Press, 1936.

City of New York Aqueduct Commission. *Report of the Chief Engineer.* New York: W. P. Mitchell & Sons, 1906.

Croton Aqueduct Department. *Annual Report of the Croton Aqueduct Department Made to the Common Council of the City of New York.* New York: 1851–54; 1856–57; 1860–66; 1868–69.

Department of Water Supply, Gas, and Electricity of the City of New York. *A Description of the Water Supply System of the City of New York.* [New York], June 1952.

———. Special Committee of Engineers on the Jerome Park Reservoir. *Reports of the Special Committee of Engineers, Upon Certain Details of Construction of the Jerome Park Reservoir, Submitted to the Aqueduct Commissioners July 28, 1903.* [New York]: E. H. Tower, [1903].

———. "The Water Supply of the City of New York." [New York], September 1950; June, 1952.

Koch, Edward I. *The Mayor's Management Report.* [New York], 1981.

Kornfeld, Robert J. *Jerome Park Conservancy Preservation Report.* [New York]: Jerome Park Conservancy, August 1998.

New York City Department of Environmental Protection. Strategic Services. "Consumption of Water in New York City Per Capita." *Drinking Water Supply Quality Report.* New York, December 15, 2000.

———. New York City, City Tunnel No. 3.

Whithall, Brinton W. Letter to Good Faith Negotiating Group, October 16, 1981.

Reports on the new Croton Aqueduct reservoirs and Dams, 1883–87; 1887–1895; 1895–1907. City of New York Aqueduct Commission.

Draft Language Re. Husdon River/Water Tunnel No.3 Authorization.

Reservoir Release Team. *Good Faith Negotiations.* October 27, 1981.

CHRONOLOGY AND INDEX OF PRIMARY STRUCTURES

1664
New York City Population: 1,500

1790
Population: 33,131

1820
Population: 123,706
Consumption Rate: 12 gallons per person per day (ppd)

CROTON DAM AND RESERVOIR, CROTON RIVER (1837–42)
Dam
STRUCTURE Masonry (stone)
DIMENSIONS Length 270 feet, height 50 feet
Reservoir (Croton Lake)
CREST ELEVATION 166 feet above mean tide
SURFACE AREA 440 acres
CAPACITY 600 million gallons
WATERSHED AREA 338 square miles

CROTON AQUEDUCT, CROTON RIVER TO MANHATTAN (1837–42)
STRUCTURE Masonry (brick)
DIMENSIONS Length 40.5 miles
CAPACITY 72 million gallons per day (raised to 100 mgd)

RECEIVING RESERVOIR, YORK HILL (1837–42)
STRUCTURE Earthen embankments with masonry core walls
DIMENSIONS Length 1,826 feet, height 38 feet
CREST ELEVATION 130 feet above mean tide (estimated)
SURFACE AREA 35 acres
CAPACITY 180 million gallons
WATERSHED None, used for storage/ balancing

DISTRIBUTING RESERVOIR, MURRAY HILL (1837–42)
STRUCTURE Masonry walls
DIMENSIONS Length 420 feet, width 420 feet, height 45 feet
CREST ELEVATION 115 feet above mean tides
SURFACE AREA 4 acres
CAPACITY 24 million gallons
WATERSHED None, used for storage/ distribution

1850
Population: 696,115
Consumption Rate: 78 gallons ppd

NEW RECEIVING RESERVOIR, CENTRAL PARK (1857–62)
STRUCTURE Earthen embankments with masonry core walls
CREST ELEVATION 120 feet above mean tide
SURFACE AREA 96 acres
CAPACITY 1 billion gallons
WATERSHED None, used for storage/ balancing/distribution

1870
Population: 942,292
Consumption Rate: 80 gallons ppd

BOYD'S CORNERS DAM AND RESERVOIR, WEST BRANCH, CROTON RIVER (1866–73)
Dam
STRUCTURE Masonry (stone)
DIMENSIONS Length 670 feet, height 57 feet
Reservoir
CREST ELEVATION 600 feet above mean tide
SURFACE AREA 379 acres
CAPACITY 2.7 billion gallons
WATERSHED AREA 20.3 square miles

MIDDLE BRANCH DAM AND RESERVOIR, MIDDLE BRANCH, CROTON RIVER (1874–78)
Dam
STRUCTURE Earthen embankment with masonry core wall
DIMENSIONS Length 515 feet, height 94 feet
Reservoir
CREST ELEVATION 295 feet above mean tide (estimated)
SURFACE AREA 452 acres
CAPACITY 4.8 billion gallons
WATFRSHED AREA 20.9 square miles

1880
Population: 1,164,673
Consumption Rate: 76 gallons ppd

BYRAM DAM RESERVOIR (1880–95) BYRAM RIVER

Dam

STRUCTURE Earthen embankment
DIMENSIONS Height 15 feet

Reservoir

CREST ELEVATION 290 feet above mean tide (estimated)
SURFACE AREA 75 acres
CAPACITY 180 million gallons
WATERSHED AREA 8.6 square miles

THE RYE PONDS DAM AND RESERVOIR (1880–85)

Dam

STRUCTURE Earthen embankment
DIMENSIONS Height 15 feet

Reservoir

CREST ELEVATION 330 feet above mean tide (estimated)
SURFACE AREA 280 acres
CAPACITY 1.05 billion gallons
WATERSHED AREA part of Byram River 8.6 square miles

KENSICO DAM(S) AND RESERVOIR [ORIGINAL] (1880–85) BRONX RIVER

Dam

STRUCTURE Earthen embankment with masonry core wall
DIMENSIONS Length 1.5 miles, height 45 feet

Reservoir

CREST ELEVATION 300 feet above mean tide
SURFACE AREA 230 acres
CAPACITY 1.62 billion gallons
WATERSHED AREA 13.3 square miles

48-INCH LINE (1882–84)

Aqueduct

STRUCTURE Cast-iron pipe
Connected Kensico Reservoir to Williamsbridge Reservoir
DIMENSIONS Length 15.2 miles, diameter 4 feet
CAPACITY 10 million gallons per day

WILLIAMSBRIDGE RECEIVING RESERVOIR (1884–89)

Termination of the 48-Inch Line from Kensico
CREST ELEVATION 100 feet above mean tide.
SURFACE AREA 10 acres (estimated)
CAPACITY 120 million gallons per day
WATERSHED None, used for storage/ balancing/distribution

1890
Population 1,441,216
Consumption Rate 78 gallons ppd

NEW CROTON AQUEDUCT (1884–93)

STRUCTURE Masonry (brick), iron lined in places
DIMENSIONS Length 33 miles (24 miles in rock tunnel, 9 miles in conduit), diameter 10 to 13 feet
CAPACITY 300 million gallons per day

SODOM DAM, EAST BRANCH RESERVOIR, EAST BRANCH OF CROTON RIVER (1888–93)

Dam

STRUCTURE Solid masonry (stone)
DIMENSIONS Length 500 feet, height 94 feet

Reservoir

CREST ELEVATION 375 feet above mean tide
SURFACE AREA 534 acres
CAPACITY 4.2 billion gallons
WATERSHED AREA 73.42 square miles
Connected to the Bog Brook Reservoir by 1,778-foot-long tunnel

BOG BROOK DAM(S) AND RESERVOIRS, BOG BROOK (1889–93)

Dams:

STRUCTURE Earthen embankments with masonry core walls

Dam #1

DIMENSIONS Length 1,340 feet, height 60 feet

Reservoir #1

CREST ELEVATION 415 feet above mean tide
SURFACE AREA 281 acres
CAPACITY 4.4 billion gallons
WATERSHED AREA 70 square miles

Dam #2

DIMENSIONS Length 1,956 feet, height 24 feet

Reservoir #2

CREST ELEVATION 415 feet above mean tide
SURFACE AREA 381 acres
CAPACITY 4.4 billion gallons
WATERSHED AREA 3.5 square miles
Connected to the East Branch by 1,778-foot tunnel

WEST BRANCH DAM(S) AND RESERVOIR (1890–96)

Dams

STRUCTURE Earthen embankments with masonry core walls

Dam #1

DIMENSIONS Length 1,800 feet, height 65 feet

Dam #2

DIMENSIONS Length 750 feet, height 45 feet

Reservoir

CREST ELEVATION 550 feet above mean tide
SURFACE AREA 998 acres
CAPACITY 9 billion gallons
WATERSHED AREA 40.94 square miles

TITICUS DAM(S) AND RESERVOIR, TITICUS RIVER (1890–96)

Dams

STRUCTURE Earthen embankments, solid masonry
DIMENSIONS Length 1,519 feet (combined dams), height 109 feet

Reservoir

CREST ELEVATION 316 feet above mean tide
SURFACE AREA 725 acres
CAPACITY 7.2 billon gallons
WATERSHED AREA 22.8 square miles

JEROME PARK RESERVOIR (1894–1905)

Dam

STRUCTURE Earthen embankments, masonry lining

Reservoir

CREST ELEVATION 131 feet above mean tide
SURFACE AREA 94 acres
CAPACITY 800 million gallons

WATERSHED None, used for storage/ balancing/distribution

AMAWALK DAM AND RESERVOIR, MUSCOOT RIVER (1889–97)

Dam

STRUCTURE Earthen embankment with masonry core wall
DIMENSIONS Length 1,280 feet, height 80 feet

Reservoir

CREST ELEVATION 390 feet above sea level
SURFACE AREA 600 acres
CAPACITY 6.7 billion gallons
WATERSHED AREA 20.5 square miles

1900
Population: 3,437,202
Consumption Rate: 105 gallons ppd

NEW CROTON DAM AND RESERVOIR, CROTON RIVER (1892–1905)

Dam

STRUCTURE Cyclopean masonry
DIMENSIONS Length 2,168 feet, height 174 feet

Reservoir

CREST ELEVATION 196 feet above mean tide
SURFACE AREA 1,962 acres
CAPACITY 19 billion gallons
WATERSHED AREA 338 square miles

CROSS RIVER DAM AND RESERVOIR, CROSS RIVER (1906–8)

Dam

STRUCTURE Cyclopean masonry
DIMENSIONS 986 feet long, 170 feet high

Reservoir

CREST ELEVATION 326.5 feet above mean tide
SURFACE AREA 899 acres
CAPACITY 10.3 billion gallons
WATERSHED AREA 30.9 square miles

1910
Population: 4,766,883
Consumption Rate: 110 gallons ppd

CROTON FALLS RESERVOIR, WEST AND MIDDLE BRANCHES OF CROTON RIVER (1906–11), HEMLOCK DAM

Dam

STRUCTURE Cyclopean masonry
DIMENSIONS Length 1,110 feet, height 173 feet

Reservoir

CREST ELEVATION 318.4 feet above mean tide
SIZE 126 acres
CAPACITY 14.2 billion gallons
WATERSHED AREA 162 square miles

CROTON FALLS DIVERTING RESERVOIR, EAST BRANCH OF CROTON RIVER

Dam

STRUCTURE Earthen embankment with masonry core wall
DIMENSIONS Length 1,185 feet, height 40 feet (estimated)

Reservoir

CAPACITY 900 million gallons

CROTON WATERSHED DEVELOPMENT COMPLETE (1911)

Supplies up to 250 million gallons per day

ASHOKAN RESERVOIR, OLIVE BRIDGE DAM ESOPUS CREEK (1907–15)

Dam

STRUCTURE Cyclopean masonry and concrete, earthen embankments with masonry core walls
DIMENSIONS Length: 4,650 feet, height 210 feet

Dikes

STRUCTURE Earthen embankments with masonry core walls
COMBINED LENGTH 3.5 miles, height average 50 feet

Reservoir

CREST ELEVATION 590 feet above mean tide
SURFACE AREA 8,192 acres
CAPACITY 132 billion gallons
WATERSHED AREA 257 square miles

CATSKILL AQUEDUCT (1907–17)

STRUCTURE Concrete (various types) and steel pipe
DIMENSIONS Length 92 miles (55 miles cut and cover, 14 miles grade tunnel, 17 deep-rock pressure tunnels, 9 miles steel pipe siphon), diameter up to 20 feet
CAPACITY 800 million gallons per day
Connects Ashokan Reservoir to Hill View Reservoir

KENSICO DAM AND RESERVOIR (1909–15)

Dam

STRUCTURE Masonry (stone), concrete, and earthen embankments
DIMENSIONS Length 1,850 feet (masonry dam), height 128 feet (masonry dam)

Dike

DIMENSIONS Length 1,450 feet

Reservoir

CREST ELEVATION 354 feet above mean tide
SURFACE AREA 2,081 acres
CAPACITY 30.6 billion gallons
WATERSHED None, used for storage/balancing

HILL VIEW RESERVOIR (1909–15)

STRUCTURE Earthen embankments, concrete-lined basin
CREST ELEVATION +295 feet above mean tide
SURFACE AREA 90 acres
CAPACITY 900 million gallons
WATERSHED None, used for storage/balancing/distribution

CITY TUNNEL #1 (1911–15)

Aqueduct

STRUCTURE Concrete-lined, deep-rock pressure tunnel
DIMENSIONS Length 18 miles, diameter 11 to 15 feet
CAPACITY 1 billion gallons per day
Connects Hill View Reservoir to the five boroughs of the city

SILVER LAKE RESERVOIR, STATEN ISLAND (1913–17)

STRUCTURE Earthen embankments with concrete core walls
CREST ELEVATION +228 feet above mean tide
SURFACE AREA 54 acres
CAPACITY 438 million gallons
WATERSHED None, used for storage/balancing/distribution

1920
Population: 5,620,048
Consumption Rate: 123 gallons ppd

SHANDANKEN TUNNEL (1917–24)

Aqueduct

STRUCTURE Concrete-lined, deep-rock tunnel
DIMENSIONS Length 18 miles, 11 feet, 6 inches; width 10 feet, 3 inches
CAPACITY 650 million gallons per day
Connects Schoharie Creek to the Esopus Creek

GILBOA DAM AND SCHOHARIE RESERVOIR, SCHOHARIE CREEK (1919–27)

Dam

STRUCTURE Cyclopean masonry, concrete and earthen embankments with masonry core walls
DIMENSIONS Length 2,000 feet, height 182 feet

Reservoir

CREST ELEVATION 1,950 feet above mean tide
SURFACE AREA 1,145 acres
CAPACITY 17.6 billion gallons
WATERSHED AREA 314 square miles

1930
Population: 6,930,446
Consumption Rate: 141 gallons ppd

CATSKILL SYSTEM COMPLETED (1927)
Supplies up to 650 million gallons per day

CITY TUNNEL #2 (1928–34)
STRUCTURE Concrete-lined, deep-rock pressure tunnel
DIMENSIONS Length 20 miles, diameter from 15 to 17 feet
CAPACITY 1 billion gallons per day
Connects Hill View Reservoir to Bronx, Queens, and Brooklyn

DELAWARE AQUEDUCT (1937–45)
STRUCTURE Concrete-lined, deep-rock pressure tunnel
DIMENSIONS Length 84 miles, diameter from 15 to 18 feet
CAPACITY 1.4 billion gallons per day
Connects Rondout Reservoir to West Branch, Kensico, and Hill View Reservoirs

1950
Population: 7,891,957
Consumption Rate: 150 gallons ppd

MERRIMAN DAM AND RONDOUT RESERVOIR, RONDOUT CREEK (1937–54)
Dam
STRUCTURE Earthen embankments with concrete core walls
DIMENSIONS Length 2,500 feet, height 195 feet
Reservoir
CREST ELEVATION 860 feet above mean tide
SURFACE AREA 2,032 acres
CAPACITY 49.6 billion gallons
WATERSHED AREA 95 square miles

NEVERSINK DAM AND RESERVOIR, NEVERSINK RIVER (1941–53)
Dam
STRUCTURE Earthen embankments with concrete core walls
DIMENSIONS Length 2,820 feet, height 195 feet
Reservoir
CREST ELEVATION 1,440 feet above mean tide
SURFACE AREA 1,480 acres
CAPACITY 34.9 billion gallons
WATERSHED AREA 93 square miles

NEVERSINK TUNNEL (1949–55)
STRUCTURE Concrete-lined, deep-rock pressure tunnel
DIMENSIONS Length 6 miles, diameter 10 feet
CAPACITY 500 million gallons per day
Connects Neversink Reservoir to Rondout Reservoir

DOWNSVILLE DAM AND PEPACTON RESERVOIR, EAST BRANCH OF DELAWARE RIVER (1947–54)
Dam
STRUCTURE Earthen embankment with concrete core wall
DIMENSIONS Length 2,450 feet, height 204 feet
Reservoir
CREST ELEVATION 1,279 feet above mean high tide
SURFACE AREA 5,178 acres
CAPACITY 147 billion gallons
WATERSHED AREA 372 square miles

EAST DELAWARE TUNNEL (1949–55)
STRUCTURE Concrete-lined, deep-rock pressure tunnel
DIMENSIONS Length 25 miles, diameter 11 feet
CAPACITY 700 million gallons per day
Connects Pepacton Reservoir to Rondout Reservoir

1960
Population: 7,781,984
Consumption Rate: 165 gallons ppd

STILESVILLE DAM AND CANNONSVILLE RESERVOIR, WEST BRANCH DELAWARE RIVER (1955–67)
Dam
STRUCTURE Earthen embankment with concrete core wall
DIMENSIONS Length 2,800 feet, height 204 feet
Reservoir
CREST ELEVATION 1,151 feet above mean high tide
SURFACE AREA 4,568 acres
CAPACITY 95.7 billon gallons
WATERSHED AREA 490 square miles

1970
Population: 7,894,862
Consumption Rate: 180 gallons ppd

RICHMOND TUNNEL (1962–70)
STRUCTURE Concrete-lined, deep-rock pressure tunnel
DIMENSIONS Length 5 miles, diameter 10 feet
CAPACITY 350 million gallons per day
Connects City Tunnel #2 in Brooklyn to Staten Island

SILVER LAKE STORAGE TANKS (1965–70)
STRUCTURE Concrete-lined tanks and covered storage facility
CREST ELEVATION 220 feet above mean tide (estimated)
CAPACITY 100 million gallons
WATERSHED None, used for storage/balancing/distribution

1980
Population: 7,071,639
Consumption Rate: 185 gallons per person per day

1990
Population: 7,322,564
Consumption Rate: 191 gallons per person per day

2000
Population: 8,00,8278
Consumption Rate: 166 gallons per person per day

CITY TUNNEL #3 (1970–2020 PROJECTED)
STRUCTURE Concrete-lined, deep-rock pressure tunnel
DIMENSIONS Length (anticipated) 54 miles, diameter up to 24 feet
CAPACITY One billion gallons per day
Connects Kensico and Hill View Reservoirs to eastern and southern areas of the city

ACKNOWLEDGMENTS

Beginning in 1995, when we embarked on the restoration of the Department of Environmental Protection archive, through the production of the "Water-Works" exhibition in 2001 at The Cooper Union, to the completion of this book, many people have recognized the merits of this project and contributed to its success.

Although dozens contributed their energies over the years, the core group that consistently devoted themselves to the project was ultimately a small one. Gina Pollara and Paul Deppe, both graduates of The Irwin S. Channin School of Architecture, have worked with me for eleven years, from the start of the internship program in the summer of 1995 through the final days of publication. Gina and Paul's hands and eyes were on every aspect of this project. Throughout the years of organizing the materials, mounting the exhibition, administering the archive, and participating in the production of this book, they have been consistently professional and passionate, and I consider myself privileged to have them as colleagues.

In the production of the book, several associates stand out. Steven Hillyer, director of the Archive at The Irwin S. Chanin School of Architecture sought support for the project, worked patiently with the various contributors, and oversaw and coordinated the texts, images and captions. He brought his excellent visual sensibilities to every page as the book took shape.

My wife, Eugenia Bone, an accomplished author and editor, helped me with every single line of text in this book. Gena coordinated and worked with all the contributing authors and gave coherence to a wide range of writing styles. She patiently worked and reworked text, providing detailed and technical editing support. Without Gena's help we would not have this book.

I also want to thank the contributors: Gerard Koeppel, Al Appleton, and Dr. Peter Gleick, who brought experience, insight and expertise to the project. I am particularly grateful to Gerard for his advice throughout the production process.

Kilian Matish, an architect who contributed his expert graphic skills to our work for over five years, helped coordinate image production techniques and handling protocols for the thousands of digital files we maintained during this project.

The photographer Stanley Greenberg worked with the project for several years. His knowledge of negatives and prints helped us properly preserve these archival materials, a priority for the project.

No aspect of this work would have been possible had we not had the unwavering support of The Cooper Union for the Advancement of Science and Art. Former President John Jay Iselin immediately saw the merits of the work and backed our early efforts to get the project going. Jay never tired in his willingness to support the project and steered us toward others who would lend support. Our beloved Dean of Architecture, the late John Hejduk, was also a champion of the project. John loved the materials and thought that architects (among others) should have access to this important archive. This initial institutional support was key to making the publication of *Water-Works* a reality. Dr. George Campbell had just begun his tenure as president of the college when we were preparing for the "Water-Works" exhibition in January 2001. Dr. Campbell helped fund the exhibition and never hesitated in his belief in the project. Dean Anthony Vidler, who took the helm of the architecture school in the fall of 2001, has also made substantial resources available, including, most importantly, the staff, energy, and facilities of the Chanin School Archive. The use of the archive freed us from the financial problems that confront a project like this and provided a home from which we could work. I am very grateful for Tony's insights and personal advice. Associate Dean Elizabeth O'Donnell has also been very supportive, for which I am grateful. I also want to thank The Cooper Union Research Foundation for administrative support, particularly Dr. Jameel Ahmad, director, and Maureen Deol.

The New York City Department of Environmental Protection was an active partner every step of the way. Through four separate administrations, the DEP recognized that the archive was not only extremely important to the effective management of the system, but also that it was a valuable part of the historic record of the City of New

York. In 1995 Commissioner Albert Appleton was the first official willing to take a chance on this project. The Cooper Union team was, after all, outsiders, not professional engineers or bona-fide archivists but architects and students. The idea that such a group would be—at least in the short term—responsible for such a body of records seemed unorthodox to some in the agency. But Commissioner Appleton and his assistant John Bennett believed that a small program, functioning outside of the ordinary channels of governmental procurement, might succeed in getting this project off the ground. They were proven correct. The next four commissioners of the DEP, Marilyn Gelber, Joel Miele, Christopher Ward, and Emily Lloyd, have all continued to support the improvement and long-term stability of the archival collection. Each stepped in to make sure that the archive had the resources it needed and that The Cooper Union had what was necessary to make this book possible. The DEP staff members who have made important contributions include Charles Strucken, Louis Tazzi (Deputy Commissioner), Alfred Miller, Carl Ambrose (whose photographs appear in the text), Joe Landau, Kim Estes-Fradis, Gino Sass, Doug Greeley, Jonathan Ridgeway, Michael Cetera, Steven Adamec, and Eileen Alter. I especially want to thank Samar Qandil, who has directed the archive since 2001, Raphael Hurwitz, an engineer with incredible passion for the system, Frederic Grevin, who personally worked through many of our requests, and Rosemarie Subasic. The New York City Municipal Archives has been helpful in the archive recovery effort since 1995, with recommendations and guidance from Ken Cobb, director from 1990 to 2005, and Leonora Gidlund, deputy director from 1992 to 2005, and now director. The expert conservators Laura McCann, Mirah von Wicht, and Ellen Chin worked on the conservation of many of the drawings in the DEP collection. I would also like to acknowledge Diane Galusha for her monumental book *Liquid Assets: A History of the New York City Water Supply*. This book is the best general overall history of the system written to date, and it was a valuable research tool throughout the making of this book. I would also like to thank Robert Kornfeld from the Jerome Park Reservoir Conservancy for sharing his materials and information about Jerome Park.

Our sincere thanks to The Monacelli Press for its commitment to The Cooper Union School of Architecture and for the quality of publication that we all believed this project deserved, to Elizabeth White, Andrea Monfried, and John Gregory, and to Abigail Sturges, the designer. And finally to Gianfranco Monacelli, whose long-standing publishing relationship with The Cooper Union School of Architecture began with John Hejduk.

I also want to thank the dozens of Cooper Union students and graduates who have contributed as interns at the DEP archive, as assistants in the preparation of the exhibition, or as interns at the School of Architecture Archive, in particular Lucy Pritchard, Kathleen Brumder, and Gia Mainiero

Finally, I would like to thank those organizations that generously provided financial support for the project: The American Council of Engineering Companies of New York, The Graham Foundation for Advanced Studies in the Fine Arts, The Lucius N. Littauer Foundation, Malcolm Pirnie, Inc., the New York State Laborers'- Employers' Cooperation & Educational Trust, New York State Laborers' Union Local 147, and Bone/Levine Architects (especially my partner, Joseph Levine) for the substantial contributions that our office has made over the years.

But maybe most of all, I would like to express my profound gratitude and admiration for the thousands of engineers, workers, and sandhogs and their families whose monumental efforts over the decades provided our glorious water supply system. It is with great humility that I offer this document to their achievements.

CONTRIBUTORS

ALBERT F. APPLETON

Albert F. Appleton is currently a Senior Fellow at the Regional Plan Association in New York City and a Visiting Fellow at the City University Of New York Institute on Urban Systems. He also works as a private consultant nationally and internationally, including water utility and watershed management projects in Canada, Mexico, Eastern Europe, and China. Mr. Appleton is a former Commissioner of the New York City Department of Environmental Protection, in which capacity he served as the Director of the New York City Water and Sewer system where his tenure was marked by the successful implementation of a wide-ranging series of management and financial reforms, the streamlining and acceleration of the Department's infrastructure programs, and the creation and implementation of watershed protection and water conservation programs that have become internationally recognized best practice models.

KEVIN BONE

Kevin Bone is a Professor at The Irwin S. Chanin School of Architecture at The Cooper Union. He edited and contributed writings to *The New York Waterfront, Evolution and Building Culture of the Port and Harbor*—The Monacelli Press in 1997. Since 1983 Bone has been a partner and principal in Bone/Levine Architects. The practice won AIA design awards in 2002, 2001, 2000, and 1999 for residential work, and the Chicago Athenaeum Award for American Architecture in 2005. He was a co-founder and Studio Director for the First and Second International Summer Workshops for Architecture at the Zenobio Institute in Venice, Italy (1994–95), Visiting Assistant Professor at the School of Architecture at Columbia University, (1988–89), and Guest Professor of Architecture and Urbanism at the Hochschule der Kunste in Berlin, Germany (1989–90).

DR. PETER H. GLEICK

Dr. Peter H. Gleick is co-founder and President of the Pacific Institute for Studies in Development, Environment, and Security in Oakland, California which addresses global environment and development problems, especially in the area of freshwater resources. Dr. Gleick is an internationally recognized water expert and in 2003 was awarded the prestigious MacArthur Fellowship for his science and policy work on water issues worldwide. He serves on the boards of numerous journals and organizations and was elected an Academician of the International Water Academy, in Oslo, Norway, in 1999. In 2001 he was appointed to the Water Science and Technology Board of the National Academy of Sciences, Washington, D.C. Dr. Gleick is the author of many scientific papers and four books, including the biennial water report *The World's Water* published by Island Press (Washington, D.C.).

GERARD KOEPPEL

Historian Gerard Koeppel is the author of *Water for Gotham: A History* (Princeton UP, 2000) and a forthcoming history of the Erie Canal; a portion of that research appears as "Andrew Bartow and the Cement That Made the Erie Canal," *New-York Journal of American History* (Spring-Summer 2005). Mr. Koeppel contributed numerous entries to *The Encyclopedia of New York City* (Yale UP, 1995), *The Encyclopedia of New York State* (Syracuse UP, 2005), and *The Encyclopedia of the New American Nation* (Scribners, forthcoming). He has reviewed books for the *New York Observer* and the *Economist* and written opinion pieces for the *New York Times*. He speaks regularly on aspects of New York City history to governmental, historical, and general audiences. Mr. Koeppel, a former radio journalist, was a reporter, writer, editor, and producer at CBS News and other radio networks and stations.

GINA POLLARA

Gina Pollara was the Associate Director of the Irwin S. Chanin School of Architecture Archive at The Cooper Union until September 2006. She contributed to *The New York Waterfront: Evolution and Building Culture of the Port and Harbor*, published by The Monacelli Press, and revised and updated the second edition. She is currently at work on a book about the architectural history of The Cooper Union's landmark Foundation Building.

MARTY POTTENGER

Marty Pottenger is a playwright, solo performance artist, director, and labor activist. She is the creator of the Obie Award–winning *City Water Tunnel #3*, a multi-media performance about New York City's 60-year-long public works project as told through the collected stories of the people building the tunnel. Performances and plays include *Winning the Peace*, *Construction Stories*, and *What It's Like To Be A Man*. She has received generous and critical financial support from the funding community including the NEA, The Trust for Mutual Understanding, Lila Wallace Arts Partners, Rockefeller MAP, Animating Democracy, NYSCA, and NYFA. Marty is a founding Member of *Heresies: A Feminist Publication on Art & Politics*. She is currently working with the city and school district of Portland, Maine, as Director of the Arts and Equity Initiative, which uses stories and performances to create policy and achieve departmental diversity goals.

INDEX

Page numbers in *italics* refer to illustrations

NOTES ON THE MATERIALS

With the exception of the images noted below, all materials (drawings, photographs, documents) reproduced in this book are from the Collection of the Archives, New York City Department of Environmental Protection. For security reasons, most of the individual line drawings were extracted from larger sheets of drawings, and identifying or sensitive information was redacted. The following hand-drafted drawings are details from a larger sheet of drawings. Otherwise, the drawings have been reproduced in whole.

THE RISE TO CROTON
Figs 22, 26, 27, 30

THE EXTENSION OF THE CROTON SYSTEM
Figs 4, 8, 9, 20, 25, 33, 35, 49, 74

PREFACE

Fig. 3 Kevin Bone, photographer. Courtesy of Kevin Bone.

INTRODUCTION

Fig. 16 Liselot van der Heijden, photographer. Courtesy of The Cooper Union School of Architecture Archive.

THE RISE TO CROTON

Figs. 4–7, 11 Courtesy of Gina Pollara.

Fig. 12 Nicolino V. Calyo, *Burning of the Merchant's Exchange*. Courtesy of the Museum of the City of New York.

Fig. 23 Courtesy of Kevin Bone.

Fig. 31 Collection of The New-York Historical Society, PR 020.

THE EXTENSION OF THE CROTON SYSTEM

The following plates are reproduced from Edward Wegmann, *The Design and Construction of Dams including Masonry, Earth, Rock-Fill, Timber, and Steel Structures*. Sixth Edition. New York: John Wiley & Sons, 1911. Courtesy of Kevin Bone.

Fig. 40 Plate LXIX
Fig. 46 Plate M
Fig. 47 Plate LXXII
Fig. 52 Plate LXXIII
Fig. 53 Plate LXX
Fig. 54 Plate LXXI
Fig. 55 Plate LXXIV
Fig. 59 Plate LXXX
Fig. 60 Plate XXIX
Fig. 64 Plate LXXXII
Fig. 65 Plate XCIV
Fig. 76 Plate LXXXIX
Fig. 82 Plate XCI

THE POST-DELAWARE ERA

Photographs of City Tunnel # 3 were taken by the following photographers:
Figs. 1, 2, 9, 10, 19 Carl Ambrose
Figs. 3, 5 F. Chaney
Figs. 4, 6, 11, 12, 17 Herb Coleman
Figs. 7, 8, 13, 14, 15, 18 J. Manger

EXCERPTS FROM MARTY POTTENGER'S PERFORMANCE PIECE *CITY WATER TUNNEL #3*

Marty's performance brought the people who currently work on the system to life and added a human dimension, a contemporary voice, that we wanted to bring to this historic body of work. We have incorporated these excerpts as moments of pause in the long history of the system and as links between the story and its contemporary counterparts.

ADDITIONAL CAPTIONS

Page 25 View of one of the mountain streams that feeds into the Rondout Creek, which flows to the Rondout Reservoir and then on into the Delaware Aqueduct for delivery to New York City. 1940.

Page 53 Core samples (or borings) taken to determine geologic profiles. November 12, 1909.

Page 105 Headframe at top of shaft with bucket for raising and lowering men into the tunnel. June 10, 1938.

Page 179 Sign delineating border of future Cannonsville Reservoir. c. 1955.

Page 215 City Tunnel #1. Crew drilling bench. May 23, 1913.

Page 235 Delaware Aqueduct. Holing through at Shaft 3. November 23, 1940.

Page 249 City Tunnel #3. Van Cortlandt Valve Chamber. Lateral lines prior to valve assembly. Bronx, New York. October 28, 1985. Carl Ambrose, photographer.

Page 250 Comparative plans and transverse sections of masonry dams. c. 1900.